Forest Bioresource Utilisation in the Eastern Mediterranean since Antiquity

A case study of the Makheras, Cyprus

Julia Ellis Burnet

BAR International Series 1243
2004

Published in 2016 by
BAR Publishing, Oxford

BAR International Series 1243

Forest Bioresource Utilisation in the Eastern Mediterranean since Antiquity

ISBN 978 1 84171 603 9

BAR Publishing is the trading name of British Archaeological Reports (Oxford) Ltd. British Archaeological Reports was first incorporated in 1974 to publish the BAR Series, International and British. In 1992 Hadrian Books Ltd became part of the BAR group. This volume was originally published by Archaeopress in conjunction with British Archaeological Reports (Oxford) Ltd / Hadrian Books Ltd, the Series principal publisher, in 2004. This present volume is published by BAR Publishing, 2016.

Printed in England

BAR
PUBLISHING

BAR titles are available from:

BAR Publishing
122 Banbury Rd, Oxford, OX2 7BP, UK
EMAIL info@barpublishing.com
PHONE +44 (0)1865 310431
FAX +44 (0)1865 316916
www.barpublishing.com

Table of Contents

List of Appendiees

List of Figures

Chapter One: Introduction

Vegetation is never permanent. It responds dynamically to a wide range of influencing factors, not least of which is human activities. A forest is frequently seen as something timeless, everlasting and unchanging or as the "climax" of a vegetative evolution (Clements 1919; Tansley 1935; Sprugel 1991), whereas, the true aesthetic of a forest lies in its internal dynamic, a measure of its life force and capacity to regenerate following disturbance.

The current paucity of information relating to Mediterranean-type ecosystems and the anthropogenic impact which may have contributed to the extensive decline in regional forest reserves prompted this study of an eastern Mediterranean forest and its bioresources. Although Meiggs (1982) and to a lesser degree Thirgood (1981; 1987) have presented aspects of the ancient Mediterranean economy in relation to forest resource exploitation, neither author fully understood the forest ecology and its relationship to the unique geomorphological and island situation of Cyprus.

Cyprus is the third largest island in the Mediterranean, lying along latitude 35 degrees north off the coasts of Turkey and the Levant. From both these shores the island can be seen in clear weather. Cyprus provides a classic example of an island as a spatially delimited "evolutionary laboratory" (Schüle 1993), where the exploration and evaluation of specific social, economic, historical, or theoretical issues contribute to a growing body of cultural knowledge (Knapp pers.com.). Against this pattern of recent research it became apparent that there was a need to place known prehistoric and historic socio-economic shifts within an environmental context. In particular, the value of the Cypriot forest resource which lay adjacent to mineral-bearing deposits had not previously been evaluated. Neither had studies been made on the anthropogenic impact on a forest's dynamic, that is, on its structure and ability to maintain its internal integrity.

The landscape of Cyprus needs to be understood as a place in which human and natural elements interact. Each interaction results in a different set of consequences. Links between settlement sites, and the environmental attitudes of the people associated with those sites, arise. Changing attitudes may lead to the encapsulation of the landscape in mythology. Traditional associations with specific places may be incorporated into the social identity of certain groups of people. Landscape may also be a human construct as real as the physical features that distinguish it. In this way a culture might combine attributes of geographic reality with a perceived reality. Together these perceptions could form an evolving tableau of ecological action and reaction in response to landscape management or occupation. Landscape as a "place" also reflects a certain idiosyncratic appreciation of locality which tends to emphasise individual attitudes and biases. Konigsson (1992) described the result as, "landscape aspect, which is dynamic, and which mirrors the presence, and so living, relations between people and nature".

Notions of ecological systems are derived in part from their history, but their history cannot be divorced from human involvement. There are very few, if any, major forests on the planet that have not been significantly influenced by cycles driven by people. Cultural influence has frequently contributed to the biodiversity of the forest resource (McNeely 1994). Ecology is rooted in natural history and the detailed study of habitats and ecosystems is frequently made without reference to a wider framework, an example of "not seeing the wood for the trees" (Lovelock 1987:126). In Nietzsche's "theory of knowledge" an attempt was made to categorise the framework of the natural sciences seen as the relative *a priori* of a world of objective illusion that has been produced for the purposes of mastering nature, and thus preserving existence. Nietzsche believed that the meaning of existence will unfold increasingly in the course of its process, that history was not in the service of pure knowledge, but of life (Neitzsche *Werke* 1:217). An echo of Neitzsche's philosophy was presented by Gould (1991) who said: "Our view of the past is compromised by our failure to recognise the uncharacteristic state of our present". For Neitzsche the entire cognitive apparatus was one for abstraction and simplification - not directed at knowledge - but at the control of things. The "end" and the "means" were as removed from the essence as the "concepts". With "end" and "means" he thought that the process was controlled as it was invented, therefore, rendering it comprehensible.

Ecological proaction toward a physical environment can be related to a community's interpretation of its immediate environment. Community actions and responses to the environment can be understood as a variable component of an ecosystem. The deduction put forward by Burrows (1923) and Roxby (1930) was that society seeks to establish an ecological balance with a natural environment through the adaptation of political institutions. The postulation by White and Renner (1936) that political institutions were developed to ensure effective control over an area to which a community laid claim may be applicable to prehistoric Cyprus. Such control over a community habitat would necessitate: (i) an internal political organisation which will sustain that control, (ii) the delimiting of satisfactory boundaries of the area, and (iii) the maintenance of effective relationships with other groups which occupy neighbouring areas (White and Renner 1936:20). The continued importance of agro-pastoralism even after the advent of a metal-based trading economy is expressed through the rise in decorative art and iconography.

The dimensions of the study undertaken for this thesis are twofold. The first broad goal of the project was to reconstruct the ecological and economic landscape of the eastern Troodos area of Cyprus in antiquity and trace its long term development. The second goal was to examine the evolution in time and space of cultural ecosystems within and adjacent to the forest reserves. Following from these two primary dimensions five secondary aims were pursued:

- to explore prehistoric and historic influences on the Cypriot forest resource in order to compile a definitive record of the evolutionary ecology of the forest environments
- to characterise short-term environments on various spatial scales which relate to particular phases of occupation at settlement sites
- to infer, using both on-site and off-site data, what resources were available to and utilised by past forest inhabitants through modelling, spatially and temporally (e.g. seasonally), the manner in which the landscape was exploited for target resources
- to consider the possible impact of such exploitation on the resource base in particular and the environment in general
- to build longer-term environmental sequences, over time scales relevant to the archaeological investigations of the island, in order to model relationships between environmental changes (both natural and anthropogenic) and changes in resource exploitation and subsistence systems.

The schematic outline of the environmental research undertaken demonstrates the complexity of the forest ecosystem. It also facilitates an understanding of the interrelations between various factors, particularly human interaction. Each facet of the study is therefore seen as an aspect of an interlinking matrix.

In order to encapsulate the principles of the investigation undertaken in this thesis, a formula published by Fanta (1982; 1983; 1986) has been adopted, as it incorporates the principles of a vegetative dynamic within an evolutionary framework. It is both succinct and comprehensive. The formula is: vegetation equals environmental factors as a function of time in its relation to stochastic elements, that is, **V = f (t,u)** (Fanta 1986). This equation serves as a theoretical basis for research into forest succession. The concepts behind the equation encompass the ecosystem theory and they are applicable to given forestry conditions translated into, and interpreted by, conventional forestry terminology. The equation expresses forest succession as a development in time (t) in its relation to stochastic elements (u), as related to forest site and forest stand mosaics.

An initial inquiry into the concepts lying behind the equation is relevant to this study as these concepts themselves have stimulated the investigation whilst providing the structural framework for the expression of the analysis.

The dynamic structure of vegetation

Natural forests are extremely complex, heterogenous societies (Florence pers.com). To examine the ecology of forest stands, a classification system which includes the range of possible variables, or characteristics, which typify forest species associations, must be established. Various systems for vegetation classification have been developed and applied to the flora of Cyprus. Holmboe (1905) devised a vertical classification for forest types observed on Cyprus, a system later adopted by Zohary (1973) and Held (1983), while Meikle (1977) based his descriptions on morphological zones. Other systems developed and applied by Cypriot foresters (Serephim 1957; Michelides 1983) relied on either natural variability factors, such as species associations, ecology, biogeography and geomorphological factors, edaphic factors, or artificial variables (Polycarpou 1956:67). The approach taken in this analysis of ecological change and human impact on the environment applies Fanta's model to determine forest dynamics, structure and potential regeneration sequences, utilising data obtained from field research and empirical observation.

The unique character of forests depends on their composition, structure and dynamics. A difficulty is encountered in evaluating the many interacting factors contributing to a forest site such as climate, soil, topography, aspect, biotic and historic circumstances. Taken together these factors constitute "site" **at a given time and place**. Site, and associated forest stands, therefore, must be assessed within the conceptual framework of "ecosystem", the site in its widest sense including vegetation, determined as a natural unit of biological evolution. In analysing the forest site, the ecosystem is treated as if consisting of individual parts, but it must be remembered that the parts are not separate, but aspects of an integrated whole.

Although this methodological approach describes the causes and processes observed in present forests, it may also be used to consider the patterns evident in the fossil or archaeological record. The fidelity of transition from the living assemblage to the fossil assemblage is uncertain. The predicability of the relationship between

Figure 1: A flow chart depicting the areas of research undertaken and presented.

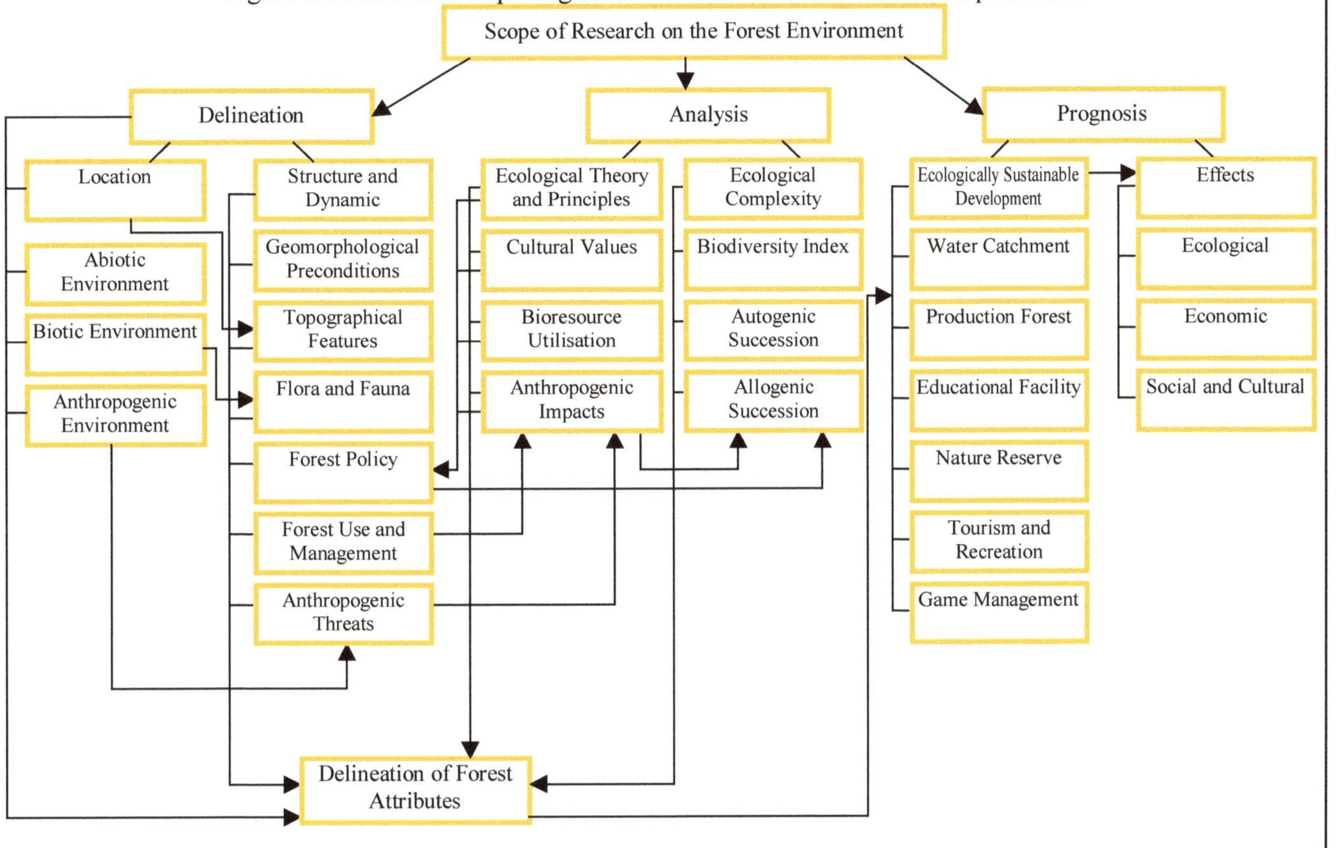

forest community composition and its external environment is bound to a temporal factor. Time influences specific ecologies and ecological processes as an element in the evolutionary process.

Factors Contributing to Forest Succession

The factors which influence the forest site can be numerous. For the purpose of this study four basic groups of factors were determined. These factors were (i) topographical and geomorphological, (ii) forest structure and dynamic, (iii) faunal associations, and (iv) anthropogenic impact. Again, each set of factors is considered separately but they represent aspects of an evolving ecosystem.

The ecosystem equation developed by Jenny (1941,1980) and formalised by Fanta (1986) makes provision for system properties such as vegetation, animals and soil which are dependent on, or are a function of, the physical factors such as climate, species pool, topography, parent geological material, and time. In this relationship, time plays a special role. The temporal factor is critical in understanding the ongoing and evolutionary character of vegetation, and it must be included as a primary element in function modelling.

Fanta used his equation to describe the development of trends in potentially natural or "climax" forest communities as related to specific sites. He believed that by using the model, communities related to each other could be analysed on one common successional line. The concepts behind the equation when applied to the climax theory require careful questioning.

The classical doctrine of the regional or climatic theory stems largely from the writings of F. E. Clements (1916,1936) and it is based on two key assumptions:

> (i) that species replacement during succession occurs because populations tend to modify the environment, making conditions less favourable to their own persistence and leading to progressive substitution; and
>
> (ii) that a terminal stabilised system, the climax, finally appears which is self-perpetuating, in harmony with the physical environment (Florence 1996:46).

The strong influence of "Clementsian" thinking ignores the contribution of the physical environment as a major element on vegetation patterning. Arguments against Clement's theory were led by Gleeson (1926) who recognised the influence of soil factors and hydrological conditions on ecosystem functions (Christensen 1989).

Foresters in particular continued to voice objection to the climax theory from the 1930's through to 1970 (Serander 1936; Graham 1941; Watt 1947; Drury 1956; Lutz 1956; Biswell 1961; Roew 1961). The impact of the climax theory, particularly on American scholarship, has been long lasting and is reflected in current literature outside environmental and forestry science. Research in landscape and environmental archaeology still applies the climax theory and the conceptual framework it endorses (Held 1993; Knapp, Held and Manning 1995). Critical studies in plant community development clearly demonstrated that climax succession rarely, if ever, developed (Elger 1954; Horn 1976,) and ecological research has widely repudiated the climax theory in favour of natural disturbance models (Loukes 1970; Dayton 1971; Hanes 1971; Heinselman 1971, 1973; Rowe and Scotter 1973; Habeck and Mutch 1973; Kilgore 1973; Loope and Gruell 1973, Biswell 1974; Levin and Payne 1974; Whitmore 1974, 1975; Bormann and Likens 1979; White 1979; Sprugel 1991; McNeely 1994).

Successional pathway theories may reflect the gradual emergence and dominance of species which were present, but were possibly inconspicuous prior to some form of disturbance. Alternatively, site occupancy by a particular species may restrict the subsequent entry of other species in direct contrast to the classical viewpoint that each species association acts altruistically to facilitate the entry of its successors (Florence 1994). Succession is not a development towards climax, but rather a change away from seral conditions (Roberts 1987).

Conwell and Slatyer (1977) and Noble and Slatyer (1981) have suggested three main successional pathway theories. A fourth theory presented by Watt (1941) and Florence (1996:47) makes a major contribution to the understanding of pathway successional vegetation models and appears, in the context of this thesis, to be the most appropriate. Described as "cyclic succession" or the "natural rotation of species", the concept was first developed by Watt (1941) who argued that in many situations (at a given time), changes in species composition may appear to be seral, but are in fact cyclic, that is, one species replaces another in a repetitive or cyclic way. Watt's example was of the "beech-birch" relationship in which short-lived, soil-improving species contribute to the amelioration of a forest site affected in some way through the occupancy of a long-lived species. Where the long-lived beech begins to degenerate, new beech may not establish within the environs the old growth trees. The soil- improving birch develops instead and, at this stage, may appear to be a successional replacement for beech. However, as the birch in turn begins to die out, the beech can regenerate and develop again to dominance. A similar process has been suggested for old growth redwood forests on some soils in the Pacific North West of the United States (Florence 1965b).

Ecological History as Human Experience

There is an increased expectation that archaeologists excavating in Cyprus will provide their sites with an environmental context together with an interpretation of short-term ecological events correlated to human history. Since the late 1960s and early 1970s the influence of the "ecosystem" concept on archaeology has become apparent, not only in defining cultural relationships to the development of complex agricultural systems, but also in determining resource distribution and land use patterning. The role of vegetation in metallurgical production in Cyprus has produced its own radical theories, many of which require careful questioning.

The components of the cultural landscape may provide limited environmental contextual evidence, particularly of the inter-relationships between early societies and their environment. Mechanised forestry and land improvement activities have led to widespread destruction of material evidence. Thus a potentially dangerous bias has been introduced into the evidence at hand, particularly in relationship to distributional data (Edwards 1989). Hence, there is increased pressure on systematic survey techniques and environmental archaeology to provide evidence of former cultural patterns within the localised forest environments.

The implementation of archaeological practices to on-site environmental data allows the recording of a set of specific modifications resulting from human exploitation in a general resource catchment of the site, rather than the preponderance of particular types of environments in the vicinity of the site (Harris and Thomas 1989). Archaeological sites, therefore, can be viewed as "palaeohabitat islands" where segments of the past ecosystem processes are preserved, or alternatively, as areas of intense human activity in which elements of past ecological systems were either artificially, and selectively, concentrated, or where new and unique habitats were created which some elements of local flora and fauna were able to exploit. The resulting pattern in the archaeological record may therefore present a bias, and it is for this reason that an understanding of the ecological dynamic in the wider, non-impacted environment, must be established as a "yard-stick" for the interpretation of on-site evidence.

People as an evolutionary component

The role of the human species as a key component of the evolutionary process needs to be assessed within the composition and structure of the ecosystem. An important aspect of ecological inter-relationships is the location of humans. People can be considered either as an identity located outside the natural structure of the ecosystem, or as a species embedded within some part of the ecosphere. Between the extreme positions of the Anthropocentric (Watson 1983) and Biocentric (Taylor 1983; Naess 1977) viewpoints, the human within the ecosystem requires recognition as a "key-stone" species (Erlich 1995) and one capable of environmental manipulation on a massive scale.

An assessment of environmental change brought about by anthropogenic impact on an island resource base and the resulting transformation of the society as it develops in response to environmental factors is frequently underestimated by regional archaeologists. Four elements need to be assessed in order to understand the complexities of a developing society within a defined location:

(i) Exploitation of the settlement catchment area
(ii) Catchment and range resource exploitation
(iii) Movement across the landscape involving natural corridors and specific site exploitation.
(iv) Energy or resource loss from the resource base through external control.

It is possible that the environment of Cyprus encountered by its first settlers dictated their mode of existence. The development of a cultural geography on Cyprus may be intrinsically linked to spatial relationships and the effect of, and dependence on, an immediate environment. It relies on a set of cultural factors which constitute a specific relationship to a defined region which is interpreted politically or analysed geographically through cultural associations. Political regions can also be delineated by variables relating to the actual spatial expression of political power, or in terms of the attitude of a population to a given political or religious ideology or issue. The dichotomy between the structural organisation of ideology over social process and the emphasis on evolving social systems over structuring must be assessed in relationship to their impact on Cypriot history. An ecological approach to political determinism can be made by linking people to their environment in a mutual relationship nexus (Patrick 1976:380). A three-staged analysis of the geo-political nexus can be seen as: (i) the extent to which geographical factors are considered in the determination of policy, (ii) the influence of geographical factors on policy implementation, and (iii) the influence which the expression of policy has on the environment (Prescott 1968:11).

Political group-earth relationships can take two forms, the relationship between a group, as a separate unit, and the landscape, and the relationship between inter-group "political contacts" and the environment. In the former situation the correlation between population and the natural resource base and its distribution within a demarcated boundary of culture is emphasised. In the latter situation the processes of co-operation or conflict between cultural groups are modified by the characteristics of intervening space across which interactions are performed. This type of assessment fits well with the current archaeological understanding of the Neolithic sites on Cyprus. Although analysis of the relationships between sites and the physical environment has yet to be undertaken, resource utilisation and cultural relationships within an environmental setting might be determined from a range of indicators presented in the archaeological record.

Time-related values and human perceptions

Philosophies of time are considered constructs of a given society. Subsequent changes in time concepts would, therefore, reflect evolving historical notions, from cycles which parallel nature to a linear progression of sequential events. Aristotle thought that motion was measured by time, but that time was likewise measured by motion. This is not to say that motion was first measured by time and that afterward motion measures time, but that they were devised for each other like relative concepts. Furthermore, movement and time were considered both perpetual, an absolute beginning or end to movement was inconceivable. Similarly, the idea of time before the beginning of time, and after its end, was seen as self-contradictory (Allan 1970:87).

The strict periodicity of the heavenly standard of time and the affiliated fixed cyclical nature of the seasons became the main reason for the belief (during Late Antiquity) in the eternity of the world . Periodicity on both a small and large scale, together with the seeming return of the identical on a cyclical basis, balanced the ancient conception of an everlasting universe. However, the debate already begun by the concepts of time, as expanded by Aristotle in *The Physics* and by his teacher, Plato, was to have a profound and lasting effect.

The asymmetry of the Aristotelian cosmos, that is, finiteness with regard to space and infinity with regard to time, lent additional weight to the classical analysis of time. The finite extension of becoming was seen to be embedded in the finite being of eternity (Sainsbury 1962:16). Aristotle also touched on the relationship between time and its experimental influence in the consciousness, identified by the ancient Greeks as the **yuch,** the *psyche* or soul. He thought that although in some contexts the body remained unaffected by change, there was an association between any movement that takes place in the mind with lapse in time (Aristotle: *The Physics* IV:11,218b21 trans.Hussey 1983:43).

Aside from the problem of time and consciousness, Aristotle challenged the perceived relationship between time and change:

"The following consideration might make one suspect that time either does not exist at all or scarcely and in some obscure way. Some part of it is past and no longer exists, and the other is future and does not yet exist; yet time - whether infinite or any time we can consider - is made up of these. One could hardly conceive something made up of non-existing things as having a share in reality (Aristotle, cited Sainsbury 1962:17).

Adding another dimension to the proposition that time is a human construct is Dummett's assertion that time is caused through involuntary action (Dummett 1964). Such a proposition provokes several immediate questions. The first, whether time exists before the event occurs that actualises it, and secondly, whether time is an event in itself. If the context of time is the event of its expression, in either a linguistic or mathematical form, it would become, by necessity, subjective. Time events are thus individual and personal. If time, however, only exists in the embodiment of an event, and the interval reality of that event does not necessarily correspond with any other, then it could be said that concepts in time are relative.

The western linguistic expression of time appears to have formalised the sequential form attributed to it. The use of tense in grammatical structure is analytical, consequential and progressive (Shallis 1982:15). Time arrows are directional and in the formulation of recognised structures they are symbolic of a shared concept.

Applying Dummett's argument to human management of natural elements, in this case forests, the relative time event pertaining to either a human concept of time or the time period for forest maturation, becomes irrelevant. Human time events are subjective and forest time events cannot be quantitatively measured within the human time scale. Therefore, the relationship between the two is an unknown, non-temporal, property. It is useful at this point to return to Aristotle:

> "...time, while it is resoluble into parts, some [parts] have been, some are to be, and none is. The now is not a part, for a part measures [the whole], and the whole must be composed of the parts, but time is not thought to be composed of nows. Again, it is not easy to see whether the now, which appears to be the boundary between past and future, remains always one and the same or is it different from time to time". (Aristotle: *The Physics* IV:10,127b32 trans. Hussey 1983:41)

Chrysippos appears to have first defined "now" to coincide with the present moment, a personal and relative event, and one which, in contradistinction to the moment that has passed and the moment to come, is coupled with the immediate awareness of reality (Sambursky.1959:103). Damascius, in an amplification of relational time says:

> "Just as the parts of separate things do not overlap because of space, the occurrence of the Trojan war does not become mixed up with the of the Peloponnesioan war because of time, nor in a man's life does the state of a new-born become mixed up with that of youth..." (Damascus, cited Sambursky 1962:14)

The idea of biological irreversibility (Shallis 1982:69) has frequently been used to maintain the notion that the human perception of time's flow is innate, and an aspect of one's biological conditioning. A geological equivalent of this mechanism would be the deposition of sedimentary rocks whose constitution and arrangement form a "memory" of some physical event which occurred at some time during the history of the planet. These changes, too, were irreversible and tend toward the maximum probability. The rocks were formed by entropy-increasing processes, and their subsequent erosion by weathering processes also increases entropy (Shallis 1982:69).

The application of time values to management encompasses much wider concepts. The subjective relationship of time from Dummett's argument distances the human objective from the properties pertaining to forests. The lack of a perceptible time link between two non-temporal objects thus defined means that their relative realities are distinct and separate and the unconscious projection of human time values onto non-human objects is therefore invalid.

The human approach to natural systems, including forests, has been historical and retrospective. By determining a definite starting point in time to identify succession, frequently only one element in a mosaic is chosen. The reconstruction of sequential events from a 'now' perspective introduces the difficulties of time projection from one dynamic system, the human, onto another, very different, dynamic system, the forest. The incompatibility between the different time values employed results in a hostile confrontation between different natural systems.

Economic rationalists have manipulated forest resources by insisting on the establishment of rapid-maturation species. "Now" time in forest management raises the difference in technical approach utilised by forest managers across the board. The regulated forest as an end to itself achieved in a minimum time frame and manipulated into an economic resource, has influenced to a large extent the policies adopted in early 20th century western national park management. As an ideal, the regulated forest reflects the economic strategies implicit in utilitarianism: "A forest enterprise should not only strive for internal capital efficiency ... but ... should be conscious of more attractive investment opportunities ... The appropriate rate of return is the alternate rate of return - the opportunity cost" (Thompson 1975, cited Beuter 1892).

Alternatively, forest rotational exploitation (that is, cyclical harvesting regimes) is frequently advocated as an economically viable form of forest management. The biological consequences of both eighty year and two-hundred year rotation were demonstrated by Boyce (1982). Although the eighty year rotation provided a larger sustained yield of sawn timber, and consequently a larger cash flow, the two-hundred year rotation provided old growth stands with the enhanced wildlife potential more typical of a "natural" forest. The economic benefits were, however, considered very "discouraging".

Time values within a forest can be expressed in another way. In 1992 the Cyprus Forestry Department felled a *Juniperus foetidissima* which was subsequently determined to be in excess of 1200 years old. If the occupation of Cyprus spans 9000 years then there would have been a natural rotation of seven for *Juniperus foetidissima.* Likewise, *Pinus nigra,* which has a maturation of nine hundred years would have experienced a natural rotation of ten, and *Pinus brutia* which lives 500, years a faster natural rotation of eighteen. Given a human life span of about 35 years in antiquity and rounding that to 40 years to compensate for periods of slightly longer life spans, a human rotation of 225 would have been experienced, about the same natural rotation as a *Cistus* species.

Human time perception and its application to the natural cycle of existence is often an inappropiate measure. Short, rapid maturation rates are frequently considered insignificant while long term maturation rates such as forest stands are seen as eternal. However, philosophies of time and duration need to be explored in order to understand the results of human environmental impacts.

Chapter Two: Philosophical Underpinning of Environmental Theories

The development of environmental studies, and its inclusion in the sciences, has been described in scientific terms as a "paradigm shift", a concept expounded by Thomas Kuhn in 1970. It has been suggested (Nix pers.com.) that environmental studies cannot be included within the scientific framework until an adequate philosophy has been formulated to underpin the range of investigations undertaken. This argument by Nix has wide-ranging implications. New forestry perspectives for ecosystem management, particularly in the United States, have sought to couch procedures within the philosophical framework of the paradigm. This has been the result of concern expressed by some that the intensive forest management model was not an appropriate vehicle to maintain ecological processes and values (Franklin 1989, Robertson 1991, Overbay 1992). The theory that enshrined ways of balancing forest commodity production, sustainable productivity, and multi-use potential, at scales that transcend the individual forest compartment stand was, therefore, termed a "paradigm". It called for a forest-stand structure objective with a landscape-level infrastructure that differed radically from the economically orientated timber harvest prescription previously in place (O'Hara *et al.* 1994).

The need perceived by ecologists, biological conservationists, and foresters to nest their discipline within the parameters of natural scientific understanding has several interesting repercussions. The arguments between extreme groups, those adhering to the principles of Ecologically Sustainable Development (ESD) (Likens 1992) on the one hand, and the opposing views of biological conservation (Soule 1994) on the other, appear to have generated, with respect to the relevant issues, a spiral of successive theories. Each new theory is in turn completely replaced by subsequent ones. However, **with each new theory the fundamental meanings of the language employed also change** (Feyerabend 1962:29). On this sort of point Kuhn (1970), operating with his novel analysis of the history of science, notes that an old scientific paradigm is occasionally displaced by a new "revolutionary" one. His principal example was the Newtonian concept of motion which differed in all aspects, including word meaning, from the Aristotelian model. The change involved a radical alteration in the way nature was viewed and the language in terms of which it was discussed - a conceptualisation so different that it could not merely be seen as constructed within the old framework by additions to knowledge or by the correction of former mistakes. Thus, the term "paradigm" appeared to provide the scientist with a new learning tool - a theory, methods and standards, usually in an inextricable mixture (Kuhn 1970).

Critics arguing against paradigm shifts claimed that they allowed scientists to create their own paradigms and then define objectivity and rationality in terms of that paradigm. In defence others have described the criticism as that levelled at any holistic, non-foundational theory of knowledge (Charlesworth 1982:38-39). Kuhn wrote: "That observation ... provides our first explicit indication of why the choice between competing paradigms regularly raises questions that cannot be resolved by the criteria of normal science ... Like the issue of competing standards, [it] can be answered only in terms of criteria that lie outside normal science altogether, and it is that recourse to external criteria that most obviously makes paradigm debates revolutionary" (Kuhn 1970). Possibly the most important point that emerges from the paradigm theory is the argument that science cannot be considered unitary and holistic and that there is not one single and distinct scientific method. Science must be seen as a human invention or construct (Charlesworth 1982).

In developing a philosophical approach to the body of this thesis Wittgenstein's concept of the visual room is utilised to provide a philosophical perspective, or window. The visual experience is independent of both physical and physiological factors for what occurs is a symbiosis of mind and matter as a personal understanding. As with the product of physical, chemical, or physiological processes, the result is the basis for all that can be sensibly said about such processes. The inorganic mystery of existence cannot be broached within the scientific framework. Therefore, even after exhaustive inventories and analysis about environmental relationships, there remains much that is imponderable (Thomas 1984).

When the natural world is approached through the understanding of personal experience, what is known may seem to lose some of its value. The reliability, and solidity of knowing or understanding may become subjective (Wittgenstein 1964). In accepting every new specification that language gives to objects the specification of the subjective world exists concurrently. Each new configuration of the objective sphere, whether spatial, temporal or numerical, produces a new picture of subjective reality and discloses new traits within the "inner" world (Cassirer 1955:250). In this situation a person can "... see nothing until we truly understand it" (Constable cited Hoskins 1976).

Seeing is contingent upon the self as "I", a fact that does not denote a particular body, for "I" cannot be substituted for a description of a body (Wittgenstein 1964:74). The fundamental characteristic of the "pure I" in contrast to all objects, is absolute unity (Cassirer 1955:262). The gender conflict currently surfacing in academic writing needs to be acknowledged, for whenever one talks of neutrality or asexuality, what may have been sought in a linguistic or grammatical form is negated. Asexual neutrality does not desexualise. On the contrary, the original positivity and power of essence is lost (Derrida 1991:387). The study itself, its title, form, research content and interpretation cannot be divorced from the personal. It has become an extension of the *psyche*, the soul or essence of being, as an intellectual activity which has given an integrity to this individual in search of self.

Therefore, rather than identify an alignment with an ecological cause, political idealism, or a new paradigm, the perspective sought through this research is a philosophical integration into the biological process, a personal journey through perception, conceptualisation, analysis, interpretation, and presentation. The meanings and connections between conceptualisation and the physical reality, are thus contained in the use of language, as a seed might be said to contain the tree (Wittgenstein 1964). In this sense whatever attributes pertain to the tree - its form or mechanics, colour, chemistry, its interaction with the environment - its entirety in time and space, becomes a reciprocal relationship, an embodiment of the tree, "for what I encounter is neither the soul of a tree, but the tree itself" (Buber 1970:59).

A theory of concept formation begins with the fact that information is received from the environment through perception. This immediately implies that there is an objective world of which humans are a part, and which is comprehensible in some fashion. A piece of information about the world is an element in the structure of the world which is encoded and transmitted in various ways within the world. The transmission of information about something that is not immediately present to the perceptual awareness requires some mediating device. Normally a word takes the form that designates the object, or objects like it. That is, an item which has no naturalistic meaning, like the sound "tree" is given meaning by being associated with a set of rules for application. Those who understand the rules will also understand anyone who makes the sound (or inscribes the symbol). Perceptual information of this nature is fundamental to scientific investigation and the concepts that arise in language are basically the product of the need to digest, encode and communicate information. One might in fact suggest that the philosophical window is not "thinking like a mountain" (Leopold 1949) but learning to be a tree (Burnet 1989).

Thus, Neitzsche conceived science as the activity through which nature could be turned into concepts for the purpose of mastering nature. The compulsion to logical correctness and empirical accuracy exemplifies the constraint of the interest in possible technical control over objectified natural processes and thereby, the compulsion of preserving existence (Habermas 1968:296).

Banuri and Marglin (1993) coined the term "systems of knowledge" in a plural form to signify the multiple ways of defining reality. They saw it as expressing a multiplicity of communities of knowledge as distinct from domains of knowledge. From this perspective, Third World indigenous and Westernised communities were not simply different political groups aiming to maximise their income or wealth, but they embodied different systems of knowledge, different ways of understanding, perceiving, experiencing and therefore defining reality, which included notions of one's relationship, not only to the social milieu, but also to the natural environment. Moreover, the expression was meant to refer not only to knowledge self-consciously explicated but also to knowledge which is implicit in all action. The "knowledge" part of the expression implicitly recognised that all forms of human activity must contain a cognitive content. The 'systems' component of the expression indicated human thought and action (or thought in action) that was not random.

Most descriptions of traditional knowledge systems emphasise the fact that they are embedded in the social, cultural and moral milieu of their particular community. In other words, actions or thoughts are perceived to have social, political and cosmological implications, rather than possessing a purely technical dimension. In contrast, the modern system of knowledge seeks to distinguish very clearly between these different dimensions. Technical questions pertain to cause-and-effect relations in the natural environment (Banuri & Marglin 1993:1-22).

The philosophical arguments called into play by various protagonists have sought to determine a shift from the Cartesian concept of the universe, one which was "essentially atomistic, devisable, isolated, static, non-relative and comprehensible by reduction" (Fox 1986:197) to one constantly in the process of replacement, a concept advocated by Deep Ecology (Naess 1977). Alternatively, on the other side of the debate there are those who assume a different philosophical stance, acting in the belief that there is an obligation to care for something, if and only if, it has intrinsic value. The objective assumed is an automatic limiting of the environmental qualities or intrinsic values, and following, there is an automatic limiting of the environmental framework. This stance may lead to an attempt to broaden the concept of intrinsic value rather than one of abandoning the concept altogether. In the extreme argument this philosophical viewpoint attempts to "articulate the feeling that one ought to care about beings/entities/'things in the world' regardless of whether or not they can matter to themselves" (Fox 1986:79).

Similarly, the Deep Ecology concept of unity in diversity is not new. It has its roots in the Pythagorean "Harmony of the Spheres", and the Hippocratic "sympathy of all things". The doctrine that everything in the universe is interconnected, either through mechanical causes or hidden affinities, has provided a base line for sympathetic magic, astrology and alchemy. It also runs as a *leit-motif* through the teachings of Taoism and Buddhism, as well as the neo- Platonists, and the philosophers of the early Renaissance. Pico (circa 1550) wrote: "Firstly there is the unity in things whereby each thing is at one with itself, consists of itself, and coheres with itself. Secondly, there is the unity whereby one creature is united with the others and all parts of the world constitute one world" (Pico 1557:40). Deep Ecology's call for a new metaphysics is said to be based on an alternative set of assumptions about the nature of reality, seeking something other than mechanisms, a concept of wholeness rather than atomistic units, process rather than a re-arrangement of parts, internal rather than external relationships, the non-linearity and unpredictability of fundamental change, and pluralism rather than reductionism (Merchant 1992). A contrary view of the Deep Ecologist movement was expressed by Gore (1992) as defining the relationship between human beings and the earth almost solely in physical terms as though humanity was nothing more than "bodies genetically programmed to fulfil a bubonic destiny, having neither free will or intellect" (Gore 1992:218).

The "Environmental Ethic" based on the writings of Naess, the co-founder of Deep Ecology, has been critically assessed by Thompson (1990) who failed to find a philosophical basis for an ethic in the premises proposed. Thompson argued that the environmental ethic fails to satisfy the requirements to which any ethical system must comply in order to be an ethic. It also failed to provide reason to suppose that the values promoted by an environmental ethic were intrinsic. Thompson's argument was that an ethic has two components. It says what states of affairs, things, and properties are intrinsically desirable or valuable (as opposed to what is valuable as a means to an end); and what action is required to promote, protect, or bring into existence that which is of intrinsic value.

According to Thompson (1990) here are three formal requirements which an ethic must satisfy:

> "1.The requirement of consistency which presupposes that the ethic in question has provided an account of the necessary differences and similarities that are relevant and why.
>
> 2.The requirement of non-vacuity must be clearly satisfied as an ethic sets out rules for what ought or ought not be done. If it is found that all things and states of affairs are equally valuable then there is no need for an ethic to bring about one state of affairs rather than another.
>
> 3. The decidability requirement through which the criteria of value offered by an ethic is such that it is possible to determine what counts as valuable and what does not" (ibid).

Callicott's (1992) interpretation of Leopold's land ethic is based on moral sentiments resting on the one single premise that the individual is a member of a community of inter-related parts. The ethic, therefore, aims to identify the boundaries and actual structural organisation of a co-operative community or society. In Callicott's understanding, "Ecological theory provides a synchronic link - the community conscience of social integration of human and non-human nature". The ultimate moral dictum of Leopold's land ethic was: "A thing is right when it tends to preserve the integrity, stability, and beauty of the biotic community" (Leopold 1949:262). Not only does the land ethic fail to satisfy the formal requirements of an ethic, it fails to consider the dynamic phenomenon of nature wherein everything is in a state of flux (Burnet 1993).

Taken at a species level each individual member has an integrity of its own which can be maintained under certain circumstances, but will be destroyed under others. At a community or system level, the defined community has a potential for change and development, and a dynamic which may be compatible with the destruction of particular populations - such as a shift in vegetational development after wildfire. So if ecological systems are entities with a good of their own, and the parts of an ecosystem also have a good of their own then all elements must be regarded as valuable.

In suggesting that environmental ethicists should simply declare that what is intrinsically valuable are living creatures, or wilderness, or ecological systems, a state of arbitrary evaluation will be implemented. If the degree of complexity for self contained ecosystems was determined as the criteria for intrinsic value the non-vacuity requirement might be circumvented. The scope and application of the value criteria must be defined and limited. When an environmental system is degraded it is assumed to be less valuable. The lower value placed on degraded environments may inhibit their ability to recover lost diversity through the cost of management or rehabilitation input. In cases where human intervention increases the diversity of a system without damaging its stability the action may be condemned by some ecocentrics who consider interference "unnatural". The emphasis placed on limiting human intervention toward preserving and protecting natural communities, suggests a hidden agenda. The value criteria depend on a covert reference to an anthropogenic viewpoint (Thompson 1990).

The major ethical principle facing an environmental philosopher is the intellectual response to a given situation. In order to respond one must be freed from conceptual fetters of either metaphysical or naturalistic interpretations of philosophy. The Tao or teaching as described by Buber (1910) emerges as a pragmatic case-by-case approach to any given problem. Each problem confronted is spontaneously, and sensitively, addressed and the continually changing patterns of inter-relationships are formed by the particulars of each given context.

A Taoist approach to environmental issues does not attempt to deduce what is correct from ethical or metaphysical first principles, which in their universality are abstractions divorced from a particular context. Neither is action reduced to utilitarianism, Kantianism, or any other single-criterion ethical system. Rather, it relies on bringing one's considered judgement into equilibrium with one's moral intuition, with one's personally and communally held beliefs. As a pragmatic, post-modern approach to environmental issues it allows one to by-pass the metaphysical dilemma. By refusing to countenance epistemic claims about the way nature is in and of itself, one is able to avoid the metaphysical bifurcation of the cosmos into human and nature (Peerenboom 1991).

The terminology and concepts introduced into the environmental debate tend to reflect the attitudes of educated, middle-class people who choose to exhibit a distaste for material production. When drawn into the conservation movement this group brought with them imagery of the 'beautiful'. The emotive language used and the contrasts depicted between polluted cities and pristine forests eventually became almost religious in its overtones. Such language when directed for political ends created a 'marketable commodity' built from non-use purposes and values (Watson 1990:95-99). So while a wide range of ideologies present paradoxical alternatives to the ecological debate, there appears to be no single unitary philosophy which can be identified, although some green extremists share a critical core of beliefs (Lewis 1992:42).

The omnipotence of complexity

In discussing natural selection Darwin wrote: "Let it be borne in mind how infinitely close-fitting are the mutual relations of all organic beings to each other and to their physical conditions of life" (Darwin 1859:80). Following the rise and acceptance of the theory of evolution little thought was given to the issue of the environment. The environment, in its past, present and future, had been an independent variable in natural selection, and failed to enter early twentieth century speculations, nor was investigation undertaken to gauge its subjection to natural law which was considered the most important factor in organic evolution (Henderson 1913:5).

Darwinian evolutions frequently are seen as a set of evolving biological systems rather than one running down. As systems they are moving from disorder to order; they are becoming more organised rather than disorganised. The direction of change over time is from simple to more complex life forms. Mechanical systems are closed systems isolated from the environment and their laws pertain to only a small part of the universe. In contrast, most biological and social systems are open, not closed. They exchange matter and energy with the environment (Marchant 1992). This seems a logical extension of Neitzsche's theory of nature.

In a contrary viewpoint the ecosystem consists of all of the organisms in an area and the physical environment with which they interact. An ecosystem is described by the quality of material and energy that is contained in each of its major compartments together with rates of flow of energy and materials among those compartments. The flow of energy is unidirectional, whereas the flow of materials may be cyclic (Ehrlich and Roughgarden 1987).

The Laws of Thermodynamics are said to govern ecosystems (ibid). According to the first law, energy is conserved. Energy is not created or destroyed, only changed in form. According to the Second Law of Thermodynamics, every spontaneous process that involves the transformation of energy from one form to another proceeds in a certain direction, as in an exothermic chemical reaction or the movement of material down a concentration gradient (ibid). Ecosystem description as an energy flow is inadequate as either a predictive or analytical tool. Energy conversion efficiency in an ecological or evolutionary context is not necessarily maximised. It cannot be assumed to be the most important ecological factor. Modelling energy flows in an ecosystem may describe certain aspects of that system, but it cannot predict how evolutionary change will occur. Analogies between non-equilibrium thermodynamics and ecological systems, for example, equating species diversity of communities with "negative entropy of chemical systems", must be discounted as untenable (Fenchel 1987:17).

One of the prominent features of modern ecology is its use of organisational hierarchies to model and explain the ways in which various levels of biotic organisation (for example, cells, organs, organisms, populations and communities) interact with one another and with the physical environment. In particular, ecology is largely concerned with populations and communities and their interactions with the abiotic environment (Odum 1983). Five levels of organisation have been defined to assist the description of ecological hierarchies: (i) population (a group of individuals of any one kind of organism); (ii) community (all of the populations occupying a given area); (iii) ecosystem (a community and its abiotic environment); (iv) biome (regional biosystems distinguished by major vegetational types or landscapes); (v) biosphere or ecosphere and all its organisms. This hierarchy represents an increase in complexity as one moves from populations to the ecosphere, and in turn raises questions concerning organisation of research into ecological structures (Steverson 1994).

The history of ecology as it has developed from a biological base, exhibits a general trend from classification of facts described by the class-concepts of ordinary language to a more definitive meaning in scientific language. Aristotle's zoological system and Theophrastus' botanical system exhibit high degrees of coherence and methodological order. Simon (1992) wrote: "If there are important systems in the world that are complex without being hierarchical, they may, to a considerable extent, escape our observation and our understanding. Analysis of their behaviour would involve such detailed knowledge and calculation of the interactions of their elementary parts that it would be beyond our capacities of memory or computation".

The most meaningful argument for the retention of hierarchal systems has been presented by Fenchel (1987) who said that the properties of ecosystems and communities must be explained in terms of the properties of the constituent species populations and their relationships to one another and to the environment. The properties of species populations, in turn, must be understood from the properties of individual organisms. Arguing against "holistic" philosophy Fenchel (1987:17) described the terminology employed as a "rationalisation of pseudo-explanations". Higher levels of organisation have properties which tend to be qualitatively different from the sum of the constituent components. Therefore, an approach which circumvents a true understanding of the phenomena may lead to false predictions (ibid).

In defence of the concept of natural complex at community or habitat level, I would argue that such a concept permits the identification of all discriminda generically, without prejudicing the pursuit or analysis of differences, of further similarities within the differences, and other differences within known similarities (Buchler 1966:2). If nature, in so far as it is distinct from the common factor of all "natures", is to be considered as a complex of complexes, then such a concept cannot be either justified or expressed at the outset. For it does not suggest a finished collectivity or an absolutely determinate whole of any kind, nor does it suggest immensity in the way the terms "universe, or "cosmos" do (ibid).

The concept of natural complex allows for generic identification and permits a range of distinctions and categorisations. It also encourages an investigation into the function of various component parts. There is, however, the implication inherent in the concept of natural complexity that the traits of any complex can never be regarded as fully ascertained or completely circumscribed. A complex, if it is accessible at all, is able to be analysed and interpreted without end, or alternatively, it is open to manipulation in an infinite number of orders (Buchler 1966:6). There is a tendency to perceive natural complexes as compounds or composites which are divisible into components of elements, that is, simples without parts. Thus the conception of complexes as clusters or bundles, or chains of traits gives rise to problems that can only be resolved by repudiating the orientation which underlies them.

One of these misconceptions is that in a "cluster of traits" the traits seem more real than the cluster, or in other words, the species is "more real" than the community. For it is supposed that: (a) a trait or species does not depend upon a given cluster for its being, whereas a cluster or community depends on given traits or individual species, or that (b) although a trait or species requires a cluster or community for its being, it is the same trait or species no matter what the cluster or community, whereas a cluster or community is not the same without the same traits or species components (Buchler 1966:12). If a natural complex, such as a tree, is considered, its constitutional traits cannot be thought of as all literally or geometrically enclosed within it, or attached to it. The tree's ability to photosynthesise (that is, its relationship to light and air) and its visual effect on the observer, are attributes equal to the colour and texture of its bark. Its maturation rate (or life span) is likewise an attribute. Together these attributes are designated either by diverse conditions that inhabit and promote its growth or abstractly by the formulation of principles. No adequate analysis of such an object could limit itself to the enumeration of components within a container or cluster.

All complexes are equally authentic as their natural parity (their ontological integrity) is the element which reveals differences between them and enables identification (Buchler 1966:33). The scope of a complex may reflect its inclusiveness. There is an advantage in perceiving any natural complex as including traits or sub-complex, or of an order of complexes as including a complex that is discriminable in another order as well. Thus, every complex is inclusive, regardless of the manner of its inclusiveness. Therefore, no natural complex can be atomistic. An atomistic theory of ultimate actualities is a type of metaphysical theory stressing the crucial role of components. Each individual element may perhaps be atomistic in the sense that there is an order to the nature of which it contributes. Its own integrity is dependent on this order within which it is located. But metaphysical atomism cannot survive in a theory of natural complexes since every natural complex includes and is included in a complex (Buchler 1966:51).

The dichotomy represented by the extremes of ecology as

a science, and ecology as a metaphysical exploration, is difficult to bridge. The complexity of the arguments are as intricate as the vital systems that each camp seeks to promote and protect. It is as well then to remember that ecology is, as expressed by Shepard (1969), "a subversive science - the basis of a social and scientific resistance movement".

Through the Gaia hypothesis (Lovelock 1979) a definition of the ecosystem was explored as: "a complex entity involving the Earth's biosphere, atmosphere, oceans, and soil: the totality constituting a feedback or cybernetic system which seeks an optimal physical and chemical environment for life" (Lovelock 1979:11). While recognising the human element amongst the many within the development of Gaia, the emphasis on contemporary ecology is seen by Lovelock to be inappropriate (Lovelock 1979:124). With reference to the Second Law of Thermodynamics Lovelock admits that the entropy of an open system must increase, as life is an open system death is the essential complement to the unceasing renewal of life.

The antithesis of organic life and matter is the answer to an evident scientific problem: that of the opposition between the increasing organisation that characterises life and the progressive disorder of a random nature which is the increase in entropy. If vital mechanisms obey the Second Law of Thermodynamics, and if their functioning is not an anti-randomness, the law may enable them to be independent of it. This dualism has been re-structured in the light of open systems (Piaget 1971).

Moreover, it is not recognised by most biologists that what is presently known about the behaviour of ecosystems is far too complex to be fitted into any man-made system, and thus is beyond simulation, except as a gross approximation within short time spans. As biologist Egler (1964) has observed, "not only is the ecosystem more complex than we think. It is more complex than we can think". What is referred to as "complexity" of ecosystems arises because in nature "everything is a part of something else" (Bearlin 1971:9). The maturity of a scientific discipline has recently been assessed through its ability to predicate. An experienced ecologist might reasonably be expected to predict for an area with a specific climate, geology and topography the kinds of plants, animals and micro-organisms present, even the numbers of species of plants and animals, the kinds of temporal change and the primary productivity. Therefore, while it is reasonable to say that a measure of the maturity of ecology is the ability to predict the result of new kinds of information, manipulation or changing conditions that have not been experienced before, a sound mechanistic understanding is necessary for success using this criterion. Behind the ecological level will lie successively physiological, chemical and physical levels of expressing mechanism (Grubb 1989).

In the personal quest for an understanding of what is perceived in the ecological matrix a return to the philosophical basis of the thesis is appropriate at this point. A question is raised by Wittgenstein: "does my visual image of this tree consist of parts? What constitutes its complexity?" The former question can only make sense if it is already established what kind of complexity is in question. To the philosophical question: "is the visual image of this tree composite, and what are its component parts?" the correct answer is: "That depends on what you understand by 'composite'". (And that is of course not an answer but a rejection of the question.) (Wittgenstein 1974:22e).

Problems within the Ecological Debate.

"If we want to undertake an investigation into the problems of truth and falsehood, of the agreement and disagreement of propositions with reality, of the nature of assertion, assumption, and question, our study is problematic due to our craving for generality" (ibid). Empirical generalisations made from the observation of ecological patterns of succession remain a descriptive approach to the discipline (Fenchel 1987:17). Generalisations in ecology are conditional and fail to represent an absolute truth. However, some may be sufficiently nearly true to be useful (Grubb 1988:). One impediment to an exact ecology, in a sense of completeness in understanding, is thus a lack of the mechanical basis for many generalisations. Even when attempting a description about the characteristics of pioneer plants whose nature depends on the substratum, a single, simple generalisation cannot be made. Pioneer plants on smooth rocks are generally lichens, and those on rougher rocks are generally mosses, while those on gravel and silt are generally herbaceous dicotyledons and grasses (Grubb 1986b; 1987). And to take one example at the level of manipulation, we may consider the effects of so-called "intermediate levels of disturbance" on species-richness in animal and plant communities. Again we cannot make a useful single generalisation. In communities with "niche control" of community function, intermediate disturbance decreases species-richness (as long as it is non-selective), but in communities with what is called "dominance control" it increases species richness. Ecologists are therefore confronted with "the ineluctably contingent nature of such rules and patterns as are to be found governing the organisation of communities" (May 1986b). One of the primary means of advancing toward a more exact ecology will be through the definition of a more complete array of generalisations that apply to specific sets of circumstances (Grubb 1988). Fenchel (1987:18) has argued that ecological systems can only be understood in terms of lower organisational levels.

Universals and Particulars.

A system of classification seeks for something in common to all the entities which can be commonly subsumed under a general term. An assumption is made here that there is an inclination to think that there must be something in common to all 'ecosystems', wherein "ecosystems" might form a family the members of which have a family likeness. The idea of a general concept being a common property of its particular instances connects with other primitive, simplistic structures. It is comparable to the idea that properties are ingredients of the things which have the properties (Wittgenstein 1964). Scientific knowledge depends on knowledge of the inter-relationships of specific characters. Each classification requires grouping by means of some character or group of characters. Knowledge of shareable qualities, properties, or relations constitutes the basis of scientific generalisation (Baylis 1953). Wittgenstein explains that a craving for generality has its main source in the preoccupation with scientific method. "I mean the method of reducing the explanation of natural phenomena to the smallest possible number of primitive natural laws; ... and of unifying the treatment of different topics by using a generalisation. Philosophers constantly see the method of science before their eyes, and are irresistibly tempted to ask and answer questions in the way science does. This tendency is the real source of metaphysics, and leads the philosopher into complete darkness. I want to say here that it can never be our job to reduce anything to anything, or to explain anything. Philosophy is "purely descriptive" (Wittgenstein 1964:17-18). The common characters of particular things when embodied or exemplified are called characteristics and resemblances between things are classed on the basis of these resemblances as a result of empirical observation. Abstract characters which are conceived but not exemplified are termed concepts. Concepts can be combined to form more complex concepts by simple conjecture. Pure universals are abstract characters which are neither exemplified nor conceived (Baylis 1953). There is a tendency rooted in our usual forms of expression, to think that once a term is learnt and understood, for example, the term "leaf", there is a conceptual expression induced. This may be a general picture of a leaf, as opposed to pictures of particular leaves, an idea which is imagined to be some kind of general image. We say that someone sees what is in common to all these leaves and this is true if we mean that someone on being asked can tell us certain properties or features which are common. But we are inclined to think that the general idea of a leaf is something like a visual image, but one which only contains what is common to all leaves. This again is connected with the idea that the meaning of a word is an image, or a thing correlated to the word. Again, the idea we have of what happens when we get hold of the general idea "leaf", "plant", etc., is connected with the confusion between mental state, meaning state of a hypothetical mental mechanism, and a mental state meaning a state of consciousness (Wittgenstein 1964:18).

Psychological and cultural associations.

The tree in the landscape in its form, dimension, structure, colour or age, creates a challenge within the intellect of the observer. What is perceived is not a single image but a complexity of psychological and cultural associations attributable to that tree, and to the viewer. In this, landscape is as much a human construct as a geographical one. It is a combination of physical reality and perceived reality, an action which creates a tableau in which soils, rivers, trees are in some way managed, manipulated or modified by changing patterns of human occupation and social attitudes. From a scientific perspective the tree in its natural landscape is seen as a biological transformer in the energy exchange between lithosphere and atmosphere, as part of the dynamic equilibrium of an interacting process. Whereas for a village farmer the tree may designate a field boundary or an inheritance. The same tree might be regarded by a green activist as evoking a reverence for nature. Each human perception has its individual bias and from the articulation of bias there is created the mythology associated with objects to explain hidden associations.

In order to approach the understanding of another ecosystem one must acknowledge that all aspects of a personal world view, structuring principles and values, in seeming to be natural, timeless and universal, have been formulated in some other society, some different place, and this will effect any interpretation of another society.

The idiosyncratic appreciation of locality emphasises individualistic attributes and these are in turn interpreted in spacio-temporal terms. Levels of appreciation for the layers and variations in landscape understanding adds a dimension not explicit in ordinary language. The aesthetic is sometimes seen as beyond science, as something binding an individual psyche to an ineffable and infinite creation. A road side sign near a mountain village in Cyprus is a striking example. It reads ΜΗΝ ΚΟΒΕΤΕ ΤΑ ΔΕΝΔΡΑ "Don't cut the trees". It represents an evaluation of a social and psychological evolution within the Cypriot community, something that rests in its aesthetic being at once both a physical entity and an abstract ideal. It is a symbolic extension of a cultural heritage in which trees signified wealth and stability.

Ecological aesthetics as a conceptual apparatus, stemming from an understanding of the unity of the integrated biodiversity, structure and multifaceted use of the forest resource, delineated by principles of management, has become a natural extension of the forest. The aesthetic quality is thus not confined to the visual but rather to the dynamic of the integrated whole. The maintenance of bio-diversity as the intended goal of regulated effort through management techniques, enhances the aesthetic value of the resource by enforcing the evolutionary processes contained within the various genetic pools. The task of conserving bio-diversity lies between extremes. It exists in accepting and retaining everything that genuinely pertains to the particular environment and by encouraging the dynamic structuring which continually transforms it as an internally cohesive creation. The use of environmental resources therefore, needs to be compared to the fulfilment of evolutionary processes, both in human and non-human terms. Expression of an ecological landscape is, therefore, both metaphysical and metaphorical. When simplistic ideology is applied to a landscape the imagery becomes a caricature.

Conservation mythologies

Mythologies are social histories explaining origins of natural phenomena whilst also acting as an expression of spiritual association with an experienced environment. Until recently a mythology about conservation, environment and nature was a presumption that it was highly structured, ordered and regulated steady-state ecological system. The "climax" theory was projected and applied universally with the assumption of uncompressed land degradation sequences occurring in the Mediterranean region. Change and the value of dynamics are now seen as intrinsic elements in spacio-temporal terms for the continuation of life forms.

Western culture has retained the Neitzschian commitment to an imperium over nature. Wilderness areas are "lock-ups". The otherness of wildland is objectified into human resource, or value categories and allocated by law to specific uses (thus bring law to the land). The US Wilderness Act (1964) states that "its purpose is to secure for all American people of present and future generations the benefits of an enduring resource of wilderness". Wilderness areas in the US are meant to serve four of the five multiple uses for public lands: watershed, wildlife, grazing and recreation. Only timber harvesting is excluded while mining is frequently permitted (Birch 1990).

Transformations between myth and myth, in this case between the North American myth of "Frontier" and "Wilderness", reveal a formal and deep structuring of relationships inherent in the culture. For instance, the natural message of myths is not carried in their surface, in natural language meanings or in morals, but rather in the meaning structures of synonymy, analogy, homology, opposition, and inversion between different but related myths. Levi-Strauss (1975) notes that the common content of sets of mythic transformation is often some paradoxical situation in social life that is in need of resolution. The continual repetition and living out of the frontier myth in various socially acceptable ways acts to resolve an otherwise intolerable state of contradiction and suffering. In this way the Green Movement acts as a myth which imposes order on a chaotic and unintelligible experience of post-industrialism. The mythology enshrined in the US Wilderness Act is an attempt to thwart destruction of an established order, and an increasing sense of chaos which endangers the equilibrium of the American ideal (Eliade 1952:38).

The relationship between myth and science, also seen as an ordering principle, is that both are extensions of cognitive concepts which parallel each other. All knowledge has the function of imposing coherence and order. Science does this with practical intent - it issues in causal regularities and aims at technical control. Conversely, myth making meets different intellectual requirements - it introduces symbolic order into the universe. It is not that myths express true experience, but that experience is lived in the context of mythologies which order experience and provide it with meaning (Arbib and Hesse 1986:211-212). Likewise, the notion that 'nature knows best' is meaningless in a world already re-made into anthropogenic contours. Pristine nature is non-existent throughout the world and has been for thousands of years. Only thorough-going human intervention can preserve remaining biological diversity until a more enlightened society can allow nature to reclaim its internal dynamic. Human impact has distorted the balance between ecological diversity in a wide range of forest types and the idea to "lock-up" these forests to conserve their naturalness fails to comprehend the necessity to redress the existing imbalance. Forests treated in this manner become stagnant and anachronistic, perhaps better seen as 'tree museums' rather than living forests. The unity of the forest, as reflected in the stand dynamics which characterise its development, must be acknowledged as that to which the intrinsic value of the forest pertains. The aesthetic value of the forest is, therefore, an appreciation of that which the forest is when the interplay between its related parts is in balance and ongoing.

From the period of Classical Antiquity the differential imperative, that is, the maximisation of species-specific differences, was achieved by humans transcending their animal and vegetative nature, and by domestication of a supposed wilderness (Plato *Timaeus* III:5 trans. J. Warrington 1965). The universe was considered to be under cosmic control which while requiring explanation, was outside human dominance. The doctrine of the mean was the final outcome of the philosophical debate wherein virtues were the mean between extremes.

In contrast to the Classical paradigm the Cartesian, developed from the philosophical understanding of Descartes, emphasised an atomistic analysis of discrete elements, mechanistic reduction and a separation of subject and object. The world became an artefact with nature as a resource to be mastered and manipulated by humans. Material growth, which favoured technological advances to further efficiency and productivity became a "natural" progression of industry with the ultimate result of specialised skills and technological knowledge being used as power. Under the Cartesian technocratic paradigm, science was seen as value-free and objective, with every problem addressed as soluble through technique and specialisation. People became evaluated in terms of their role, an extension of a mechanical dimension and in need of control (Dringson 1980).

The current dominant western world view was largely influenced by the Cartesian and Hegelian philosophies. In assumptions about the nature of human beings it is held that they are fundamentally different from all other creatures on the planet over which they have domination. Furthermore, people are considered masters of their own destiny capable of choosing their own goals and learning the skills necessary to achieve their end. The context of human society is within a vast world which provides unlimited opportunities for the human species.

An alternative to the dominant western world-view is termed the Human Exemptionalism paradigm. It emphasises the cultural heritage of the human species in addition to, and distinct from their genetic inheritance, which distinguishes them from all other animal species. Social causation is seen as determined by social and cultural factors including technology as major determinants of human affairs. The social and cultural environments are the crucial context for human affairs, and the biophysical environment is largely irrelevant. Moreover, culture is considered cumulative, and thus technological and social progress can continue indefinitely with the ultimate result that all social problems will be resolved (Calton and Dunlop 1980).

Between 1984 and 1987 the environmental attitudes of groups in the United States; the United Kingdom and West Germany marked a dramatic shift in ecological perception. In the New Ecological paradigm humans, while exhibiting exceptional characteristics such as culture and technology, remain merely one species among many interdependent in a global ecosystem. Extending this concept, it is seen that human affairs are not only influenced by social and cultural factors, but also by intricate linkages of causal effect and return, a pattern onto which the impact of human action may have profound and unintentional consequences. Within the New Environmental paradigm humans are seen as dependant parts of a finite biophysical environment which imposes potential physical and biological restraints on human affairs. The restraints thus imposed on human activities are perceived as being the extent of set ecological laws beyond which human intervention cannot extend.

An appreciation of the nature and pattern of environmental relationships is the starting point in determining complex forest mosaics. A description of the species patterning in response to natural succession and anthropogenic modification of ecosystem processes must be a structured procedure with specific objectives.

Chapter Three: Field Methodology

The Makheras Forest is comprised of 146 forest compartments covering an area of 4,054 hectares. The compartments were devised by Chapman (1953) and follow natural landform features. A comprehensive register and renumeration of standing stock was undertaken by Polycarpou (1954). This original work has provided the basic guidelines for forest policy and utilisation to date.

The pattern of the Holocene forest covering the area now defined as the Makheras Forest was that of a complex, uneven aged forest with tolerant and intolerant tree species competing for the available soil nutrient and moisture supply. The pillow lavas of the foothills carried a dominant canopy of *Pinus brutia* with a rich and diverse sub-strata of *Olea europea*, *Crataegus azarolus*, *Pistacia terebinthus* and *Quercus alnifolia* and *Quercus coccifera*. The *Pinus brutia* is light-demanding and intolerant of species competition, particularly co-dominant canopy cover, thus rendering it an excellent species for secondary forest re-establishment.

At higher altitudes the forest carried a cover of co-dominant species represented by the deciduous hardwoods, *Quercus lusitanica* (Royal oak), with *Juniperus oxycedrus* (Prickly Juniper) and *Juniperus excelcus*, two species now severely depleted through raised soil temperatures after the removal of sub-strata species. The *Cedrus brevifolia* may have grown on the diabase geological formation found at higher altitudes. Utilisation of the sub-strata for fuel wood and the warefare technique of wild fire are believed to be the major factors in the destruction of the former forest structure (Burnet 1993). Forest depletion was further compounded later by free range goat grazing.

Following the establishment of British colonial rule, the forests were annotated by Wild (1879) as a starting point for forest management. He described the Makheras Forest in the following terms: "Leaving Lythrodonda on the way to the Macheria Monastery somewhat older but still immature growth of *Pinus brutia* intermixed with *Quercus alnifolia* was encountered. From there higher and older growth was entered though it was scattered and open canopy of *Pinus brutia* of 20-80 years mixed on northerly and north-westerly aspects with *Quercus alnifolia*. This pattern continued up to and around Machera and Chawni Hill and further with no variation in either quality, age or composition ..."

Between 1880 and 1927 the area was planted to a monoculture of *Pinus brutia* at a spacing of ten feet. The current forest is composed of the remnants of this planting with a gradual natural recovery of the sub-strata species.

During the First and Second World Wars the forest was extensively harvested to meet the Allied war demand. A final heavy clear felling of standing stock occurred after the Turkish invasion of 1974 to meet the needs of economic reconstruction. The clear felling policy has since been abandoned and replaced by selection felling.

Wild fire has contributed to the current character of the Makheras Forest which has been burnt at regular intervals. In 1986, 13 compartments were destroyed by fire, the heat intensity of which has caused vitrification in some areas of soil (Swiny pers. com.).

The dominant species type is the *Pinus brutia* with *Quercus alnifolia*, *Arbutus andrachne*, *Quercus alnifolia*, *Quercus coccifera*, *Pistacia terebinthus*, *Acer obtusifolium* and *Crataegus azarolus* occurring in the sub-strata. Open areas carry *Cistus* as the dominant ground cover while the *Lavandula stoechas* is also widespread.

The current condition of the forest appears very poor. The forest is under stocked with wide canopy gaps and areas of sheet erosion. The increment potential of the standing trees is limited and very little natural regeneration of *Pinus brutia* was observed. The monoculture plantation of *Pinus brutia* is in decline and without immediate silvicultural treatment the forest will disappear when the current trees reach senescence.

An understanding of the complex nature of the forest reserve and its associated bio-resources has been compiled from observations recorded during a systematic survey of the Makheras Forest and the Panagia Bridge Division of the Adelphi Forest, Cyprus, over a period of seventeen months between September 1994 and February 1996. During that time, parallel north-south transects one kilometre apart were walked across the eastern Troodos Range. In each forest compartment crossed by the transect line a 30 metre square quadrat was established to record the forest's stand structure and dynamic[1], together with an evaluation of attributes pertaining to its current biodiversity (see Figure 2). In total 181 quadrats were established across diverse terrain (see Figure 3). All trees with a girth at breast height (GBH) greater than 56 cm were measured and recorded[2]. Sub-strata species were mapped and the position of canopy cover species was plotted. A list of the ground cover flora was compiled and faunal activity at macroscopic level was recorded. This

[1] Forest dynamic is an evaluation of the consequences of competition between components of the forest, and their relationship to each other over time (Florence 1992)

[2] This measurement of girth is in accordance with standard forestry practice

Figure 2: Map of Cyprus Showing Forest Boundaries

information, together with the anthropogenic evidence observed, forms the basis for the analysis presented.

The methods employed to determine the evolutionary ecology of the Cypriot afforested landscape depend on a number of factors. In part the purpose of the field research was to establish a contextual relationship between archaeological sites and their contemporary environment, and to define and describe the creation of a cultural landscape in relation to exploitation of the Cypriot forest resource. The features and characteristics of the vegetation were described within the parameters of these aims and objectives.

Methodology must be more than the establishment of significant data bases. Although data bases can sometimes be related, they can appear to be merely randomly connected depending on the nature of the data collected. The methodology, therefore, represents the cohesive pattern of the investigation's internal structure. It also reflects the complex inter-relationships within the ecosystem being studied. The methodology employed is not a theoretical study of the ecology as a mathematical model, for it encompasses both the physical and chemical environment which has influenced the evolution and evolutionary shifts of the species described. Moreover, it aims to discover the random fluctuations which have occurred in response to anthropogenic modification of the environment. Large and unprecedented perturbation imposed by human activity in the past has been considered more traumatic for complex ecosystems than for simple units (Balon 1993). A critical aspect of the information collated was to examine the former structural interdependence of the Cypriot forest ecosystem. The result of this analysis will gauge the forest's resilience to, or internal destruction by, anthropogenic impact. Arguments for the fragility of more complex ecosystems with rich biodiversity and structural variations must be seen as a mathematical generality (Soule 1987). The tenacity of endemic species observed during the survey, however, may be indicative of the stability of complex ecosystems which survive although much of the forest dynamic has been lost.

Environmental Setting.

The scale of the study has been determined by the size and nature of the afforested areas. In 1992, State designated forests of Cyprus covered an area of 175,420 hectares, that is, 18.96% of the island's total area. The areas covered by the five major forests are considered "natural" and occupy about 78,000 hectares, about 58% of the total afforested land (Christodoulou 1990:7). Four of the forests are situated on the steep and frequently rocky slopes of the Troodos massif. These are the Makheras forest, the Adelphi forest, the Troodos forest and Paphos forest.

The Makheras Forest, located on the easterly extreme of the Troodos massif, was, at the outset of the field season, considered the poorest in both structure and dynamic, with large areas appearing relatively depleted. The most

Figure 3: Map of the Makheras Forest Showing Transect Lines and Quadrats Established

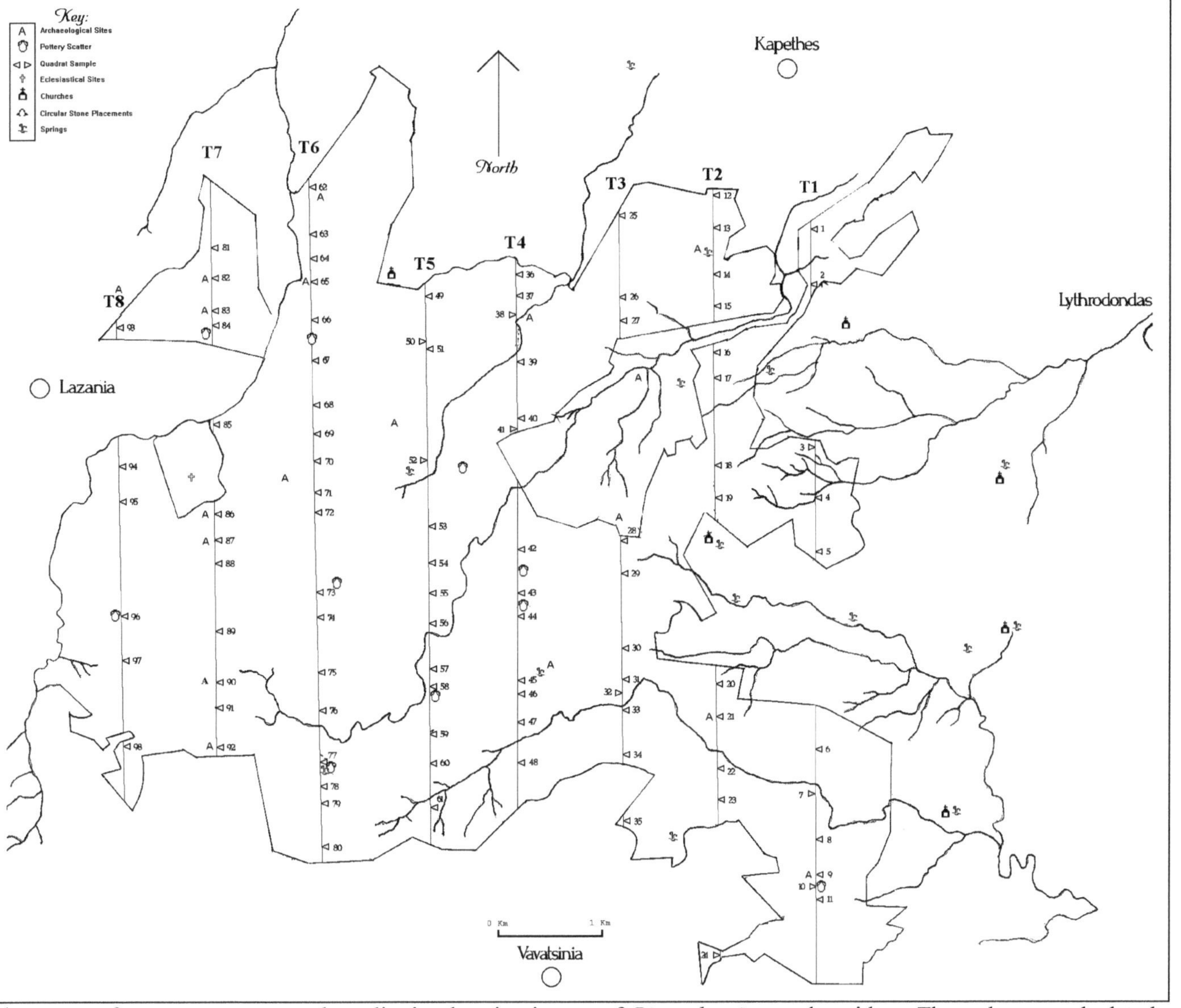

important forest areas appeared to lie in the riverine regions where soils are deep red or brown loams, derived from the pillow lavas. During a preliminary field survey of the forest in 1993 *Platanus orientalis* were observed climbing high into the range along river and gully lines. *Quercus alnifolia* in pure stands occurred on northerly slopes. Some dense stands of *Pinus brutia* were found on lower slopes with gentle inclines. A rich sub-strata of *Crataegus azarolus*, *Olea europea* and *Quercus alnifolia* were also noted. Ground cover of *Lavandula stoechas*, *Sarcopoterium spinosum* (thorny burnet) and *cistus* was in most cases continuous. At high altitudes *Acer obtusifolia* was most noticeable in recolonising the steep, inaccessible slopes. An immediate conclusion from the observation of increased riparian cover was that its recovery was a result of grazing exclusion and the preferred harvesting of non-hardwood species.

The adjacent Adelphi Forest, in comparison, seemed to be in a spectacular state of recovery with good examples of *Pinus brutia* stands evident. The only areas depleted are those on which clear felling was previously practised. The variety of species and the growth pattern of those encountered at high altitudes were considered indicative of a mixed species forest dominated currently by *Pinus brutia* in its optimum range. The Troodos forest located in the central, mountainous area of the massif, is greatly enhanced by the pure stands of *Pinus nigra* at high altitudes. The *Pinus nigra* stands represent the remnants of old growth forests defined as ecosystems distinguished by senescent trees and related structural features characteristic of the later stages of stand and successional development (Florence 1992). In comparison with North American, European or tropical rainforest old growth examples, the *Pinus nigra* forests do not exhibit many of the old growth characteristics such as multiple canopy layers, large accumulations of fallen timber or multiple species. However, they do function as old growth forests indicated by the attributes that are present (Everett *et.al* 1994). The Adelphi Forest was thought to be responding

Figure 4: Map of the Adelphi Forest Showing Transect Lines and Quadrats Established

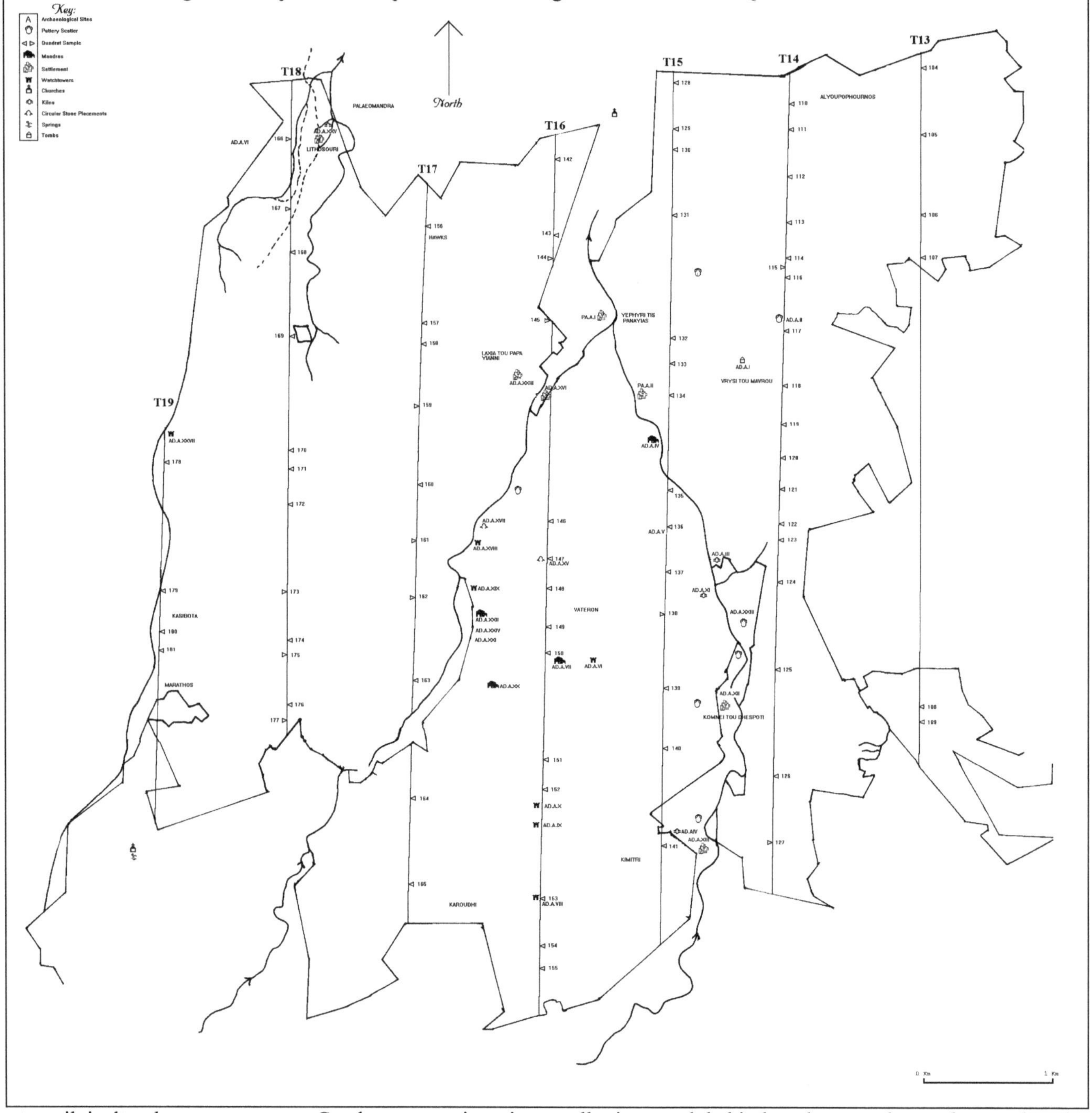

to silvicultural management. Good regeneration in uneven aged stands was noted during a preliminary investigation (Burnet 1993). The reality of the forest dynamic after seventeen months intensive field research has produced a very different appreciation of both forests.

The Paphos forest on the western extreme of the Troodos massif supplies the highest percentage of locally produced timber. The Cedar Valley, where pure stands of *Cedrus brevifolia* occur, supports good regeneration. A sub-strata of *Quercus alnifolia* grows in association with the *Cedrus brevifolia* though ground cover species appear sparse. *Cedrus* occurring on the north-east aspect appear taller in growth habit than those on the south-west though the latter seem more compact, denser and older in stand structure. In the *Pinus brutia-Cedrus brevifolia* association *Quercus alnifolia* understorey is more stunted and *Cistus* is more apparent. Even aged stands of *Pinus brutia* occur in the Paphos forest, largely as a result of post-fire regeneration. In these stands no understorey occurs though a light covering of ferns as ground cover is present. Trees appear in excellent condition and self thinning of the lower limbs occurs (Burnet 1993).

The Akamas forest occupies the central range which runs from north to south along the Akamas Peninsula, and the

Figure 5: Aerial Photograph of mid-sections T5 and T6

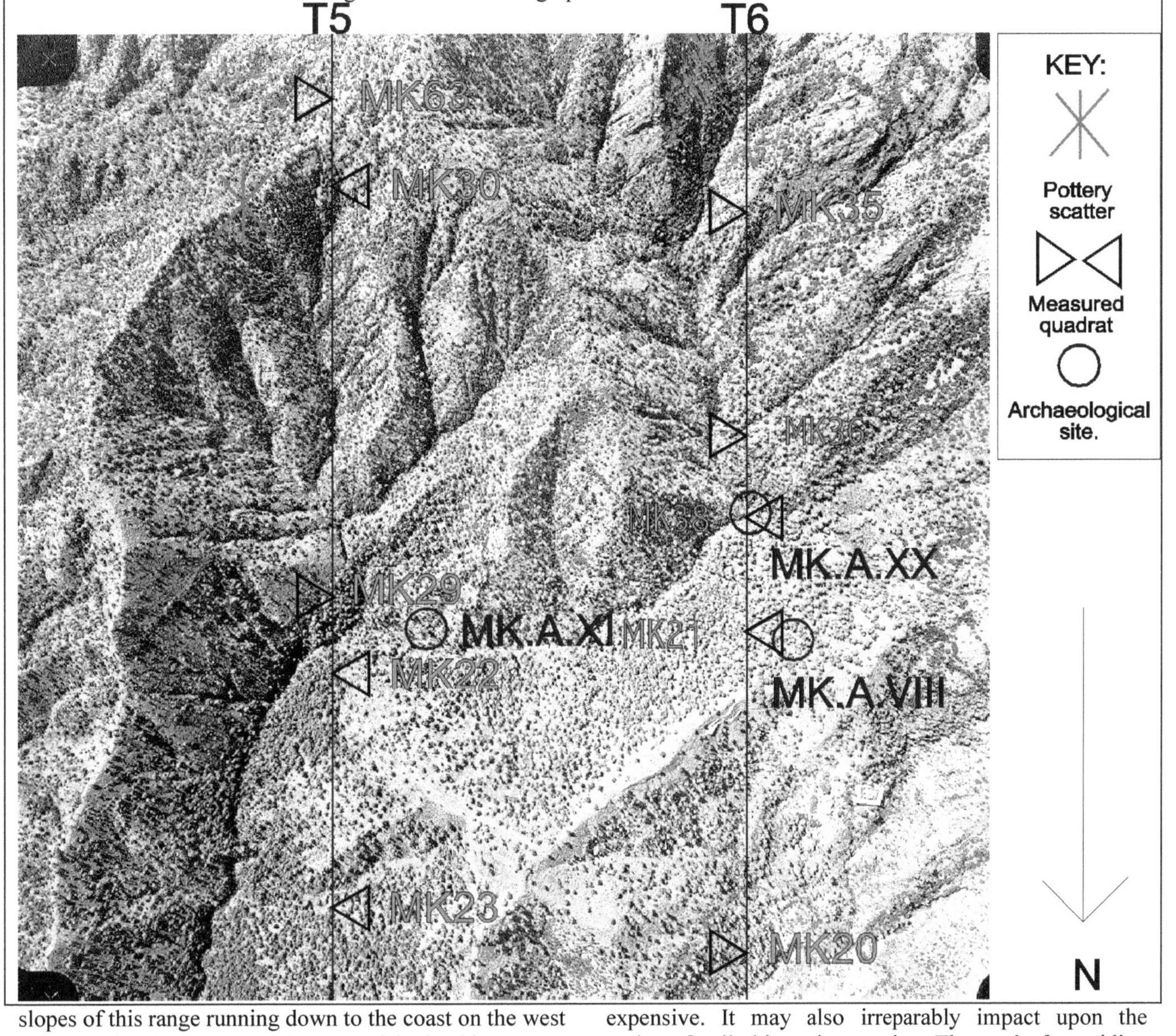

slopes of this range running down to the coast on the west of the peninsula. The general aspect of the land is western and it is deeply dissected by numerous ravines, gorges and gullies running westward. Along the coast there are flat coastal plateaux and similar plateaux, although smaller in extent, are found inland and along the central ridge. The forest lies in the area of Mamonia complex with a mixed geology of Diabase and Basal groups that form ragged outcrops, giving way abruptly to the gentle seaward slope of pillow lava. The vegetation is dominated by *Pinus brutia* and *Juniperus phoenicia* with *Ceratonis siliqua* and *Olea europea* occurring as scattered trees (Chapman.1934). The widest range of endemic species occurs within this forest region.

The logistical constraints incurred by the field research made it impossible to produce a census of every individual standing tree within the compartment community. Such a strategy, while being scientifically desirable, would be prohibitively time-consuming and expensive. It may also irreparably impact upon the various fragile biotas in question. The goal of providing descriptions of flora, fauna, forest stand composition and dynamic, floristic associations, human impact, and floristic indicators necessitated the establishment of a plot sampling technique (quadrats) on parallel transects which cross the Troodos massif at one kilometre intervals. These transect lines were drawn from the designation of 33° longitude which crosses the island from north to south in roughly the centre. The quadrats were laid out on a north-south axis and situated in each individual forest compartment lying along the transect path. Random site location does not guarantee that all the pertinent biological and environmental information at a site is captured most efficiently (Mackay 1991). The use of parallel transects established within forest compartments is essential in mapping vegetational change along environmental gradients. The transects, therefore, cover quadrats on northern and southern slopes where attributes such as slope angle, drainage and altitude combine, and across major changes in the geological parent material (Kent and Coker 1992:54). The time frame and terrain inhibited an east-west transect survey, although quadrats located on the southernmost sectors of each transect line have been recruited to provide an east-west perspective along the high range line.

Site history involving research into spatial patterning of plant populations, forest dynamics, and vegetation processes indicates the historical impact on the sample area through the current behaviour or location of species. The temporal dimension in ecological studies is a prime object of palaeoecology and must be defined at the outset. For the analysis of human impact on the Cypriot forest resource, a time frame of ten thousand years Before Present (BP) has been accepted as the maximum scale. The evolutionary questions concerning ecological responses to historical and pre-historical events are thus confined to the period of known island occupation by humans. A central tenet of ecology is that eruptions of ecologically dominant species reduce diversity and lead to decline in ecosystem health and integrity (Leopold 1949). Forest health is, therefore, described within the parameter of an ability to recover following disturbance, while ecological integrity is the maintenance of a co-evolved diversity of life form.

While acknowledging the contribution made to the understanding of the Cypriot environment (Holmboe 1905; Christodoulou 1959; Zohary 1973; Meikle 1977; Held 1983; 1987) an investigation into the Cypriot afforested landscape and an evaluation of anthropogenic modification which has shaped the environment since first settlement, cannot be based on singular attributes. Each forest stand has its own unique character reflecting its composition, structure, and dynamic.[3] In one forest stand several territorially-adjacent stands may be assessed as a compartment if the differences between them are not very marked, if they have the same average forest characteristics, and if the same forestry measurements can be carried out in them (Tyurin 1945). Difficulty in evaluating the many interacting factors, such as climate, soil, topography, aspect, biotic and historic circumstances is encountered, for together these factors constitute "site" at a given time and place. Site, and associated forest stands, must be assessed within the conceptual framework of "ecosystem", the site in its widest sense, including vegetation, determined as a natural unit of biological evolution.

The paucity of information on Mediterranean-type ecosystems has previously limited the extent to which hypotheses on ecosystem function were made, particularly in relation to biodiversity, and the acceptable limits of ecological change. Changes in functioning were linked to changes in ecosystem properties such as biomass or structure, where reductions affect ecosystem function. Biotic effects (e.g., reduced numbers of phenotypes and genotypes per species, species per genus, individuals per population, population by community, communities per region) could influence function *via* ecological effects (e.g. reduced variation in age structure and levels of sharing of resources between taxa, fewer species in pathways, and fewer pathways for resource transfer) (Richardson and Cowling 1993).

Ecological Methods and Applications.

Pertinent data relating to the observation and interpretation of the forest dynamic process requires the recording of forest stand composition and frequency of tree species, the characteristics of the understorey, and the way these vary with the local topography. The forest stand composition needs to be interpreted in terms of vegetational gradient and environmental relationships for the forest as a whole.

The requirements for forest regeneration must be determined. Assessment needs to include species type, site conditions, understorey, the presence of advanced growth, and the nature and intensity of site disturbance. The listing and mapping of all overstorey and understorey species on a $30m^2$ grid laid out along north-south parallel transects over the five southern forests will record the stocking rate on each plot, the basal area of the stand, the condition of the tree crowns, and the vertical structure of the stand. While an intensive description was attempted through the collection of specific data from each quadrat, the wider environmental setting within the forest compartment was assessed, as it is understood that distinctive variations may occur within a few metres. The biodiversity of each quadrat was, therefore, determined on an integrated basis.

The relative tolerance of tree species and the effect this had on ecological associations was considered. Past disturbances, whether natural or anthropogenic, must likewise be assessed, with particular note made of factors such as fire frequency and intensity. The tendency of fire behaviour in Cyprus is to advance uphill, using deep gullies as chimneys. The rate of fire spread is increased by factors such as forest cover, fuel moisture content, wind movement, topography, temperature and atmospheric humidity. The principal cause of fire is human activity (Peonides 1993).

Forest Compartment Description Criteria.

To facilitate the range of data relating to the factors (f) necessary to compile a statistical description of the forest's biodiversity, a form for forest compartment description was devised (see Appendix 2). The criteria

[3] A forest stand is a forestry term rather than an ecological term and is best described as a "parcel" of trees which can be evaluated as a unit.

description, a three paged inventory of typological data, was compiled at each individual quadrat plot. The importance of establishing the limits of data in terms of environmental precision and comparability was thus defined. In conjunction with the compartment description criteria, a log book was kept to note all fine variation over the forest structure as a whole. Each log book description indicates the same transect and quadrat number as the Forest Description Criteria form together with the date and hour of day (this was to ascertain light quality and visibility). Each description contained information on plant cover and ecological conditions. All species of plants taken as specimens for identification were included in the description with a special symbol or note. A question mark was placed against all species names, the accurate identification of which was uncertain, to be later confirmed. Observations of flora or fauna which fell directly outside the quadrat, but within its immediate area, were noted in the context of the compartment definition. A note of the map number, compartment map number and aerial photograph from which the quadrat was located were also recorded in the log book (Viktorov *et al.* 1964:123). After data collation and a quantitative analysis of the ecological information completed, the pattern of species, community, and environmental relationships was revealed.

In addition to the recording of data, specimens of flora and fauna were collected. These included herbarium material which has been identified by forest officers in Nicosia. A range of faunal material including birds' nests and fox castings was also collected. The faunal material has provided a wealth of information on habitat and ecosystem elements that were not otherwise readily identifiable.

Topographical and Geomorphological Recording.

Subjective assessments on attributes such as tree vigour, wood quality, tree condition, and potential response to release are pertinent to specific topographic sites within a defined region. *Pinus brutia* can attain heights of 45m on valley floors whilst on exposed ridges trees might only grow to 5m. Thus changes in topography and microtopography are important variables and a list of topographical situations is provided on the description criteria form.

The description of a particular area of forest must begin with its location. Accurate placement in latitude, longitude, and altitude was essential in being able to locate a map position. The Cyprus Forestry Department supplied one in five thousand cadestral maps of the forest areas studied. These maps were complimented by one in one hundred and twenty five thousand contour maps prepared by the Department of Land Survey. Aerial photographs from the 1961 series were also used when available.

Although access roads cross the forests at frequent intervals, in order to place the grids on parallel transects most of the field work was carried out on foot. In all, over 300 kilometres of rugged terrain were walked utilising forest roads, spur and ridgelines as well as abandoned ancient forest foot paths. The latter were overgrown and only visible to the practised eye. Field and logistical support for equipment cartage, the handling of a range of collected specimens and as a safety measure was provided by students and friends, predominantly Australian[4]. Reconnaissance of the region under study preceded the first stage of the transect survey and consultation with forestry officers was maintained at a local level to facilitate exchange of field experience, particularly local knowledge of the terrain. It was believed that the control over the quality of the fieldwork and the safety with which it was accomplished were enhanced as a result (Viktorov *et al.* 1964:144).

Vegetation patterns are often correlated with patterns in the underlying geological formations and, through this, the physical and chemical properties of the soil (Florence 1996:100). The geological formations of the Makheras and Adelphi Forests are basic or ultra basic and of igneous origin. The geological parent material in each sample plot was assessed and recorded. This information was then later crossed referenced with detailed geological maps. Unusual or intruding rock forms were sampled for specialist identification (see Appendix 3).

The soils derived from the igneous formation are frequently shallow, skeletal or otherwise immature, and are characteristic of ridges and steeper exposed slopes. They are mainly brown earths and vary in depth. Soils at high altitudes are slightly podsolised. Irrespective of aspect such distinctive soil conditions will have a significant effect on the composition and productivity of forest communities. Good physical soil conditions may enhance exploration by tree roots of a large soil volume, and compensate for marginal nutrient status for a particular species. Physical soil conditions and the microclimate may affect the activity of harmful soil organisms and hence the ultimate composition of the community (Florence 1996:101).

It is necessary to appreciate "soil fertility status" in terms of forest requirements and forest ecosystem processes.

[4]Field assistance over the 17 month field seasons was contributed by Dr. Michael Given (Lever Hulme Scholar 1994-95); Dr. Edvard Polak (Australian Bureau of Mineral Resources, retired); Dr. Stuart Swiny (Director, Cyprus American Archaeological Research Institute); Cameron Loundes (undergraduate student Wollongong University); Michael Kean (undergraduate student Macquarie University); Allasandra Swiny (undergraduate student Oberon College); Gabrielle Hart (undergraduate student Macquarie University) and Helenia Swiny .

There is frequently a marked distinction between forest fertility and agricultural requirements. For the forest system, soil fertility status must take into account not only the absolute amounts of available nutrients in accessible soil horizons, but also the wide spectrum of nutrients which may be important for complex associations, the volume of soil on which tree roots can draw, the "tightness" with which nutrients are held within the soil, and the rapidity with which nutrients can be recycled through the litter-soil-tree uptake pathway (Florence 1996:79-83).

Soil nutrient status may also play a critical role in plant successional processes. Soil depth and structure influence early tap root development of seedlings which in turn affects the survival of the plants, whilst soil nutrients are frequently related to composite gradients. The ability of *Pinus spp.* to dominate recently established sites, based largely on its rapid growth, light tolerance, and drought tolerance, swamps the impact of nutrients in determining the composition of immature stands (Christensen and Peet 1981). These patterns may be evident in observations of stand development. Increased competition along an age gradient may result in an increase in niche breadth and hence lead to greater species fidelity to site condition factors. A stand development sequence which moves from widespread *Pinus spp.* dominance in recently abandoned or severely disturbed sites to one of finely articulated patterns of mature, complex forest compositions governed by site nutrient and, to a lesser extent, by soil moisture status, may be enhanced by improved silvicultural methods.

Soil moisture is essential for seed germination and subsequent seedling survival. The amount of soil moisture available is influenced by a range of factors: amount and distribution of rainfall, water holding capacity of the soil, and water loss through evaporation and transpiration. Evidence suggests the critical importance of soil nutrient levels, soil physical properties, and soil water status in creating patterns within the forest. It is also possible that a nutrient and water interaction can have a profound influence on the composition of the forest within a community mosaic. The loss of biodiversity may threaten the continuation of many ecosystem functions such as water production from catchments, primary production and carbon recycling, regulation of climate, biochemical cycles, and the production of soil. Ecosystem functions at this level must, therefore, be seen as the set of processes that maintain systems through the transfer of matter, nutrients and energy (Richardson and Cowling 1993).

Approximately 90% of the permanent water resource of Cyprus is derived from forest areas or the immediate vicinity of the forest reserves. Clean water is a crucial resource. The hydrological patterning of streams leading into the catchment from any particular grid site must be fully understood, and the river drawing from the catchment identified. A water pattern is commonly recognised in drainage and river systems. The action associated with water catchment areas is usually the collection, concentration, or distribution of energy, nutrients, or other materials. The concentration of water from a dispersed state to progressively more concentrated central river results. The Troodos drainage is principally meridional and has been linked with the trend of the folding. On this view the drainage has adjusted itself to a structure which precedes many periods of denudation and one of sedimentation of the region. The valleys have been recently rejuvenated and streams are down cutting vigorously with waterfalls and rapids are common. Everywhere the valley longitudinal profiles are steep, many in a course of thirty kilometres falling over one thousand five hundred meters. The drainage is well-integrated and dissection minute (Leontiades 1963). Streams of adjacent drainage frequently abut one another near ridge lines. More commonly however, rivers from one drainage system are frequently juxtaposed with rivers from a different drainage. Protective forest corridor strips along these drainage lines may facilitate fauna movement and enhance gene flow between populations (Harris 1984:148).

The degree, concavity and convexity of a slope are important factors relating to drainage and microclimate. Steep slopes increase surface run-off and hence water loss, and contribute to soil erosion. This is more severe at higher altitudes as lower slopes are usually soil and water receiving situations and are subsequently more moist. Gentle slopes tend to retain more moisture and are less open to soil erosion. The slope incline also affects the quantity of heat received and is reflected in soil temperature. The gradient was assessed to gauge the degree of slope encountered.

Forest Stand Assessment.

The range of forest types is numerous and to a large degree reflects varied silvicultural practices, amongst the most important of which is artificial regeneration, which has become an integral aspect of forestry management in Cyprus. The result of anthropogenic and environmental factors and their impact on the ecosystem is such that the forest in many parts is unable to reproduce naturally. Reliance on human input in order to rehabilitate and repair the structural diversity, and biodiversity lost or endangered, continues the established association of cultural elements within the ecosystem. On the other hand, areas where natural regeneration occurs are important. The recording and mapping of abandoned, formerly cultivated areas re-established by the invasion of natural forest species is another important objective in understanding natural regeneration processes. Three site classes were determined by Chapman (1956) to facilitate forest management and harvesting regimes. Site

"A" represented the most favourable combination of all site factors, with the influence of soil moisture predominating. Apart from abundant soil moisture, soils were usually deep, stable and little effected by erosion with gentle slopes, or on valley floors. In this situation other site factors are of less importance. In these localities *Pinus brutia* reaches its maximum growth. The riverine hardwood species are confined to site "A" situations.

Site "B" influences the greatest part of the forest cover for it represents all sites intermediate between the very best and the very worst. Its general characteristics are slopes which are frequently steep, soil surfaces more or less unstable, soil depth medium (average 45 cm), and aspects and altitudes various. Two broad sub-divisions are discernible, and have been designated B+ and B-. Aspect plays the most important role in distinguishing between the two; unfavourable aspects, that is SE to SSW, giving a very marked bias to the B- sub-type. The most marked difference between the two sub-types is their capacity for natural regeneration. B+ sites usually carry far more seedlings than the B- type on which regeneration is either absent or slow in establishing due to hotter and drier conditions, which render B- sites more exposed and sparsely covered.

Type "C" are the poorest sites with soil again the predominant factor. Shallow, dry, loose soils are characteristic, rock outcrops frequent, and slopes steep to precipitous. Chapman (1934) noted that hill or ridge slope vegetation varies in response to any change in aspect, being generally denser on favourable aspects.

Aspect, therefore, is the major influence in determing the microclimate of specific sites. Southerly aspects receive direct insolation and are considered hotter and drier than northern aspects, although northerly aspects are exposed to the desiccating effect of the warm, dry winds which tend to be hotter at a higher altitude than those a little lower. Regeneration appears to be favoured on aspects facing from the south-west to north-west (Chapman 1936).

Seed bed condition is again a critical variable. Seed falling onto disturbed surfaces facilitates germination, while seed falling onto an ash layer results in regeneration failure due to the inability of radicals to penetrate ash-covered soil. Ash covered soils compact into a hard pan which inhibits moisture absorption and this combined with the higher temperature of burnt areas greatly reduces the germination potential.

Understorey vegetation can either favour natural regeneration by providing shade during the early growing period, or retard it by inducing weak and deformed seeding stems. Tall dense grasses are serious competitors, while short, thin grasses present almost no obstacle except on poor, shallow soils.

The accumulation of fallen pine needles, cones, leaves and branches acts to inhibit soil moisture loss and can stabilise slopes. However, accumulated ground litter also influences the intensity of wild fire. The depth of ground litter must, therefore, be accurately measured.

Stand composition for the major canopy dominant species has been calculated at 32 various stand types (Burnet 1993), either of pure conifer, pure hardwoods, mixed conifer or mixed conifer and hardwoods. Frequently the composition is determined by altitude though aspect and geological parent material are relative factors.

Dominant age considers the range of components or growth stages present in the stand and the silvicultural condition of each component. The normal development sequence for *Pinus brutia* is sapling stage, pole stage, mature and old mature or senescent trees. An uneven-aged forest or stand is one in which all stages of development are represented.

Stand structure patterns the forest mosaic. It is the spatial relationship of individual trees or groups of trees to each other, both in a vertical and horizontal plane. Structure is the consequence of competition between plants within the ecological community. Structure can be simple - a single tree canopy dominating a forest floor covered by only litter, or a minimal cover of grasses, ferns and small shrubs. A complex structure is one in which tolerant and intolerant species compete.

Crown classification is an understanding of the relative position and condition of tree crowns which indicates the struggle for existence within the forest stand. Position in the canopy provides an objective and readily accessible means of classifying trees in a stand in terms of their likely contribution to continuing stand development. Crowns can be segregated into canopy classes in even aged stands as : dominant, codominant, intermediate, overtopped or suppressed. The crown cover and position was mapped for each quadrat.

Diameter growth rates of trees in even-aged stands will vary with species, site, stand age, rainfall and dominance class. The development of an even-aged mixture of species will depend on the inherent growth characteristics of the species, their frequency, and their adaptation to the particular sites on which they are growing. The characteristic difference between species, for example *Pinus brutia* and *Pinus nigra*, are important factors when measuring the basal area of a stand.

The density of the canopy cover influences regeneration in various ways. It determines the quality of light penetrating to the forests. It affects soil temperature and available moisture, which in turn influence the decomposition of forest floor organic litter. Closed canopies intercept a considerable amount of rain, particularly of the light shower type, and may prevent it reaching the forest floor. Alternatively, it prevents insolation and reduces surface evaporation.

Five sub-strata vegetational groups further characterise the forest stands:

Quercus alnifolia association is mainly found on steep slopes and screes, generally in sheltered locations. It appears to be more prevalent on aspects facing north. *Arbutus andrachne* is the most common associate species. It forms a dense canopy and ground vegetation is usually absent.

Olea europea associations with the main species the wild olive and associated species *Certonia siliqua* and *Pistacia terebinthus* are found in the lower to middle reaches of the Makheras and Adelphi forests.

Quercus coccifera associations are similar to the *Olea europea* association with the exception that *Quercus coccifera* is frequent and *Olea europea* is rare or absent. *Crataegus spp.* are found in both associations.

Pteris acquilinium associations occupy the moister valley floors and slopes where the water table is high in the soil. *Pteris* is the main ground cover vegetation species and its growth is very dense precluding other species apart from mosses. It indicates deep, moist soil conditions.

Above 800m the *Cistus villosus* associations are most common. The association is dominated by *Cistus villosus*, some *Cistus monspeliensis*, and in certain locations *Cistus landifera* occurs.

Forest Injuries.

The sclerophyll forests which dominate the Cypriot landscape are highly flammable, a characteristic which is compounded during summer by prolonged hot, dry conditions. Carbon deposits within the soil horizon can be utilised to identify periods of wild fire. Carbon sinks in all forest ecosystems are associated with sequences in growing vegetation, soils, and peatlands. Sources of carbon in forest biomes are primarily associated with forest fires and anthropogenic disturbances. In many instances the development and functioning of forest ecosystems are closely connected with forest fires. Forest fires and logging activities are the main driving forces that create forest age structure, a critical factor which influences the rate at which carbon accumulates in forests. Forest fires and wood harvesting influence the carbon flux between forests and the atmosphere. Forest fires cause an immediate carbon efflux to the atmosphere from burned biomass, forest floor litter, and soil organic material. In the post-fire period, increased soil respiration and decomposition of killed (but not burned) trees may also result in a carbon efflux to the atmosphere. The carbon loss or gain in an ecosystem depends on the intensity and location of forest fires and the type of site. An intense fire may deplete the volatile nutrients in the soil and inhibit vegetation growth in the post-fire period. Even a long time after an intense fire, the soil profile may contain 50% less carbon than an undisturbed plot (Kolchugina and Vinson 1993).

Goat grazing was determined detrimental to the forests due to direct and indirect influence, habits and browsing preferences. Studies indicate the progressive loss of biodiversity in the forest composition due to the preference displayed by goats for broadleaved species and junipers. Riverine species were particularly depleted prior to the implementation of restricted grazing (Goat Laws) during the 1930's (Wild 1879; Unwin 1929; Chapman 1938; Serephim 1961; Wertime 1978; Thirgood 1981, 1987). Illegal grazing still occurs with sapling trees, particularly *Pistacia terebinthus* regeneration being severely affected.

Resin tapping has ceased in the forest but many senescent trees bear scars as witness to its former economic importance. Resin tapping weakens the tree over a period of years and exposes the trunk to greater than normal risk during periods of wild fire. Long periods of tapping result in bole weakness and early decline.

There is a close relationship between forest cover and soil protection. The major forests are mostly situated on steep to precipitous land, much of it exposed outcropping rock or very shallow soil, poor in nutrients. The shallow soil layer only remains *in situ* while the vegetation cover is intact. Extensive erosion and gullying occurs after wild fires which destroy the canopy protection.

The rotation of harvest in a production forest, the type of felling sequence exercised and the date of the last felling are of importance in ascertaining recovery and regeneration potential. This information was obtained from studying the cutting scars observed of tree stumps *in situ*, and from an examination of Forestry Department working files.

Forest Ecology.

Ecological succession was described by Polycarpou (1953) as occuring in four broad stages. The periods involved are roughly 5-10 years for grasses and other herbs, followed by shrub species for about 30 years before *Pinus brutia* begins to recolonise a depleted area. Scattered pine seedlings may appear earlier if there is a seed source, but 35-40 years are required before re-establishment occurs, particularly after the impact of a hot wild fire. Polycarpou's assessment of ecological succession generalises many factors and does not represent an accurate overall picture of succession within the forest mosaic.

Factors which may inhibit the development of a forest stand, or enhance its potential, are relative to site disturbance. Cast-strip planting for reafforestation, habitation, either current or abandoned, aqueducts or water pipe lines, terracing for former agricultural purposes, roads, mines, or high tension lines contribute to site disturbance.

The biodiversity of the forest, and the role of wildlife habitat in maintaining the complex characteristics of the forest, requires sensitive monitoring. Diversity consists of two measurements: variety, and relative abundance of species. No community consists of species of equal abundance, and it is normally the case that the majority of species are rare while a number are moderately common with the remaining few being quite abundant. In species-poor communities, analysis is usually made through mathematical modelling using a geometric series (Magurran 1988).

Biological diversity may be segregated into three components: species richness, rarity, and vulnerability. Species richness is a simple measure of species diversity, and is calculated as the total number of species in a habitat or a community. In coniferous forests the nutrient and moisture level in the soil and the rate of habitat diversity, seem to be the major factors governing diversity patterns. Habitat diversity reflects the number and abundance of different kinds of forest stands within an area, and may be used to evaluate alternatives with respect to species richness. The stand composition and structure, with reference to senescent or stag trees, is an important variable in habitat determination. Silvicultural techniques that create variable stand structures, depending on the intensity and type of cutting regime employed can also enhance habitat potential. Generally slight, small-scale cuttings increase diversity by creating additional successional variation, particularly if, at the same time, a sufficient area of old growth forest remains intact. Rarity is a characteristic that depends on the interaction of local (or global) abundance, habitat specificity, and geographical distribution. Rarity is evaluated in this study by defining the abundance of habitats specific to locally rare plants and animals. Vulnerability is defined as the number of endangered species which are known to persist in the research area (Kangas and Kuusipalo 1993).

Reference to each specific compartment by former officers and managers was essential in understanding the development of the stand dynamic and the subsequent changes which have occurred over time. Comprehensive files on each compartment have been compiled and a range of archival material in the form of files and internal departmental memos were available for research purposes. This material was collated in sequence to follow the transect gradient lines and quadrant plots, providing valuable historical information on the development of the forest mosaic since the implementation of silvicultural practices.

All these elements, when combined, constitute a study of the forest aesthetic. They are the criteria through which the dynamic structure of the forest environment can be ascertained in a holistic and integrated manner. They record the ecological evolution of particular forests. An important addition to the material was the establishment of a flora data base which combined species identified from ground cover within the plot areas, pollen taxa and macrobotanical remains.

Plants and animals represented at any particular site provided a range of information relating to the site's history. The species of plant life observed and the overall density of growth may reflect a history of disturbance or integrity. The population structure of selected species and the growth history of individual specimens may record natural or anthropogenic impact. By listing species which are naturally confined to primary forests and those which have a progressively weaker association with undisturbed habitats, it was possible to establish a data base that predicted with accuracy the degree of disturbance a site has experienced. It was believed that ancient, undisturbed forest was generally richer in species than secondary woodland, and contained a more diverse assemblage of vascular plants than secondary forests. More significantly, the richness in non-impacted forestry situations was not severely reduced by fragmentation of the habitat, thus rendering them identifiable on the basis of their flora.

Chapter Four: Existing Forest Products.

Forest resource utilisation is usually seen as being determined by the harvesting of standing trees. A more perceptive understanding of the wide ranging resource base pertaining to a forest can be ascertained when examining the forest as an ecosystem. The productivity of the forest reserve must be ascertained not only in terms based on subsistence economies or economic return, but also in the assessment of non-wood values and objectives. The forest represents a multi-purpose resource encompassing a wide range of community values.

Table 1 clearly demonstrates the importance of sub-strata species, particularly the *Olea europea*, *Arbutus andrachne*, *Crataegus azarolus* and *Pistacia terebinthus*, in providing a range of forest products utilised by the people of Cyprus.

Timber Resources.

Tree classes may be based on one or more artificial variables, including diameter and height, age, growth stage, position within the canopy, bole and wood quality, crown condition and vigour, the degree of competition to which a tree is exposed, and the likely response of a suppressed tree if given a greater growing area (Florence 1996:185).

The forest mosaic is patterned by structure, the consequence of competition between plants within the ecological community. Structure may be simple or complex, a simple structure being one of a single tree canopy dominating a forest floor covered only by litter or a minimal cover of grass, ferns and small shrubs. Stands composed of intolerant species, such as *Pinus brutia*, often have a simple structure though in time other more tolerant species may invade and a more complex structure will be established (Florence 1996:193).

Dominant Tree Species Distribution.

Monoculture stands within the forest mosaic are termed "pure" by most foresters. Pure conifer stands can be segregated into species groups that, with the exception of the *Pinus brutia* plantations, tend to be site-specific.

Pinus halepensis* subsp. *brutia Elwes & Henry (*P. brutia* Tenore). The Calabrian pine (sometimes referred to by American writers as the Aleppo pine) of Cyprus is the most common of the conifers. It was considered by early twentieth century forest administrators to be the "climatic climax" species of the pine association. It has a wide range in altitude and soil type, from about 1,600 metres to sea level, and occurs on both plutonic and sedimentary formations. Optimum development appears to be on the slightly acid soils of the Troodos massif, derived from andesitic rocks, and between 650 and 1,300 metres where the rainfall varies between 400 and 900 mm. On alkaline soils, however, there is a marked decline falling off in maximum height and in form. Seed is produced abundantly every year and is shed from mid-August to mid-September, though sometimes cones remain closed through winter and only shed their seed on the approach of hot weather in the spring (Chapman 1967:27-28).

Figure 6: *Pinus brutia*

Table 1: Bioresource Utilisation and Species Presence.

Key: M=medical; D=dyes; P1=perfumes; R=resins and gums; P2=preservatives; T=timber; E=roots; P3=poisons; W=wines and spirits; P4=pitch and resin; E2=edible leaves and shoots; F=fuel; T=tannin; G=gum and glue; D=dunnage; A=aromatic oil; B=browse; F2=fruit (Christodoulou 1957, Forbes 1965, McLaren 1983, Hansen 1989, Manniche 1989, Haldane 1990, Demetriou pers.com.)

Taxa	M	D	P1	R	P2	T1	E1	P2	W	P4	E2	F1	T	G	D	A	B	F2
Acer obtrusifolium		1				1						1					1	
Achillea sp.	1																	
Allium ampeloprasum	1						1											
Allium trifoliatum	1						1											
Arbutus andrachne	1					1						1						1
Arisarum vulgare	1							1										
Asparagus acutifolius	1										1							
Asphodelus aestivus	1	1					1											
Astralagus lusitanicus spp.orientalis																	1	
Biscutella didyma																	1	
Capparis spinosa var. canescens	1										1							1
Carlina involucrata spp. cyprica																	1	
Cedrus brevifolia			1	1		1										1		
Cistus salviifolius		1	1	1														
Cistus creticus var.creticus		1	1													1		
Clematis cirrhosa	1	1		1												1		
Crataegus azarolus	1					1						1						1
Cupressus sempervirens	1	1		1		1												
Cyclamen cyprium	1																	
Equisetum ramosissimum	1																1	
Fumana arabica	1																	
Galium aparine	1	1																
Inula viscosa	1																	
Juniperus oxycedrus	1			1		1											1	
Laurus nobilis	1		1		1													1
Lavandula stoechas	1		1															
Mentha longifolia	1																	
Myrtus communis	1	1	1	1														
Olea europa	1		1			1						1				1	1	1
Origanum majorana var. temuifolium	1																	
Papaver setigerum	1																	
Pinus brutia				1		1				1								
Pistacia lentiscus	1			1	1													
Pistacia terebinthus	1		1	1	1						1						1	
Plantanus orientalis						1											1	
Ptilostemon chamaepeuce																	1	
Quercus alnifolia	1					1						1	1					1
Quercus coccifera	1	1				1						1						1
Rhamnus palaestina		1																
Rhus coriaria	1	1			1								1					
Rubia tenuifolia		1																
Rubus sanctus	1	1																1
Salvia fruticosa	1																	
Sarcopoterium spinosum	1														1			
Sinapis arvensis	1																	
Smilax aspersa	1																	
Sorghum sp.																	1	
Styrax officinalis			1	1														
Teucrium creticum	1																	
Thymus integer	1		1								1						1	
Trifolium campestre																	1	
Trifolium stellatum																	1	
Valeriana italica	1																	
Vicia cassia																	1	
Vicia sp.																	1	
Ziziphus spina-christi	1																	1

Figure 7: *Pinus nigra*

***Pinus nigra* Arnold var. *caramanica*.** *(P. laricio* Poir. var. *pallasiana).* This pine occurs in Cyprus only on the higher regions of the Troodos massif, where it forms a pure forest above the 1,600 metre level. It is also found growing mixed with *Pinus brutia* down to about 1,300 metres in cool northern aspects. It is confined mainly to the gabbro, serpentine, peridodite and pyroxite rocks. Chapman (1967:26) described the *Pinus nigra* stands as "stark and sombre, with gaunt drooping boughs and crown characteristically tabular in maturity". These attributes evoked a romantic response in British paintings of the Cyprus mountains. The *Pinus nigra* is shade tolerant; the saplings send up straight sturdy leaders and yield good quality poles. Seedlings are extremely slow growing in the early years and may grow barely 60 to 100 centimetres in ten to fifteen years after germination, while an extensive root system is being formed. The tree, however, is frost resistant, drought hardy and wind fast. The timber is highly resinous and inferior in strength to the *Pinus brutia* (Chapman.1967:26).

Juniperus oxycedrus L. A small bush or tree growing to a height of 10m. It is comparatively rare in Cyprus and is confined to the forests of the Troodos massif above an altitude of 1,200 metres (Chapman.1967:25).

Figure 8: *Juniperus foetidissima*

Juniperus foetidissima L. A small tree up to 15m high, occurring only on Khionistra and on the peak of Madari, in the Adelphi forest, where in certain limited areas of exposed sheeted dykes, it forms a dominant constituent of the forest. As a true alpine species, it is not found below 1,300 metres in elevation and presents a stunted appearance. Its timber is fragrant, red and smooth-grained.

Juniperus phoenicia L. A small bush or tree up to ten metres in height, common in lowland scrub forests where it often forms dense thickets. It grows on all soil types, but it does not occur not above 700 metres in altitude. Its wood is fragrant, close-grained with yellow sap-wood and dark-brown heartwood. It does not grow to timber size (Chapman 1967:25).

Figure 9: *Cedrus brevifolia*

Cedrus libantica subsp. ***brevifolia*** Hooker (***Cedrus brevifolia*** Henry). The Cyprus cedar is endemic and in its natural distribution grows only on diabase rock at elevations of between 900 to 1,400 metres. The soils are frequently of the poorest quality, although the rainfall is high, averaging about 900 mm per annum. Good seed years occur approximately every six years and there are often two or three intervening years when no seed is borne. Seeds germinate freely but their viability declines rapidly after about nine months from the time of maturation.

Figure 10: *Cupressus sempervirens*

Cupressus sempervirens L. This is an evergreen tree attaining heights of 28 metres and girth of 2.6m. The cypress is the second most common tree on the island forming dense forests on the Kyrenia range. It grows on all soil types but prefers those of limestone origin, and in a natural state is not found commonly on soils of igneous origin. The *Cupressus sempervirens* is deep-rooted, wind fast and drought resistant. Slower in growth than *Pinus brutia*, it has a straighter growth habit and produces quality timber and poles. The wood is scented, with a light-brown sapwood and darker-brown heartwood. The *Cupressus sempervirens* produces seed abundantly with the cones ripening in late summer (Chapman 1967:24).

Hardwoods.

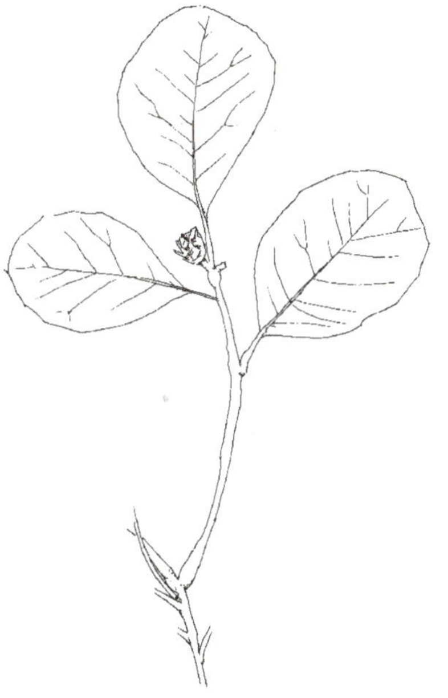

Figure 11: *Quercus alnifolia Illustration drawn by Benjamin Wright*

Quercus alnifolia Poech. (***Q. infectoria*** Gaud., ***Q. cypria*** Jaub et Spach., ***Q. ilex*** Sibth.). Evergreen bush or small tree found commonly on the igneous mountains of the Troodos massif, not below 450 metres, and occurring as pure stands in parts of the forest. It coppices readily from the base and thus re-establishes after low grade forest fires. The wood is hard and durable and it has been a principal fuel species, it also produces a high grade charcoal (Chapman 1967:33). Thirgood (1987:243) notes that the growth rate for the *Quercus alnifolia* was approximately 80 years to each 2.5 cms. diameter increment. Rotational coppicing was practiced on a forty year cycle.

Figure 12: *Quercus lusitanica*

Quercus lusitanica Lam. (***Q. infectoria*** Oliv.). A large tree with busy crown and thick grey bark with deep vertical fissures, leaves not falling until late winter or early spring, but before new foliage appears. The trees reach a very great age and the trunk of solitary trees may reach a girth of seven metres. The tree is found singly or in groups in fairly deep soils in the middle regions, particularly in the west of the island, though it never forms a forest stand. Acorns are used as fodder (Chapman 1967:32).

Figure 13: *Platanus orientalis*

Platanus orientalis L. A large deciduous tree to twenty metres in height commonly growing in river beds and near streams up to 1,200 metres in elevation The wood has a beautiful and characteristic grain, the medullary ray being very prominent when cut in radial sections (Chapman 1967:40).

Figure 14: *Alnus orientalis*

Alnus orientalis Decaise. Large deciduous tree ranging in height from four to twenty metres. It is common in valleys and near water sources in lower and middle hill regions. The wood is soft and dark-reddish in colour (Chapman 1967:31).

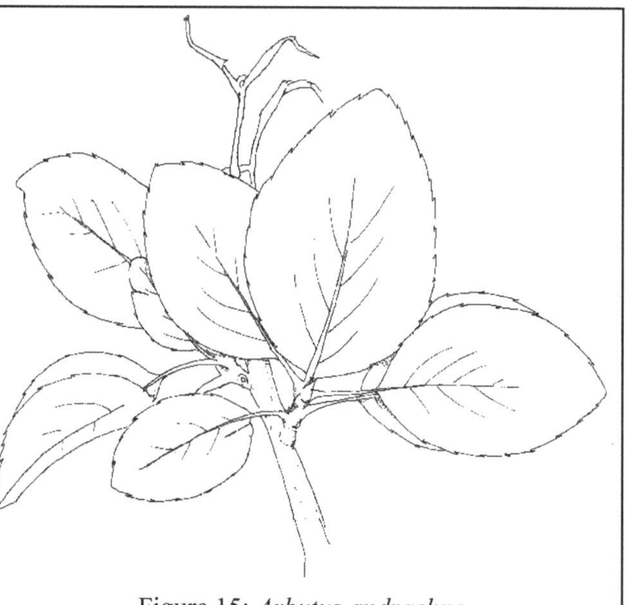

Figure 15: *Arbutus andrachne*
Illustration drawn by Benjamin Wright

Arbutus andrachne L. (***A. integrifolia*** Lam., *A.* ***sieberi*** Klotzsch). This species was believed to be a large bush or evergreen small tree common in the mountain regions up to 1,300 metres. An uncoppiced specimen was recorded in the Makheras as over 10m. in height. It coppices freely from the root and is therefore quickly re-established after cutting or fire damage. It was formerly one of the major fuel trees on the island (Chapman 1967:59).

Figure 16: *Olea europea ssp. sylvestris Illustration drawn by Benjamin Wright*

Figure 17: *Olea europea var. oleaster Illustration drawn by Benjamin Wright*

Olea europea L. The "wild" olive is generally identifiable by its shrubby, spinose habit, small leaves and small, stony fruits. Wild olives classified as *Olea europea* var. *sylvestri* and *O. europea* var. *oleaster* are merely convenient labels for distinguishing uncultivated olives from cultivated varieties. The olive is found on hillsides as a major component of the sub-strata species, and extensively in the cultivated lowlands between sea-level and 1000 metres (Meikle 1986:1095).

Ceratonia siliqua L. An evergreen tree growing to ten metres in height with a broad hemispherical crown sometimes spreading to 14 metres wide. Its natural habitat is dry hillsides in gargue, and in coastal and sub-maritime maquis, occurring frequently as a relic of cultivation. The carob is restricted in altitude to under 600 metres, however, appears suited to a range of sites particularly on sedimentary formations. It is widely cultivated in the lowland areas except on the Mesaoria (Meikle 1985:590).

Figure 18: *Acer obtusifolium Illustration drawn by Benjamin Wright.*

Acer obtusifolium Sibth. et Sm. A small evergreen tree or large bush growing to about ten metres in height. It is found on rocky hillsides, in pine forests, along field borders, and frequently along the margins of rivulets and streams to which it appears confined at lower altitudes (Chapman 1967:49; Meikle 1985:363).

Figure 19: *Styrax officinalis*

Styrax officinalis L. A decorative, deciduous bush or small tree growing to between two and seven metres producing fruit with certain medicinal properties. It is a member of the sub-strata community and is found in moist sites up to 1,300 metres (Chapman 1967:60).

Figure 20: *Quercus coccifera*

Quercus coccifera L. var. ***calliprinos*** (Webb) Boiss. (1879) An evergreen tree or shrub 2.5 to 4 metres in height with a semi-globular crown and a trunk which in senescent trees can exceed one metre in diameter. Its natural habitat was described by Zohary (1966:33) as "maquis or forest on terra rossa, rendzina, sandy clay and rock fissured soils from sea level to 1,200 metres". In Cyprus it has an affinity for the region within and adjacent to the Adelphi Forest growing as the principal substrata species on the gossan. Chapman (1967:33) described the *Qucercus coccifera* as an evergreen shrubby bush to small tree commonly found in the middle and lower regions in most parts of the island, occasionally growing into medium sized trees. The wood is used for fuel and the acorns as pig fodder.

Pistacia lentiscus L. A shrubby evergreen bush one to two metres high growing on dry, rocky slopes and hillsides, on sand dunes, in lowland pine forests with an altitude range from sea level to 800 metres (Meikle 1985:365).

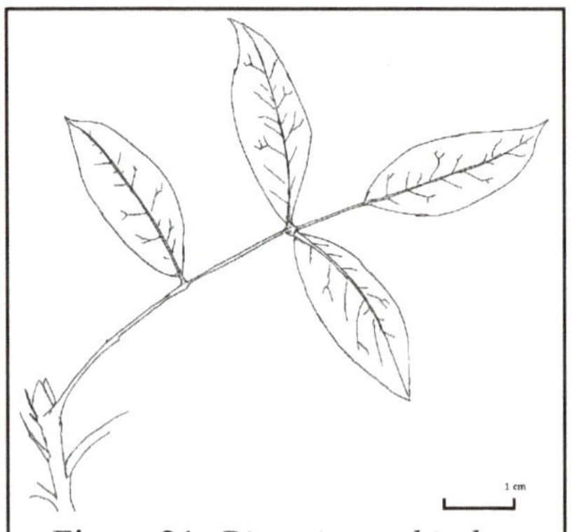

Figure 21: *Pistacia terebinthus Illustration drawn by Benjamin Wright*

Pistacia terebinthus L. This deciduous tree or small bush is common in forests from 300 to 1,200 metres (Chapman 1967: 48). The *Pistacia terebinthus* can grow into a spreading tree up to 6m. in height (Meikle 1977:368). Only three mature trees were recorded on the island during the survey.

Laurus nobilis L. An evergreen bush or tree growing to seven metres on preferred sites. Considered one of the most characteristic components of the "mesic maquis" variants it grows in remote valleys of the Paphos Forest (Chapman.1967:38; Zohary.1966:190).

Figure 22: *Crataegus azarolus*

Crataegus azarolus L. A small, rounded tree growing to six metres high which dominates the tree cover on the Mesaoria. It is frequently found growing to altitudes of 1,200m in forests, vineyards and along cultivated fields (Chapman.1967:42).

Mixed Conifers.

A mixed conifer stand contains conifers of more than one species. Mixed conifer stands tend to be classified by altitude. The *Pinus brutia/Pinus nigra* association occupy a zone between 1,200 and 1,500 metres above sea level on the Troodos. The *Pinus brutia/Cupressus sempervirens* association occurrs on the Kyrenia range and certain situations on the Troodos massif. The *Pinus brutia/Cedrus brevifolia* association is confined to defined localities in the Paphos Forest. The *Pinus brutia/Juniperus foetitissima* is an alpine forest occuring only at high altitudes in the Adelphi and Troodos Forests. The *Pinus brutia/Juniperus phoenicia* association forms lowland and coastal scrub forests in the south and east of the island while the *Cupressus sempervirens/Juniperus* sp. occur as lowland forests along the northern coastline. The *Pinus nigra/Juniperus oxycedrus* association is confined to the Troodos assif.

Mixed Hardwoods.

The *Platanus orientalis/Alnus orientalis* association occupies riverine moist beds. Other species included in the association are *Acer obtusifolia, Styrax officinals, Laurus noblis* with *Nerium oleander* and *Tamarix tetragyna* occurring at lower altitudes. *Pteridium acquilinui* and *Rubus alnifolius* are common on the banks, whilst *Hedra helix* is frequent (Polycarpou 1956). The *Quercus alnifolia* association is mainly found on steep slopes and screes, generally in sheltered localities. It appears to be more prevalent on aspects facing north. *Arbutus andrachne* is the most common associate species and *Smilax aspera* and *Clematis cirrhosa* may be present. It forms a dense canopy and ground vegetation is usually absent. Where the *Quercus* growth is open, cistus is present and *Pinus brutia* invades to form an overstorey. Occasionally the reverse occurs and *Quercus alnifolia* invades pine stands to form an understorey thus reducing light to seedlings and inhibiting regeneration. *Quercus alnifolia* reproduces from acorns but also coppices very well. It usually invades grass communities. After fires it coppices and recolonises sites. However, it is restricted in range and is confined to the igneous formations of the Troodos massif.

Mixed Conifers and Hardwoods.

The mixed conifer and hardwood stands are the dominant group in the older forests. There are three main categories: *Pinus brutia/Quercus alnifolia, Pinus nigra/Quercus alnifolia* and *Cedrus/Pinus brutia/Quercus alnifolia* (Polycarpou 1956).

Fuel.

Wood has provided the major source of heat energy for pyrotechnology and domestic purposes. Charcoal samples excavated from the Neolithic site of Ayios Epikitos *Vrysi* included *Quercus* sp., *Olea europea*, *Myrtus communis*, *Laurus nobilis*, *Styrax officinalis*, *Ulumus* sp., *Crataegus* sp., *Rhus coriaria*, *Pistacia* sp. and *Pinus* sp. (Cutler 1982). Although all but the *Ulmus* sp. current grow on the island, it is possible that some specimens may have been carried by ocean currents from the mainland of Asia Minor. The hardwood species were those most suitable for providing a long hot fire. The *Quercus spp.*, *Olea europea*, *Crataegus azarolus*, *Certonia siliqua*, and *Pistacia terebinthus* appear to have been extensively used from an analysis of charcoal associated with some ancient kiln sites (Schoch 1988). These species also coppice well (Wertime 1982). Contrary to the cyclical deforestation theory suggested by Constantinou (1982), which is based on an eighty year harvesting cycle for *Pinus brutia*, archaeological examination of charcoal from the smelting site of Ayia Varvara *Almyras* clearly indicates a preference for *Olea europea*. It is probable that woodland management was exercised during the Iron Age to provide industrial areas with coppiced logs of suitable dimensions for optimum combustion. Although hardwood tree species may have been used as they were immediately available within the forest sub-strata, it does seem apparent that certain species were more favoured than others. The lack of fuel wood does not appear to have affected industrial activity further supporting the hardwood coppice theory (Mighall and Chambers 1993).

During the Colonial period and before, Cyprus was a regional supplier of quality charcoal harvested primarily from the coppice of the *Quercus alnifolia*. Evidence for large kiln sites are obvious in many forest areas, some dating to as late as 1944 (Demetriou pers. com.)

Dyes.

Pigments used to provide colours for dyed material are plentiful from a range of species growing in the forests. The mordants[5] listed are suggestions only, as some plants, such as the *Quercus*, will exude a brown colour without a mordant but, when alum is used, the bark will give a golden shade. Most lichens will not produce a dye unless boiled or fermented. The traditional method of fermentation was to leave the dye plant in urine for a prolonged period. Table 2 lists known dye plants, the parts used and the colour produced.

Name	Parts Used	Colour Produced
Alnus sp	bark	yellow / brown / black
	roots	grey
	twigs	yellow
Rubus sp.	berries / shoots	pink / purple
	shoots / old leaves	yellow, green
	flowers	orange yellow
	twigs, berries	yellow, brown
Cistus	leaves	yellow
Cupressus sempervirens	cones	tan
Gallium spp.	roots, top	yellow, red
Acer sp.	bark	tan
	whole plant	yellow
Quercus sp.	inner bark	gold, brown
Asphodel	skins	yellow, orange
Asphodel	flowers	red, yellow
Rhus coriaria	berries, leaves	tan, brown
Rhamnus	unripe berries	yellow
Myrtus communis	leaves, berries	black
Xanthoria parietina	whole plant	black

Table 2: Dye plants recorded in the Makheras and Adelphi Forests

Mordants used in antiquity included chemicals leached from copper or iron mines, urine and potash (Forbes 1964).

Medicinal Plants.

A large proportion (over 30%), of the flora identified in the forest can be used medicinally. The pharmaceutical constituents of the Cyprus flora were known to Galen, Discorides and Pliny. A major export trade in many species was suggested by Dioscorides (*Materia Medica* trans. Sadek 1983) and a quantity of plant remains have been preserved in archaeological sites (Manniche 1989).

Edible Shoots, Fruit, Tubers and Browse Plants.

The sub-strata species within the forest provide an abundance of seasonal plants harvested for human consumption or available almost all year for animal browse. In spring, the shoots of the *Pistacia terebinthus*, *Capparis spinosa, Asparagus acutifolus* and the *Sinapis arvensis* are plentiful and nutritious. During summer wild artichokes are harvested. In winter *Asphodelus microcarpus*, *Allium ampeloprasum*, acorns from the various *Quercus spp.* and mushrooms contribute to a varied diet.

[5]A mordant is a substance used to fix a colouring matter.

Browse and fodder varies depending on the animal. Goats and sheep prefer the leaves from the deciduous hardwoods while acorns provide forage for pigs. Donkeys and goats graze on the flowers of the *Calina coryrubosa* and *Ptilostemon chamaepeuce* during summer while hares eat the leaves of the evergreen *Equisetum ramossimum*. Wild fruits are also readily available from the *Arbutus andrachne*, *Ficus caria*, *Capparis spinosa*, *Vitis vinifera*, *Crataegas azarolus*, *Ceratonia siliqua*, *Olea europea* and *Rubus sanctus*. Macrobotanic remains from almost all these species have been recovered from archaeological sites and they are still harvested annually by mountain villagers.

Resin and Pitch.

Extraction of sap for resin, pitch and tar production has been, in the past, of major economic importance in Cyprus (Christodoulou 1957: 109; Thirgood 1981:127). The *Pinus brutia* produces high quality pitch and tar used in proofing vessels from amphore to sea going ships. The *Pistacia terebinthus* produces a yellow resin burned as incense by the Egyptians during the 18th Dynasty (1500 B.C.). A large *P. terebinthus* can yield up to 2 kg of resin (Harfield and Harfield 1990). Resin tapping over a period of years weakens the trees and induces early decline. Trees that survive into maturity cannot be harvested for any other commercial use due to the extensive damage inflicted on the tree bole. At present, the only senescent trees in the Makheras and Adelphi Forests are those with resin tapping scars.

Faunal Reserves.

The range and complexity of the faunal species associated with the forest are critical factors in determining the forests' viability. In Europe, a forest represented wealth or status through its recreational facility. In Cyprus, small introduced mammals such as the hare, provide food and sport, while other species, such as the *Vulpes vulpes* (red fox), act as natural predators. Species introduced through trade and commercial links, such as the *Rattus rattus*, have naturalised into the forest environment. Many bird species from the native pheasant to the migratory robin have been a traditional food source and hunted, in some instances almost to extinction. The faunal resource has provided not only a food supply, but bone and horn for implements, hides for clothing, fluid containers and industrial use, feathers for weaponry and a range of secondary products.

Bird observations.

An important aspect of biodiversity measurement is the identification of local bird life. Identification of species was made by Dr. Michael Given from bird characteristics observed through binoculars and later compared with textual information provided by Bannerman and Bannerman (1971). Numerous nests were recovered from the ground for identification so that the ecology of birds within the wider forest ecosystem could be better appreciated. Birds were observed in all but two quadrats in the Makheras Forest.

Figure 23: Chukar nest under pine branch fellings.

The bird assemblage as an element in the ecological community of the Makheras and Adelphi Forests interacts with the population dynamics of many forest insects, consuming up to 80% or more of a population of herbivorous insects, and may also act as vectors of disease for forest insects (Ford 1989:53).

The diversity, structure, succession and stability of the Cyprus bird population are governed largely by the migratory bird pattern of the Northern Hemisphere, the stability of the forest canopy cover, and the action of seasonal hunting practices which tend to be indiscriminate. The range of bioresources available within the forest community will determine the number of species that can be supported at any location. The diversity of species may be governed by the height of trees within a stand and the total canopy cover within the habitat. The complexity in composition and age classes of the stand frequently act as the most critical element in species diversity. Succession in vegetation development, particularly after wild fire, impacts upon the diversity of bird life within the forest. Forest harvesting, particularly clear felling, also affects habitat and species density.

The stability of the bird assemblage is an unknown factor, although Bannerman observed an increase in the chukhar population during the Eoka Campaign when hunting activity was limited. The foraging behaviour of birds can be distinguished by their main food requirement. Categories include seeds, fruit, nectar, insects and predators.

Figure 24: Map of Makheras and Adelphi Forests showing Faunal observations excluding Foxes.

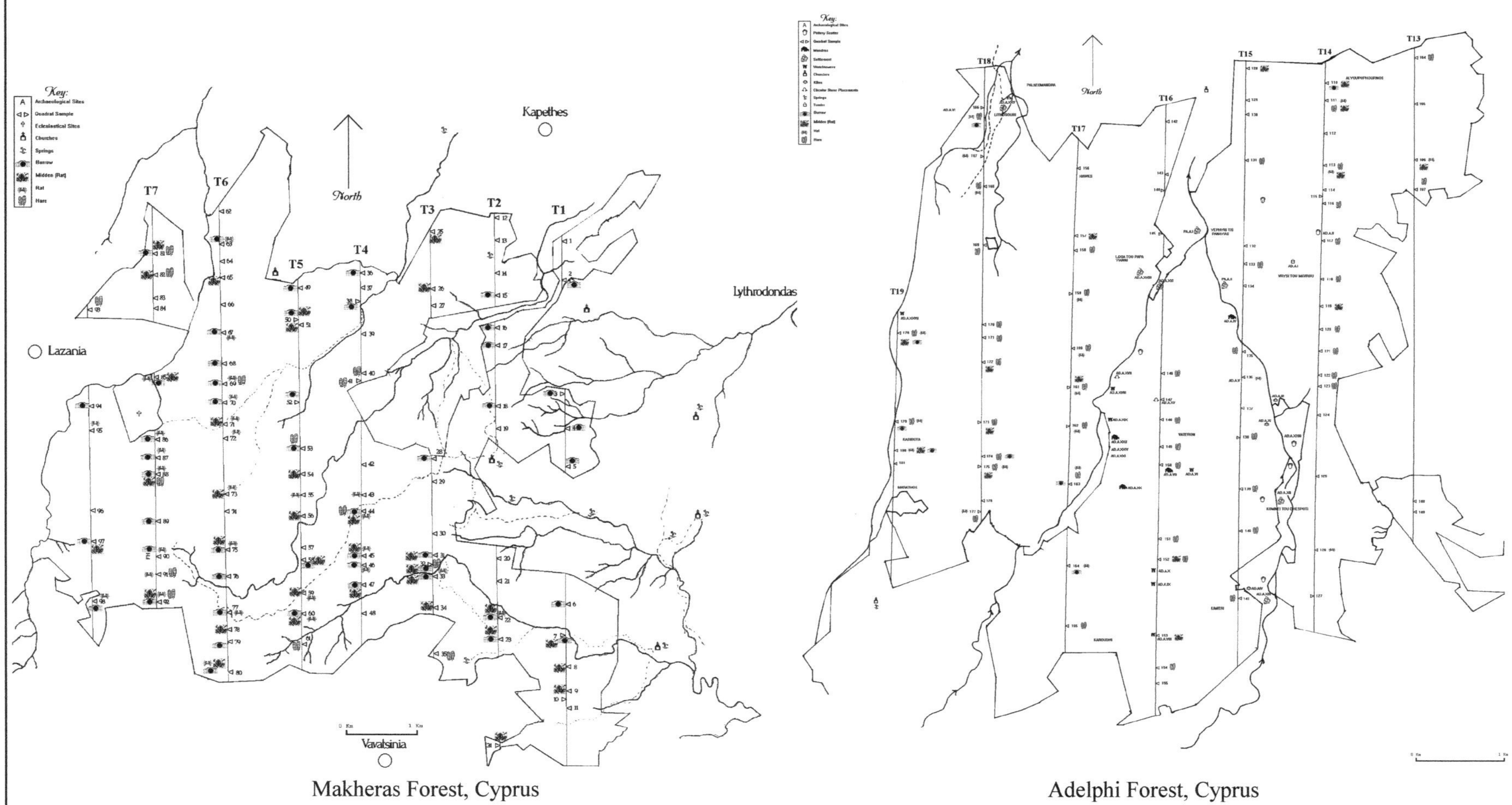

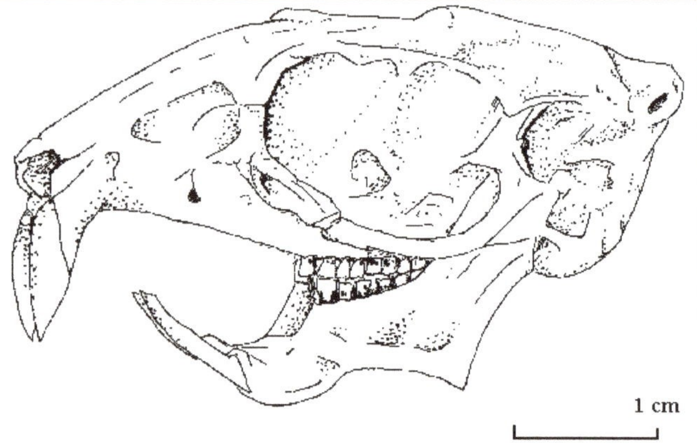

Figure 27: Skull of Rattus rattus
Illustration drawn by Benjamin Wright

Small mammal observations.

Throughout the forests, small mammals may be significantly reducing the forest regeneration potential through the direct destruction of the seed crop (Smith and Aldous 1947; Spencer 1955; Golley *et al.* 1975). Mice and shrews can consume large quantities of seed. Their mobility can also affect regeneration through the transport of organic material, seeds and chemicals within the ecosystem. The creation of rodent middens in the forest was a notable occurrence with seeds of several different species collected for consumption in one specific location (See Figure 26). The cache of seeds was not discerned, although on a few occasions a single *Quercus* sp. was recorded without an obvious seed source nearby. Small mammal activity can also alter the soil substrate through digging, burrow construction, tunnelling and other constructions. This activity can influence the microrelief of the soil which in turn can affect the soil moisture absorption rate. Soil movement through small mammal activity can also influence the distribution of soil chemicals frequently increasing the decomposition process on the forest floor. The main predator for the small mammal assemblage in forest areas is *Vulpes vulpes* (red fox) which, through an analysis of castings recovered, demonstrates a distinct food preference for the small mammal community. The two major small mammals recorded during the survey were the *Rattus rattus* (black rat) and *Lepus europaeus* (hare).

Figure 25: rat burrow
Photograph by Michael Given

Figure 28: Rattus rattus midden in pine forest
Photograph by Michael Given

Figure 26: Pine cone stripped by *Rattus rattus.*
Photograph by Michael Given

Rattus rattus is native to Asia Minor and the Orient and has been presumed to have naturalised on Cyprus during the Crusades (c. 12-14th century). Fossil rat droppings have been recovered from deposits at the Neolithic site of *Kandou*. *Rattus rattus* weighs between 115 and 350 g and it is an agile climber. Rat shelters occur in a range of habitats including burrows, under rocks or nests in trees. It is an extremely adaptable and inquisitive animal but tends to avoid new objects in a natural environment. At times, the population of the *Rattus rattus* can attain plague proportions, mostly due to human intervention through the removal of natural predators (Walker 1964:904). A study into the ecology of the *Rattus rattus* was conducted by J. S. Watson between March 1947 and April 1948 over vegetation zones ranging from arable land to forest (Watson 1951). The population of rats appeared to be the largest during December-January and the smallest during June. Mobility of individuals was small generally averaging between 50 - 100m, although males travelled further than females and six animals moved over 200m. Population density varied with environment - areas carrying a thick sub-strata carried the highest populations (Watson 1951:7).

Food intake varies seasonally. Between February to March, rats feed on the seeds of the *Pinus brutia,* frequently creating large middens of cone material. The frequency of middens varied but were larger and more frequent in the Makheras Forest than the Adelphi. From a total of 99 quadrats in the Makheras, 34 contained rodent middens. This compares with 15 middens over 73 quadrats in the Adelphi. Acorns from the *Quercus alnifolia* were also eaten and were often included in the pine midden debris. During the summer months, Asphodel and other bulbs were dug and consumed *in situ.* The *Rattus rattus,* therefore, appears to be taking advantage of an ecological niche which in Europe is filled by the squirrel.

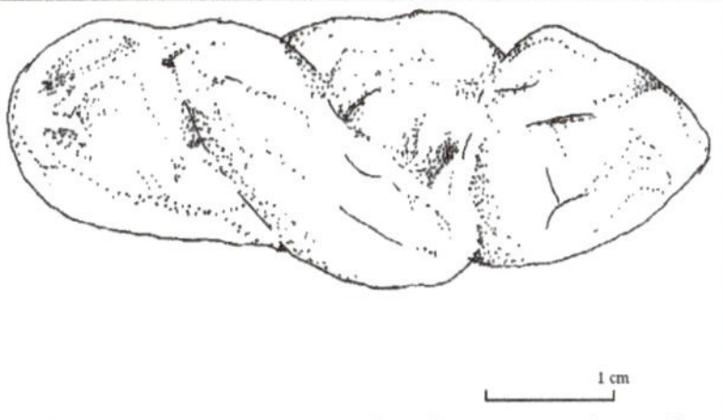

Figure 29: A typical example of the twisted turd characteristic of fox scats.
Illustration drawn by Benjamin Wright

Predators include the *Vulpes vulpes* for whom the rodent population constitutes an important, and accessable, part of the diet. The snakes such as *Zamenis gemonensis* and *Vipera libetina* are both large enough to prey on rats. However, there are very few raptorial birds in the Cyprus assemblage. Large owls that are important rodent predators elsewhere, do not occur on Cyprus (Watson 1951:51).

Lepus europaeus are generally nocturnal and neither hibernate nor estivate. They are adapted to open country and generally occupy a large home range of approximately 2.0 to 5.5 km^2. *Lepus* appear to be solitary, but research indicates the possibility of a well-defined social hierarchial system and the establishment of nonexclusive territories (Hewson 1977, Marsden and Holler 1964). Lagomorphs do not store food and they are strictly herbivorous. They are strong runners and use forms, ie, depressions dug into the soil often underneath low growing shrubs, for resting (Diersing 1984). Lagomorphs constitute a necessary component in the trophic relationships of biotic comunities serving as a principal food source for predatory birds and mammals. Lagomorphs were introduced into Cyprus during the Medieval period as a sporting animal and have naturalised into the environment

Fox observations.

Vulpes vulpes may have a home range of between two and five km^2 depending on the terrain and available water resources (Ables 1969). (See Figure 32: Fox Locations in the Makheras Forest) Food consists mainly of small mammals and birds. However, insects, eggs, fruit and carrion are eaten in smaller quantities, the last being most important during the winter months. Skeletal material from one large adult *Vulpes* was recovered from the Makheras Forest and from two individual sub-adult and one adult *Vulpes* from the Adelphi Forest. British research (MacDonald 1986:313) demonstrates a positive impact by *Vulpes* on small mammal pests of young pine plantations. The *Vulpes vulpes* is regularly shot during the open hunting season, as it is traditionally believed to prey on game birds.

Bone material identified from 20 fox scats in the Makheras Forest included 15 individual rats, 2 hares, 3 individual mice, bone material from 10 individual birds, four carion carcasses predominantely *Caprine* and a variety of different fruit seed including grape and fig (Payne pers. com.). (See Appendix F:2)

Corridors used by foxes follow ridgelines. Tracks were also observed across grassy slopes and open areas. An inhabited den complex was located on a south facing bank above a stream line on T3 adjacent to Q26 in the Vromonera area. Approximately ten entrances were recorded although several appeared to be disused.

Figure 31: Fox runs across a grassy slope above *Spring of the Black Man*

Fresh paw prints and scats were observed nearby. A large scatter of goat bones, presumably dumped from a mandra on the forest perimeter, were located nearby. The bones covered an area of over 10m^2 and included four skulls of fully-grown, but immature animals.

Figure 30: Fox dens

Figure 32: Maps of the Makheras and Adelphi Forests showing Fox Locations.

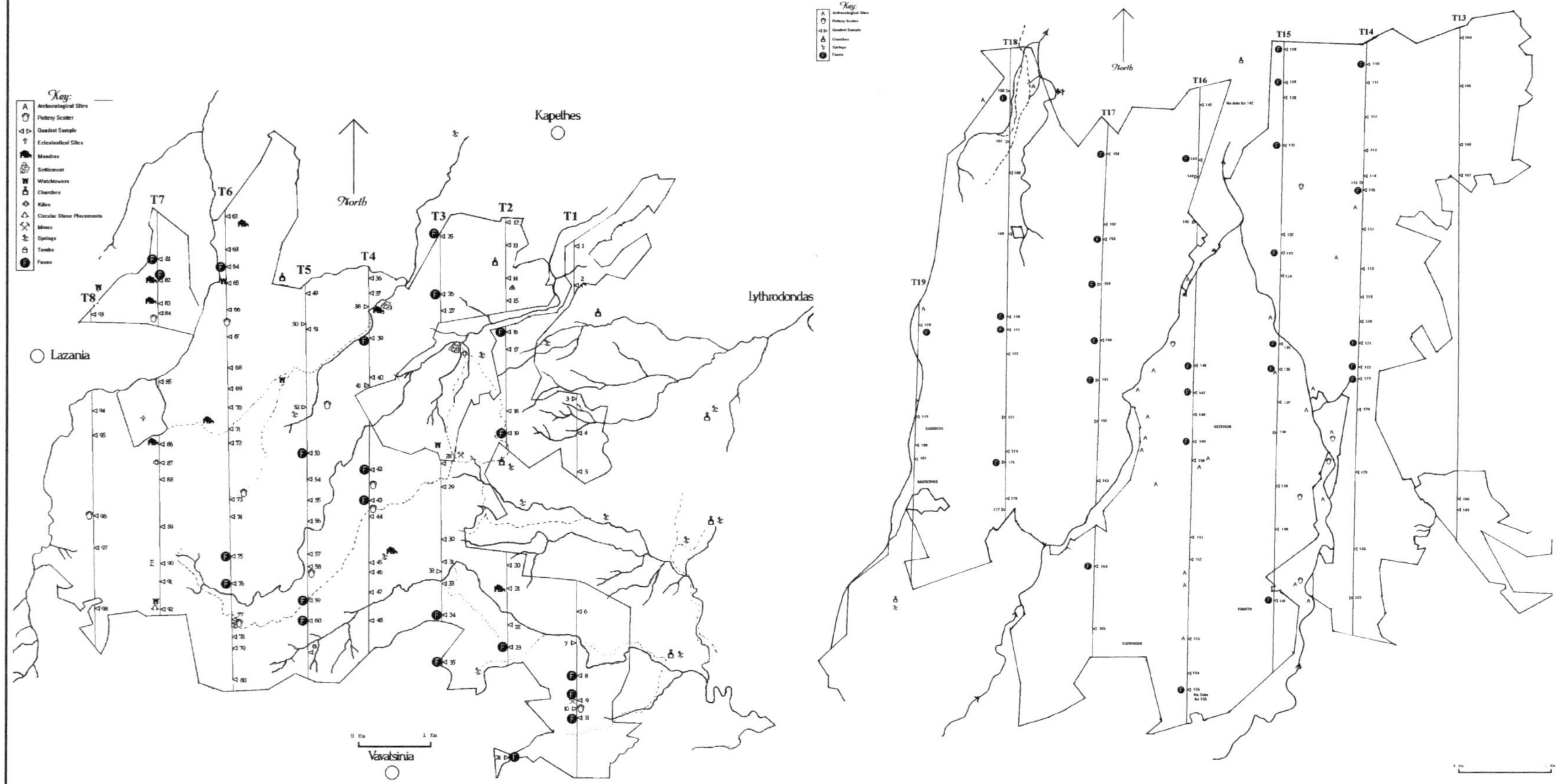

Makheras Forest, Cyprus

Adelphi Forest, Cyprus

Fish

Native fish are known to exist in the larger streams and rivers of Cyprus. However, the extensive environmental degradation that has occurred over the past two thousand years inhibits research into the scope and extent of a former supply of fish in the mountain forests.

Bees

Two species of bee have been observed in the forest areas. Bee activity is often an essential part in plant reproduction through cross pollination. The importance of honey production by bees and its use by people as a sweetener needs to be classed as a secondary forest product. The export of honey from Cyprus has continued since antiquity as Cyprus remains a major producer. The forest area is suitably located for honey production and has excellent access for bee keeping activities. The preference for the aromatic ground cover and sub-strata plants enables the utilisation of most species. Wax as a sealant, for candle production and as a component in medicinal compounds, is also of great economic importance.

Geological Extractions from the Forest Reserves.

The geological parent material associated with the high altitude forest reserves has maintained economic importance since the Bronze Age. Hellenistic and Roman mines are found within the forest and the extraction of gold and silver from quartz seams in the Makheras has also been recorded (Bear 1959). Medicinal substances leached from the sulphur deposits associated with copper mining were recorded by Discorides (Riddle 1985). Vitrol collecting on copper spoil heaps was of importance in the production of ink (Shopin pers. com.) and pigments for paint, glass colouring and glazing. Gypsum was of economic importance during the Roman period for plaster manufacture, used to waterproof cisterns, and also during the Colonial period when old gypsum mines were reopened to produce domestic plaster. None of the geobotanical indicators cited by Brooks and Johannes (1990:16-18) occured within the perimeters of the forest.

The combination of resources available for exploitation within a limited area appears remarkable. This concentration of wealth in natural resources makes the political geography of Cyprus and its trading connections assume an incomparable position.

Passive Forest Reserves.

The importance of clean water to the current expanding urban and tourist economy cannot be overlooked. Over 92% of all water resources on Cyprus rise in the forest regions (Gresham 1958). With the reduction of forest canopy cover and the subsequent increase in soil temperatures, less soil water is retained. The impact on the river catchment areas of increased insolation has increased following continuing and widening gaps in the forest canopy cover. Subsequent erosion and water pollution reduces the quality of the water catchment area.

The Forest Policy advocated by the Government of the Republic of Cyprus constitutes a broad perspective of objectives, covering the role of the forests in the provision of both material (direct and indirect), social, and aesthetic benefits. Under legislation, the forests are protected against adverse forces, both human and physical, with fire constituting the gravest danger. The forest resource is managed for sustainable wood production to meet current and future demands. The forest resource is likewise managed to provide employment in remote villages, whilst providing raw materials to local industries and fuel wood to rural communities. The heavy emphasis on non-wood values is reflected in the use of the forest resource as a soil erosion control, and as a recreational facility in response to public demand. The forest expansion into lowland areas is likewise in response to non-wood values (Christodoulou 1993:4). The forest resource is managed to provide the maximum possible social benefit through multiple use, conservation, protection, recreation and timber production. The extension of the resource onto land adjacent to the forest reserves, and the reafforestation of marginal land widens the species types and range so as to encompass a variety of ecological habitats. The application of forestry management for desertification control and environmental conservation is proving to be an innovative strategy.

Extensive areas lying outside the Main Forest reserves which have degraded through abusive use in the past, continue to degrade. These areas have, since 1976, been located, surveyed and mapped. After consultation with other Departments concerned with aspects of land use, they have been declared State Forests.

The main objectives covered in the current five year plan include: forest protection with special reference to fire prevention and control, reafforestation of burnt and debilitated areas, nature conservation and protection, timber production and allied industries, professional silvicultural management of the forest resource, maintenance of public recreational facilities, forest research, and the provision of employment for forest villagers. The Development Plan currently in place has an increased emphasis on recreational use of the forest resource and nature conservation (Christodoulou 1990:36).

Trees felled under current silvicultural treatments are selected on either negative or positive criteria. Negative criteria include trees selected exclusively on the basis of undesirable attributes without any attempt to directly favour selected individual stems growing in association. Positive criteria involve as an initial stage, the identification of selected stems which are favoured by the removal of the main competitor. Silvicultural treatments are directed *inter alia* at securing natural regeneration of the forest, so that the newly established stands maintain their natural character. However, the natural reproduction of forests cannot be an objective in itself. Rather, it is considered an appropriate means, selected to attain maximum benefit from the forest resource including wood production as well as non-wood services and values. Artificial regeneration of forest stands is applied where natural regeneration fails (Peonides 1993).

Special Provision for Ecological Enhancement.

Elite trees are selected and marked for retention as seed trees and for other forms of genetic improvement of the species. Trees declared "Nature Monuments" are also retained under special provision as part of the National Estate. Silvicultural treatments and forest management techniques in adjacent areas are to be sensitive, causing minimum disturbance to the landscape setting.

For each hectare of forest, one or two stems are retained to become senescent examples of the localised species and habitat trees to meet faunal requirements. Trees already mature, exhibiting defective growth or occurring on remote rocky areas, are preferred for retention in this instance. Trees utilized as nesting sites for birds of prey are also retained. In the case of eagle nests, a protective forest zone is to be maintained around the site. Fellings over the remainder of the compartment is planned for periods when no eggs or chicks will be present in the nest. Trees are also retained along road verges to provide summer shade and for added landscape value. Stands in the vicinity of churches, historical sites or sites of cultural value are to be included in a protective zone and treated accordingly.

If the ultimate goal of silvicultural practice is to provide:

> *the maximisation of the total benefit to the community from the forests, through sustainable and multiple use - production, recreation and protection - maintaining at the same time the natural character, and improving the structure and productivity potential of the forest* (Peonides 1993),

then it may be inappropriate to silviculturally treat regeneration areas to ensure the continuing dominance of *Pinus brutia,* where other species would ultimately be more vigorous and highly competitive (Florence pers. com.). In 1976, the Council of Europe requested a prohibition on the replacement of natural forest with single species plantations. The motion read:

> *A natural forest has advantage from the economic as well as the ecological point of view. Renewal of plant cover according to ecological criteria is therefore essential* (Council of Europe 1976).

A critical consequence of the use of single species in regeneration of degraded areas is the lack of opportunity for subsequent "ecological sifting" of species within the regrowth stand. The possibility of a regeneration site being "sub-optimal" for a monoculture of the preferred species is reflected in vigour reduction and an expression of dominance which may be relatively weak, and the natural thinning process being unduly slow. Under more extreme conditions, a "locked" or stagnant stand could result, predisposing it to disease (Florence pers. com.).

Chapter Five: Cypriot Wood Anatomy - Timber structure and Potential Use.

An examination of Cypriot tree species from a study of the internal structure of the timber, may define the use made of the forest resource and its importance as a trade commodity. This factor is particularly pertinent for species selected for ship building (Theophrastus *HP* 7.1).

In the living tree, wood performs a threefold roll of support, conduction and storage. Sapwood transports sap to the upper areas of the tree and acts to conserve or store food. Sapwood lumber is light coloured and always nondurable in character. Heartwood contributes support to the living tree and functions as a repository for metabolic waste products. It frequently has a dark appearance and is in some species naturally resistant to biological degradation (Wenger 1984:567).[6]

Pinus brutia

Figure 33: Radial thin section: *Pinus brutia*

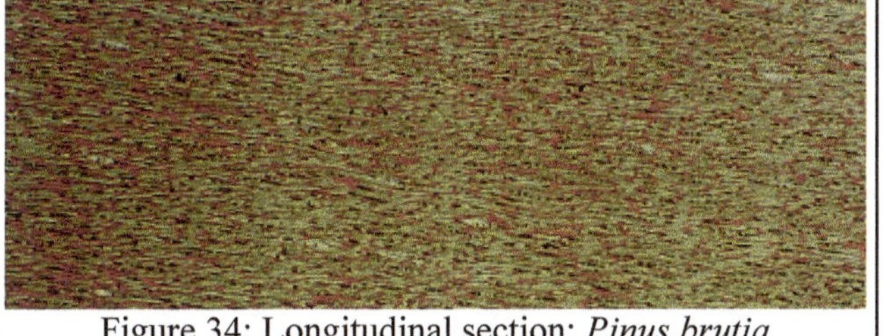

Figure 34: Longitudinal section: *Pinus brutia*

Figure 35: Tangential section: *Pinus brutia*
Brightfield light microscopy photograph by Jenny Norman

Growth rings in the Pinus brutia are distinct and very rarely vague. The transition from early to latewood is usually abrupt although sometimes it can be gradual. Tracheids have an average range of 2,450 (1,270-3,870) µm in length, with uniseriate bordered pits with round apertures in the earlywood and elliptic to slit-like apertures in the latewood. They are confined to the radial walls, except for some very rare tangential pits in the latest formed latewood. Crassulae are often present. Trabeculae are occasionally observed while axial parenchyma are restricted to resin ducts. Rays 30-40/mm^2, except for fusiform ones that are uniseriate and very rarely biseriate, (1)3-16(22) cells high, composed of marginal (rarely also intercalated) tracheids in one or several rows and fairly thick-walled parenchyma cells with nodular end walls. The walls of ray tracheids are sometimes weakly dentate. Cross-field pits range between 1-3(4) (Fahn, Werker and Bass 1986:4).

Cedrus libani

Figure 36: Radial thin section: *Cedrus brevifolia*

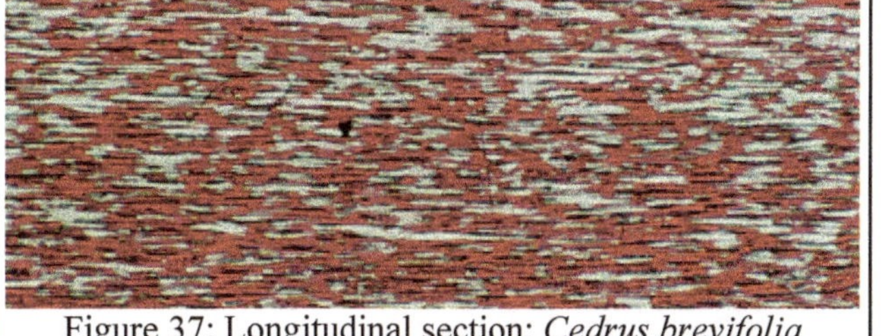

Figure 37: Longitudinal section: *Cedrus brevifolia*

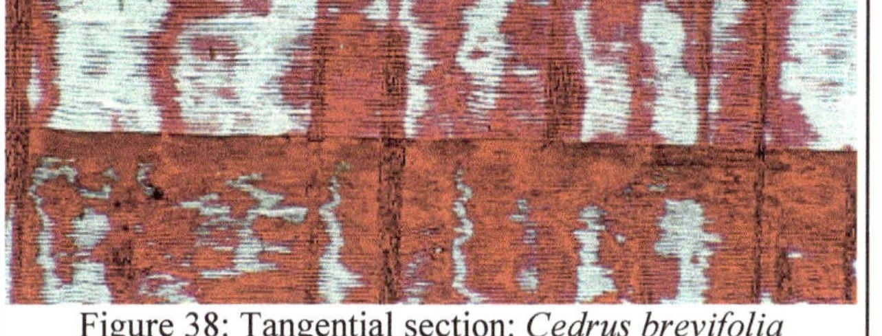

Figure 38: Tangential section: *Cedrus brevifolia*
Brightfield light microscopy photograph by Jenny Norman

Growth rings are distinct with the transition from early to latewood gradual, although in some seasons it can sometimes be abrupt. Tracheids are between 1,600 (940-

---

[6]Wood specimens of Cypriot tree species were obtained for wood anatomy analysis. Unfortunately, time restraints prevented a full description being made. Published material of regional species has, however, been included for comparison together with photographs prepared from thin section slides made available by the Cyprus Forestry College, Prodromos.

2,230) μm long, with uniseriate (very rarely biseriate, alternate) bordered pits with fringed torus margins confined to the radial walls, except for some tangential pits on and near the growth ring boundary in the latewood. Pit apertures appear circular to elliptic in the earlywood and slit-like (included to extended) in the latewood. Faint crassulae are sometimes present. Axial parenchyma appear with pitted horizontal walls and are confined to the growth ring boundary. Rays average 25/mm2 and are mostly uniserate and rarely biserate, (1)3-20(35) cells high, composed of marginal ray tracheids in single rows, and fairly thick-walled ray parenchyma cells with nodular end walls. Ray tracheids appear with smooth walls. Cross-field pits occur with 2 to 4 taxodiodid pits per field in the earlywood and piceoid in the latewood. Resin ducts are absent or occur as the traumatic type and close to each other. These are mainly thick-walled and lignified epithelial cells. Axial ducts are more common than horizontal ducts (Fahn, Werker and Bass 1986:3).

Cupressus sempervirens

Figure 39: Radial thin section: *Cupressus sempervirenes*

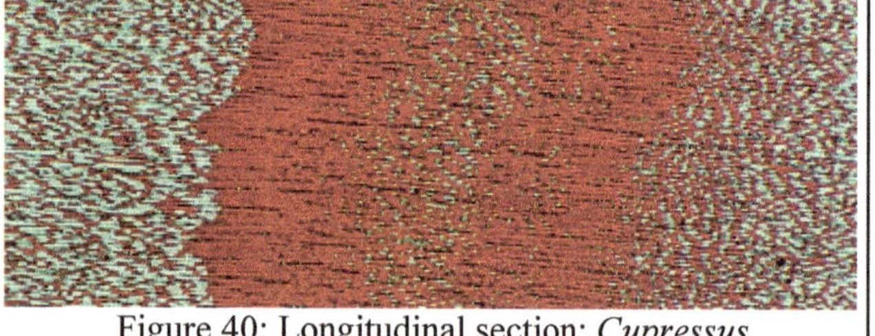

Figure 40: Longitudinal section: *Cupressus sempervirenes*

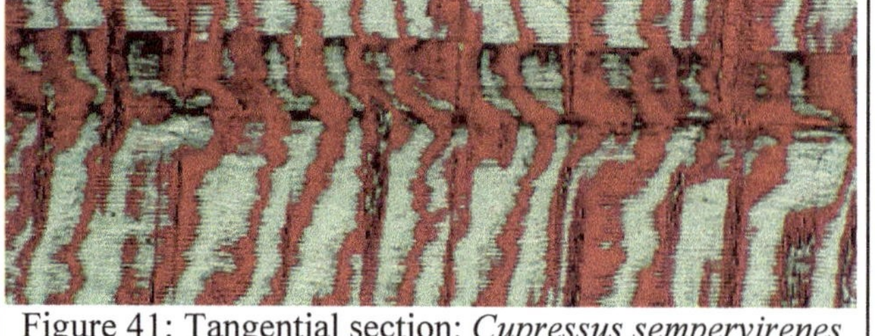

Figure 41: Tangential section: *Cupressus sempervirenes Brightfield light microscopy photograph by Jenny Norman*

Growth rings in the *Cupressus sempervirens* are distinct. The growth transition from early to latewood is gradual. Tracheids average 1,700 (700-2,800) μm in length, with uniseriate (rarely biseriate) bordered pits confined to the radial walls. However, these are also common in the tangential walls in the latewood near the growth ring boundary. Pit apertures in the earlywood are mainly circular in *Cupressus horizontalis* and elliptic to spindle-shaped in *Cupressus pyramidalis.* Latewood pits have spindle-shaped apertures in both varieties. Faint crassulae are sometimes present.

The axial parenchyma diffuse or in tangential bands, with smooth to faintly nodular horizontal end walls. Rays are 18-40/mm^2 in mature stemwood, 40-55/mm^2 in branchwood. They are mostly uniseriate and rarely biseriate occuring (1)3-20(40) cells high in stemwood, 1-8(16) cells high in branchwood. They are entirely composed of fairly thick-walled parenchyma cells with smooth to very faintly nodular end walls. Cross-field pits range between 2-4 per field, and appear cupressoid with elliptic to slit-like. Apertures are included in both early and latewood. Resin ducts are absent (Fahn, Werker and Baas 1986:1).

Pinus nigra

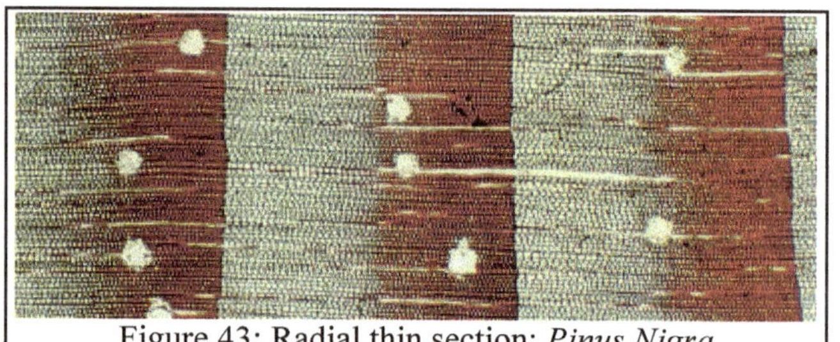

Figure 43: Radial thin section: *Pinus Nigra*

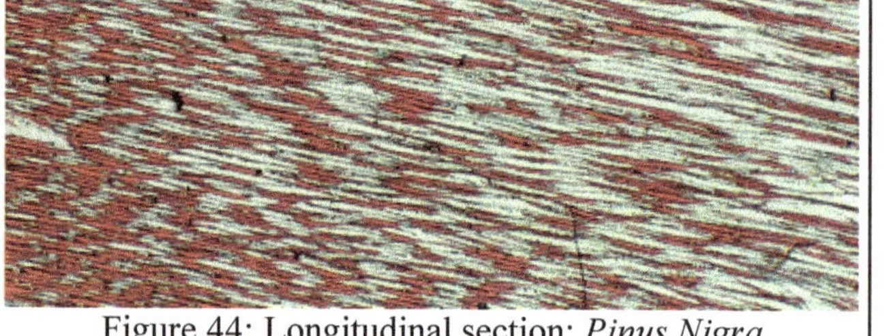

Figure 44: Longitudinal section: *Pinus Nigra Brightfield light microscopy photograph by Jenny Norman*

Pinus nigra is typical of Gymnospermae softwood in that vessels are absent while resin ducts are present. These ducts for the greater part occur in the late wood although they can be present against the growth-ring boundary in early wood. The width of early-wood zone is slightly wider or equal to the width of the late-wood zone with solitary resin ducts. The average diameter of resin ducts is about 100μm. On the radial surface there is a large difference in colour between early-wood and late-wood (Outer, Veenendaal and Versteegh 1988:46)

Juniperus phoenicea, J. oxycedrus and J.excelsa.

Figure 42: Radial thin section: *Juniperus foenicea*

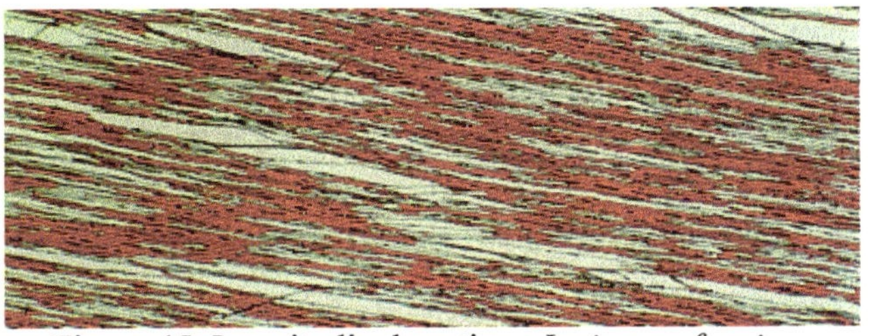

Figure 45: Longitudinal section: *Juniperus foenicea*

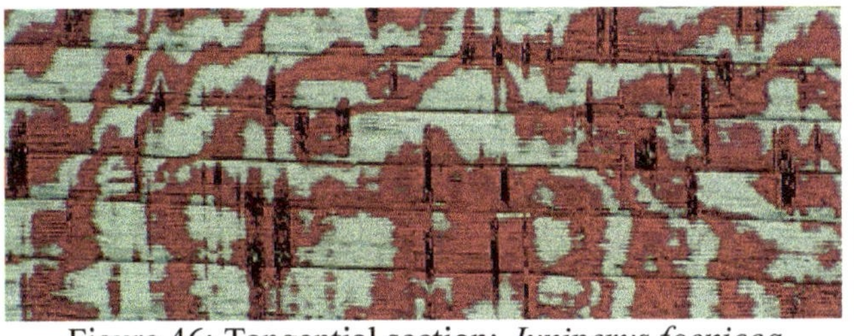

Figure 46: Tangential section: *Juniperus foenicea Brightfield light microscopy photograph by Jenny Norman*

Growth rings in these species are distinct. The transition from early-wood to late wood is gradual. Tracheids have a mean average of 1,760 (760-2,700)µm length in *J. phoenicea;* 810 (490-1,140) µm long in the thick branch wood of *Juniperus excelsa* and 1,450 (700-1,700)µm in the branchwood of *Juniperus oxycedrus* with uniserate bordered pits confined to the radial walls in the earlywood, but also commonly appearing in the tangential walls of the latewood. Pit aperatures are round to eliptic and rarely slit-like in the late wood. Axial parenchyma are difuse or in short tangential bands (especially in *Juniperus excelsa* and *Juniperus oxycedrus*) occuring throughout the growth ring and appearing with pitted to smooth horizontal end walls. Rays of *Juniperus oxycedrus* and *Juniperus phonecia* are 55-85/mm^2 in the mature stemwood. Cross-field pits are between 2 and 4 per field in *Juniperus oxycedrus* and *Juniperus phonecia* and 1 to 3 in *Juniperus excelsa.* These are cupressoid with elliptic to slit-like and include apertures in both earlywood and latewood. Resin ducts are absent.

Juniperus wood is anatomically very similar to *Cupressus*, especially in the branchwood or juvenile stewmwood. Ray height and frequency are the best diagnostic attributes for seperating mature stemwoods (Fahn, Werker and Baas 1986:56).

Platanus orientalis.

Growth rings are distinct with vessels diffuse, ca. 75/mm^2, solitary (ca. 50%) or in tangential to radial multiples or clusters of 2-4(6). These are angular in cross-section with a tangential diameter of between 40-100µm and radial diameter of up to 130 µm. Cell walls are ca. 2µm in thickness. Vessel member length averages 570(390-780) µm. Perforations are simple and scalariform occurs in oblique end walls, the latter

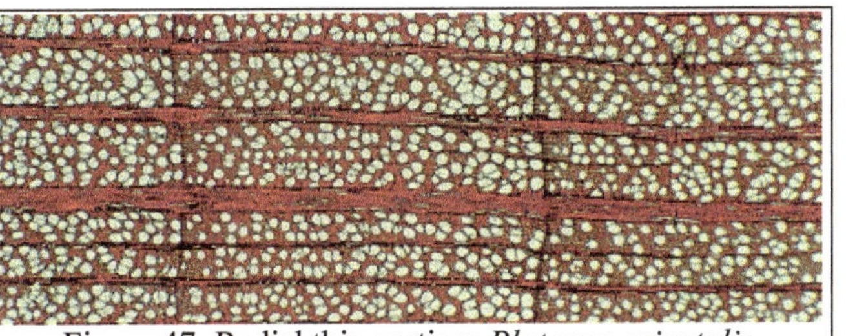

Figure 47: Radial thin section: *Platanus orientalis*

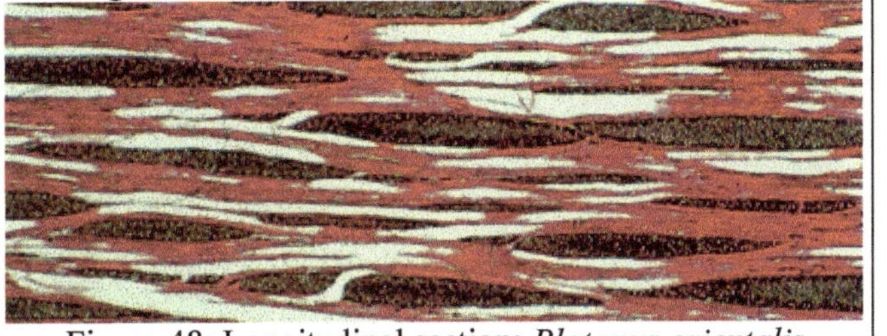

Figure 48: Longitudinal section: *Platanus orientalis*

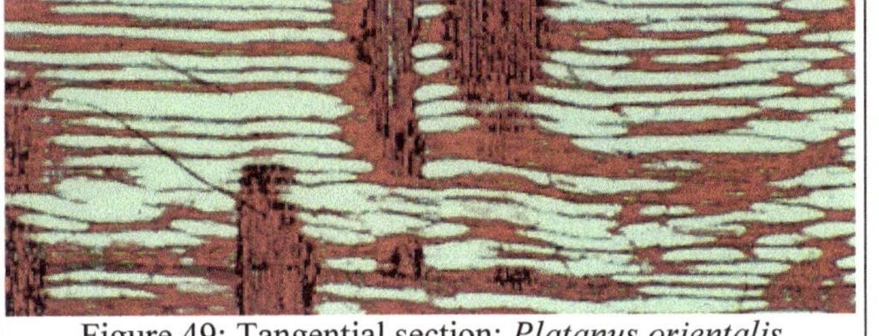

Figure 49: Tangential section: *Platanus orientalis*
Brightfield light microscopy photograph by Jenny Norman

exhibiting up to ca. 20 bars. Inter-vessel pits are opposite to diffuse and round to elongate ranging between 6-10 µm in diameter, with slit-like apertures. Vessel parenchyma and vessel-ray pits are similar but characteristically half-bordered. Fibres range between 1,450(840-1,800) gm in length. They are medium thick-walled, with distinctly bordered pits in radial and tangential walls.

Parenchyma are scanty while paratracheal and apotracheal occur either diffuseor in aggregates and in short uniseriate tangential bands; in 2-8-celled strands. Rays 2-6/mm, rarely 1(2)-seriate, mostly multiseriate up to **14** cells wide, distending tangentially at the ring boundaries. Cells are 3 mm high; homocellular, composed almost exclusively of procumbent cells, sometimes with one marginal row of square cells. There are many chambered crystalliferous. Crystals are solitary and prismatic. They are most abundant in chambered ray cells (Fahn, Werker and Baas 1986:53).

Arbutus andrachne

Growth rings are distinct. The wood is diffuse to weakly semi-ringed and porous. Vessels average between 100-130/mm^2, mostly in radial multiples of 2-6(12) although they can occur more rarely solitary (10-20%) or in clusters. They are angular to rounded in cross-section with a tangential diameter of 25-80 µm. Radial diameter can range up to 120 µm. Narrow vessels are sometimes intergrading with vascular tracheids. Vessel member

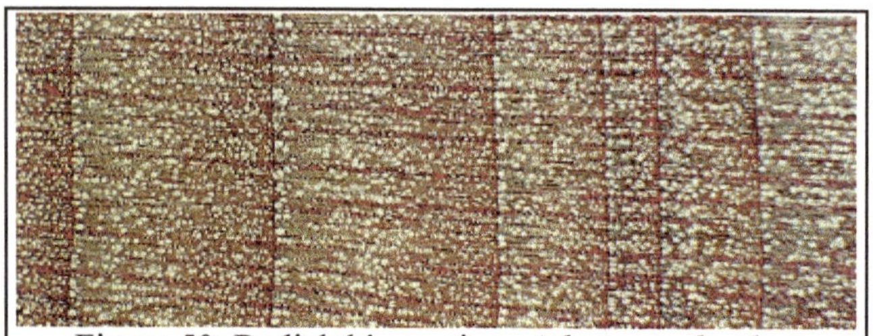

Figure 50: Radial thin section: *Arbutus andrachne*

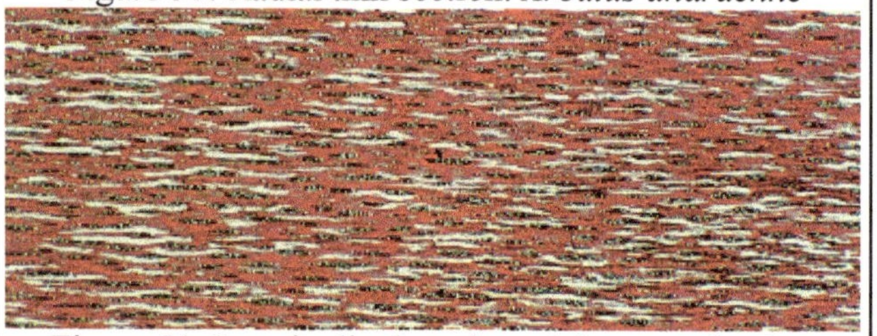

Figure 51: Longitudinal section: *Arbutus andrachne*

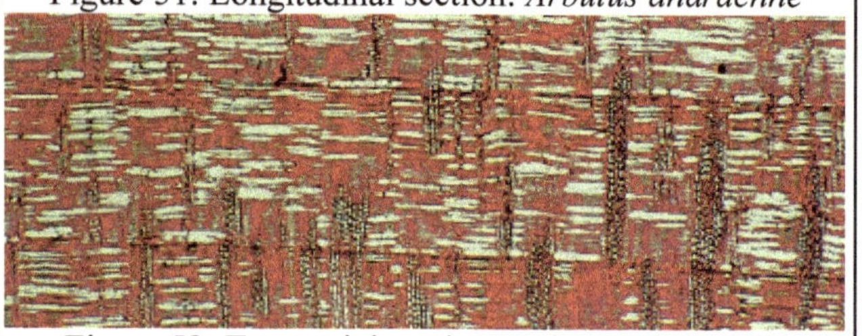

Figure 52: Tangential section: *Arbutus andrachne*
Brightfield light microscopy photograph by Jenny Norman

length is generally about 420(230-660) µm. Perforations are mainly simple although partly scalariform especially in the small vessel members, with 1-4(6) bars or even reticulate, in oblique end walls. Inter-vessel pits are opposite to alternate, rounded, 5(9) µm in diameter, with slit-like apertures. Vessel-parenchyma and vessel-ray pits are similar but half-bordered. Some vessels have coarse and distinct spiral thickenings. Fibres 610(350-830) µm long, thin- walled to medium thick-walled, mostly with simple pits confined to radial walls, partly septate and without spiral thickenings (libriform fibres); partly with distinctly bordered pits in radial and tangential walls and with spiral thickenings (fibre-tracheids) often associated with vessel multiples, especially in the latewood; rarely chambered crystailiferous. Parenchyma is very sparse with scanty paratracheal and diffuse apotracheal occuring in 2-5 celled strands; sometimes with chambered crystalliferous. Rays ca. 8/ mm, 1-4(5)-seriate, 1-25(43) cells high; heterocellular and tending to be of two distinct sizes (Kribs' heterogeneous type Il). Some of the upright cells are chambered crystalliferous. Crystals are either absent or prismatic, solitary or in aggregates together with minute ones in chambered axial parenchyma (in 3-6-celled chains) and upright chambered ray cells, rarely in chambered fibres (Fahn, Werker and Baas 1986:30).

Olea europea
Growth rings are distinct to faint. Vessels are diffuse, ca. 80/mm^2, solitary (ca. 20%), in radial multiples of 2-4(6) or occasionally in clusters; angular to rounded in

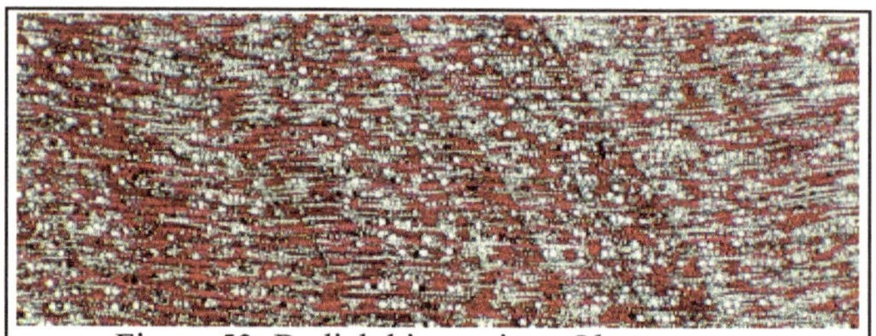

Figure 53: Radial thin section: *Olea europea*

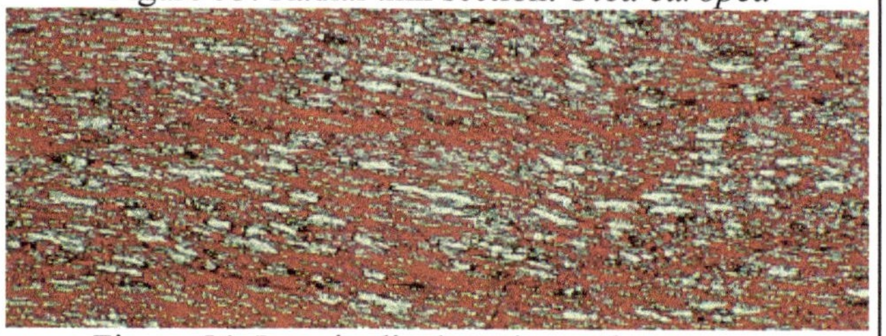

Figure 54: Longitudinal section: *Olea europea*

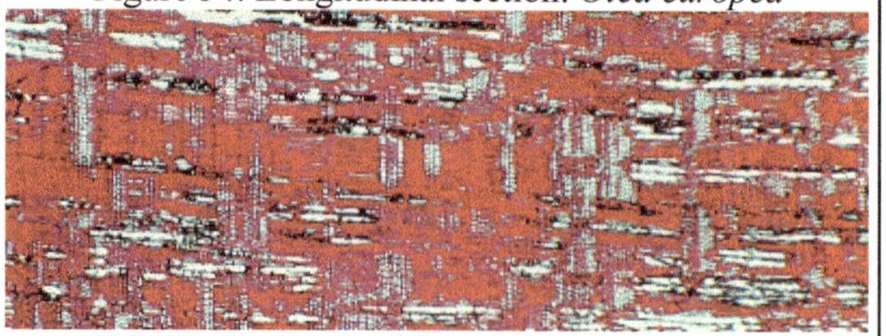

Figure 55: Tangential section: *Olea europea*
Brightfield light microscopy photograph by Jenny Norman

crosssection, tangential diameter 30-60 µm, radial diameter up to 70 gm, walls 3-6 µm thick. Vessel member length averages 310(190-380) µm. Perforations are simple in horizontal to oblique end walls. Inter-vessel are alternate and rounded, 3-4µm in diameter. Vessel-parenchyma and vessel-ray pits are similar but half-bordered. Some vessels contain gummy substances.

Fibres are 790(520-1,030) µm in length and they are medium-thick to very thick-walled, with minutely bordered pits in the radial walls. Parenchyma are mostly vasicentric although sometimes confluent, and sparse apotracheal are diffuse; fusiform and in 2-4-celled strands.

Rays 8-13/mm, 1-2(3)-seriate, up to 12(20) cells high; heterocellular (Kribs' heterogeneous types I and 11); often crystalliferous. Crystals are acicular, minute, numerous in ray cells (Fahn, Werber and Baas 1986:52)..

Acer obtrusifolia
Growth rings are distinct with vessels diffuse, ca. IOO/mm^2, solitary (40-50%) and in radial multiples of 2-5, up to 16 in latewood, very rarely in small clusters, rounded to somewhat angular in cross-section, tangential diameter 25-60 µm, radial diameter up to 80 µm, walls 2-4 µm thick. Vessel member length 280(230-340) µm. Perforations simple in oblique end walls. Inter-vessel pits alternate, polygonal, ca. 9 µm in diameter, with slit-like apertures. Vessel parenchyma and vessel-ray pits are

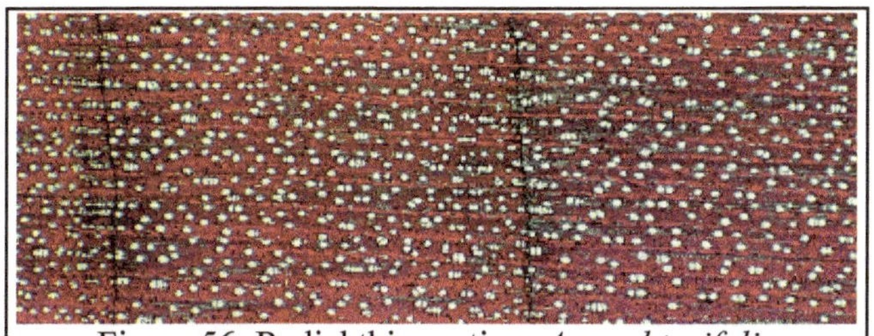

Figure 56: Radial thin section: *Acer obtusifolia*

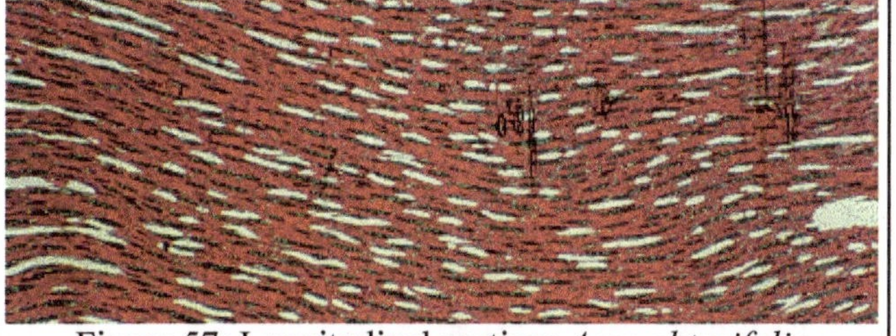

Figure 57: Longitudinal section: *Acer obtusifolia*

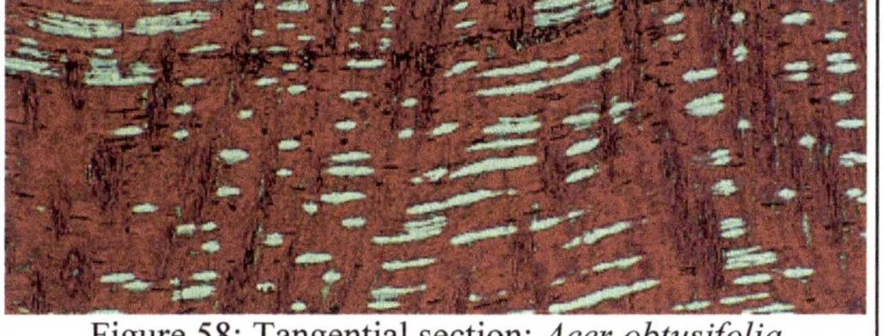

Figure 58: Tangential section: *Acer obtusifolia*
Brightfield light microscopy photograph by Jenny Norman

more rounded and half-bordered. Vessel walls with prominent spiral thickenings. Fibres 730(510-920) µm long, medium thick-walled, with numerous simple pits mainly in radial walls; some fibres gelatinous. Parenchyma sparse, scanty paratracheal and diffuse apotracheal, in 2-4-celled strands; often chambered crystalliferous. Rays ca. I I /mm, 1-4-seriate, up to ca. 50 cells high, almost always homocellular, composed of procumbent cells, rarely with some square marginal cells; infrequently crystalliferous. Crystals are solitary, prismatic, in chambered parenchyma and ray cells (Fahn, Werber and Baas 1986:6).

Crataegus azarolus

Growth rings are distinct. Vessels are diffuse, ca. 250/mm^2 appearing as solitary mostlty, but occurring occasionally in variously directed multiples of 2(3), or very small clusters In appearance they are angular to round in cross-section with a diameter range of between 20-70 µm. Walls are ca. 1.5 µm thick. Vessel member length 440(250-640) µm. Perforations simple in oblique end walls. Inter-vessel pits are diffuse, rounded, 5-9 µm in diameter, with slit-like, sometimes coalescent apertures. Vessel-parenchyma and vessel-ray pits similar but half-bordered. Some vessels with very faint spiral thickenings. Fibres 740(530-960) µm long, thick-walled, with distinctly bordered pits in radial and tangential walls; some with fine spiral thickenings. Parenchyma scanty paratracheal and diffuse apotracheal, sometimes diffuse-inaggregates; in 2-4-celled strands; occasionally

Figure 59: Radial thin section: *Crataegus azarolus*

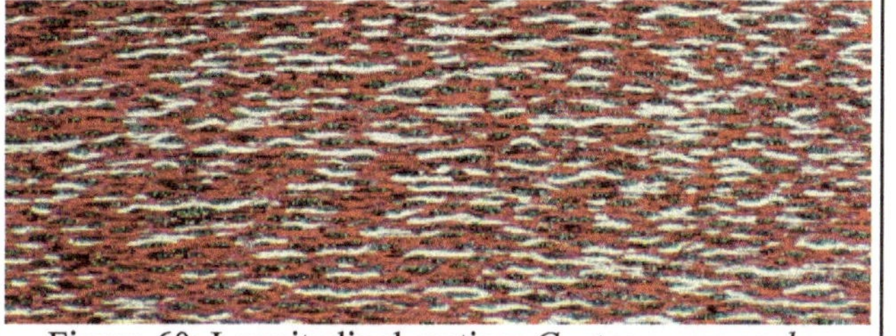

Figure 60: Longitudinal section: *Crataegus azarolus*

Figure 61: Tangential section: *Crataegus azarolus*
Brightfield light microscopy photograph by Jenny Norman

crystalliferous. Rays 11-15/mm, 1-3(4)-seriate, up to 25(35) cells high; homocellular, composed of procumbent cells; marginal cells less strongly procumbent than the central cells. Crystals sometimes in clusters or prismatic, in enlarged chambered parenchyma cells (Fahn, Werber and Baas 1986:61).

Characteristics of the genus Quercus

Vessels almost exclusively solitary, in a radial, flame-like or dendritic pattern. Perforations simple. Vasicentric tracheids present. Large and simple vessel-parenchyma and vessel-ray pits present. Fibers thick to very thick walled, with minutely boredered pits. Parenchyma canty paratracheal, and apotracheal diffuse, diffuse-in-aggregates and in short narrow tangential bands. Rays of two distinct sizes, uniseriates and very wide and high multiseriates; the latter partly compound. Crystals solitary, prismatic, in chambered and non-chambered ray and parenchyma cells. (Fahn, Werker and Baas 1986:109)

Quercus infectoria (Oliver) (*Q. lusitanica* Lam.; Boiss) (*Q. boissieri* Reut.)

Growth rings distinct. Wood ring-porous. Vessels almost exclusively solitary, rarely in pairs, forming a flame-like or dendritic pattern together with parenchuma cells and casicentric tracheids; 5-75/mm^2, earlywood vessels rounded, latewood vessels angular to slightly rounded in cross-section, tangential diameter 30-200 µm, walls 2-3 µ m thick. Vessel member length 450(250-590) µ m. Perforations simple in horizontal to oblique end walls.

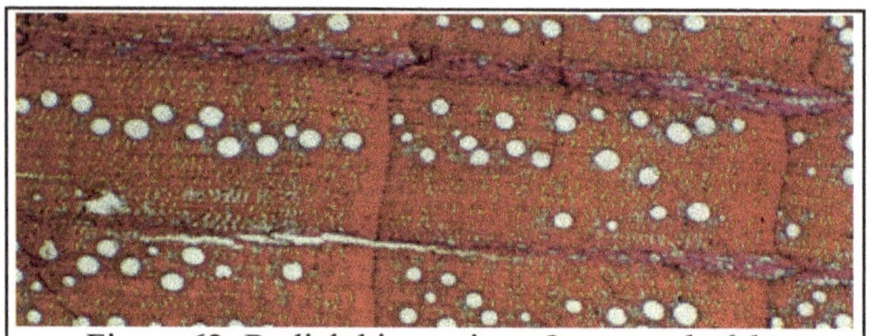

Figure 62: Radial thin section: *Quercus alnifolia*

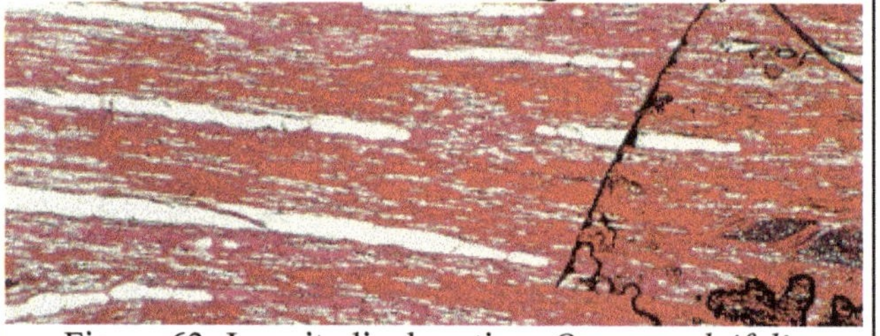

Figure 63: Longitudinal section: *Quercus alnifolia*

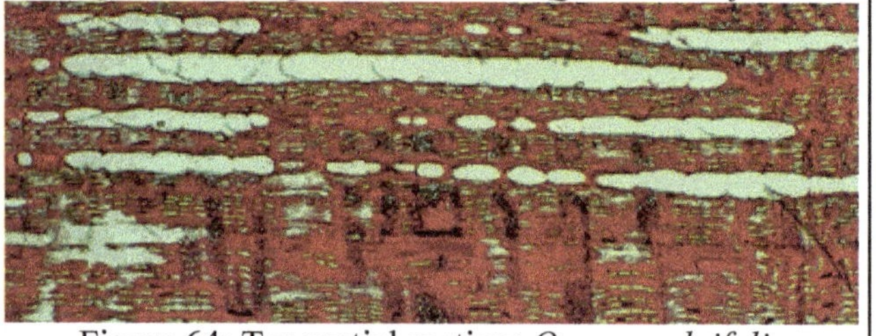

Figure 64: Tangential section: *Quercus alnifolia*
Brightfield light microscopy photograph by Jenny Norman

Vessel-tracheid pits widely spaced or alternate, round, 7-9 μm in diameter, with slit-like or oval apertures. Vessel-parenchyma and vessel-ray pits large and simple or with reduced borders, round to horizontally or vertically elongate.

Vasicentric tracheids abundant, forming the ground tissue associated with the vessels together with axial parenchyma.

Quercus lusitanica

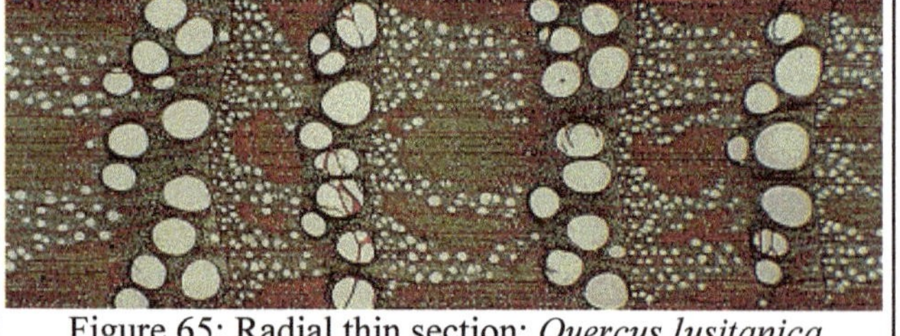

Figure 65: Radial thin section: *Quercus lusitanica*

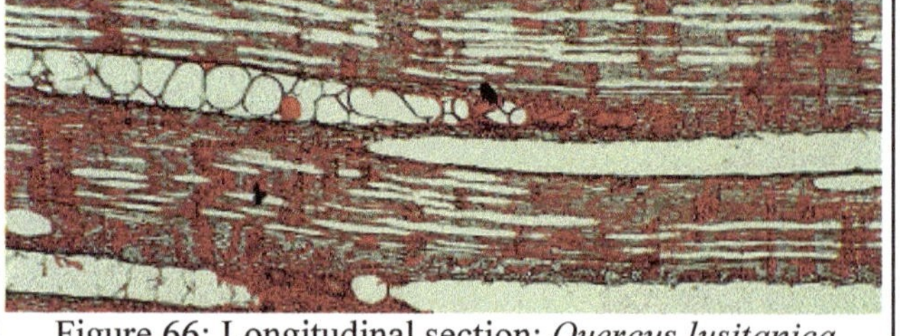

Figure 66: Longitudinal section: *Quercus lusitanica*
Brightfield light microscopy photograph by Jenny Norman

Fibres 1,020(670-1,380) μm long, thick to very thick walled, with minutely bordered pits mainly in radial walls; sometimes gelatinous.

Parenchyma scanty paratracheal and apotracheal diffuse-in-aggregates and in 1-seriate tangential bands; in many-celled strands; often chambered crystalliferous.
Rays ca. 10 / mm, of two distinct sizes, 1 (2)-seriate rays, 2-11 (18) cells high, and multiseriates 24-40 cells wide and up to 11 mm high; typically homocellular, but some upright and square cells scattered in the multiseriate rays; multiseriates partly aggregate; some cells chambered crystalliferous.

Crystals solitary, prismatic, in chambered axial parenchyma and ray cells; some crystalloferous cells enlarged and / or sclerified. (Fahn, Werker and Baas 1986:106)

Quercus coccifera ssp. calliprinos (Webb) Holmboe (Meikle 1985:1485)

Growth rings fairly distinct. Vessels diffuse, almost exclusively solitary, occasionally in pairs, forming, together with casicentric tracheids and parenchyma cells, an irregular flame-like or dendritic pattern; 5-20 / mm^2; rounded in cross-section, tangential diameter 25-130 μm, radial diameter up to 160 μm, walls ca. 5 μm thick. Vessel member length 360 (140-600) μm. Perforations simple in mainly oblique end walls. Vessel-tracheid pits round, diffuse, opposite or alternate, ca. 7 μm in diameter, with slit-like apertures. Vessel-parenchyma pits simple or with reduced borders, round to elongate. Vessel-ray pits large and simple, mainly vertical elongate.

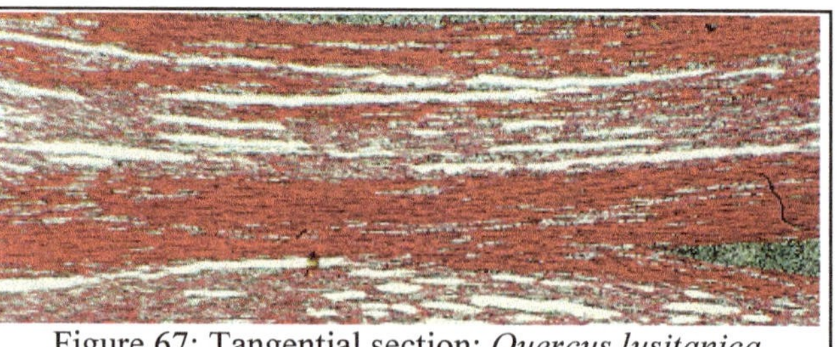

Figure 67: Tangential section: *Quercus lusitanica*
Brightfield light microscopy photograph by Jenny Norman

Fibers 810 (580-980) μm long, thick to very thick walled, sometimes gelatinous; with simple to minutely bordered pits more numerous in radial than in tangential walls (Fahn, Werker and Baas 1986:107).

Characteristics of the genus *Pistacia*
Vessels are mostly in multiples and clusters, often including vascular tracheids. Some vessels, especially wide ones, appear solitary. Perforations are simple occuring in horizontal end walls in the largest vessels.

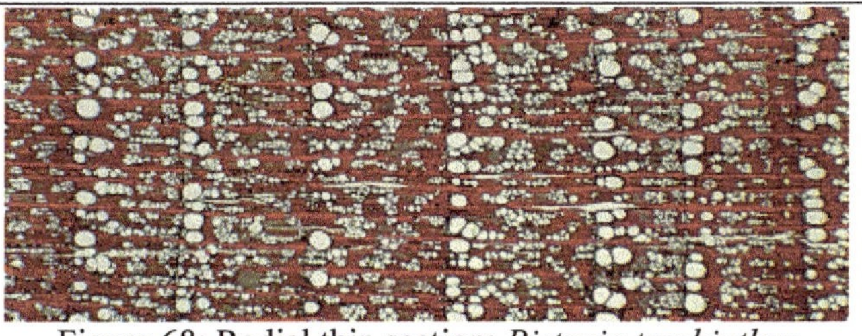

Figure 68: Radial thin section: *Pistacia terebinthus*

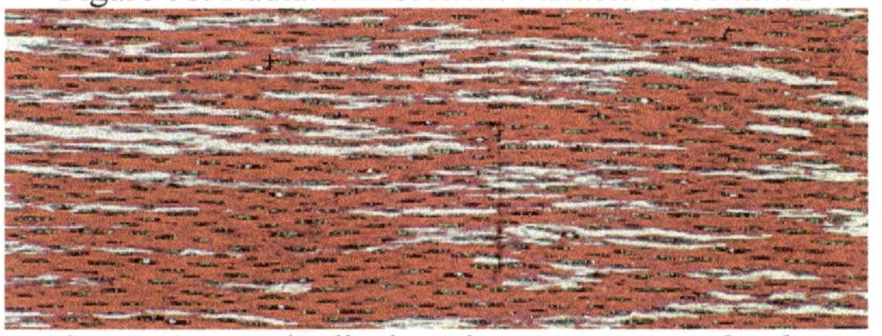

Figure 69: Longitudinal section: *Pistacia terebinthus*

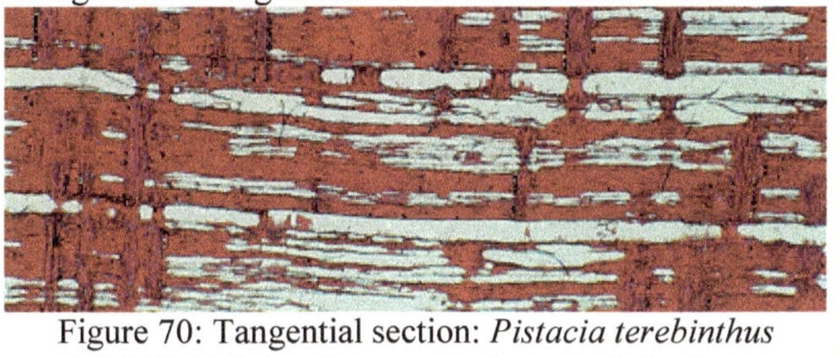

Figure 70: Tangential section: *Pistacia terebinthus*
Brightfield light microscopy photograph by Jenny Norman

Vessel-ray pits are mostly large and simple. Vessel walls, except for the widest, and vascular tracheids occur with prominent spiral thickenings. Tyloses are often present. Fibres with infrequent, simple pits are mainly or soley in radial walls. These are partly gelatinous. Parenchyma are sparse. Rays are both uni- and multiseriate, heterocellular, with short margins of square, upright and sometimes slightly procumbent cells and procumbent central cells. Resin ducts are present in some of the multiseriate rays.[7]

Timber Structure.

Sources of variation affecting wood products are: density; specific gravity; static bending; proportional limit; modules of rupture; modules of elasticity; and stress in compression. The specific gravity of timber is the ratio of the density of oven dried wood to the density of water at four degrees Celsius where the density of water is $1.0 g/cm^3$. Density is the weight or mass of a unit volume of wood. In the metric system, density and specific gravity are numerically identical. Density, therefore, relates to the basic wood structure of tree species and their variable moisture content. The presence of extractives in wood cells can have a major effect upon the density of the wood. This is an important factor relating to some of the pine species indigenous to Cyprus, as they are very high in mineral and extractable substances, such as resin, which add to the weight of a timber length, but reduce its strength.

Mechanical properties of wood define its action as a natural engineering material under load or stress, such as, for example, the performance of ship keels or ribs in heavy seas. "Static bending" is measured by loading a beam at a slow, constant rate, typically resulting in deflections of fracture of 2.5 cm per minute. "Proportional limit" is the limit of proportionality between load (stress) and deformation (strain). With an increase in load beyond the proportional limit, deformation increases by a greater percentage than the load. Stressing beyond the proportional limit results in permanent deflection. "Modules of rupture" is the measure of the breaking strength of a beam expressed in terms of stress per unit area. It is determined by computing the fibre stress in the outer fibres of a beam at maximum load and serves as a practical strength comparison for different species of timber. "Modules" of elasticity are a measure of the stiffness or rigidity of a beam or column. It is the ratio of stress to corresponding strain below the proportional limits. Other mechanical properties that are frequently measured are maximum stress in compression parallel to grain, compression strength perpendicular to grain, sheer strength parallel to grain, compression strength perpendicular to grain, sheer strength parallel to grain, impact bending strength, tensile strength perpendicular to the wood grain, and hardness (Wenger 1984).

All wood substance, regardless of the species from which it comes, has about the same density. The difference in density that occurs between species arises from the cellular structure of wood and depends on the size of the cell and the thickness of the cell walls. (See Figures 4-17)

Wood in the green state contains a substantial quantity of moisture which varies considerably between heartwood or sapwood, and again with great differences between species. The amount of moisture in a piece of wood accordingly affects the weight of the wood. Most species of wood, even when green, float in water. The buoyancy of wood is due to air contained within the cells. When wood is kept submerged, the air within the cells is gradually expelled by water, with the result that it becomes heavier and finally sinks.

The rate at which wood will absorb water when continuously exposed to it varies with: species; whether it is sapwood or heartwood; the length of the piece; size of cross section; and other factors. Whereas one species may become waterlogged after a relatively short period of exposure, another may remain under water many years

[7] Ongoing research into the description of the wood atanomy of the endemic Cypriot species is being carried out by Dr. P. Evans, Forestry Dept. Australian National University, using photographic material from slides prepared by the Imperial Forestry Institute, Oxford in 1953, and fresh material supplied by the Cyprus Forestry Department.

without absorbing so much water that it becomes heavy enough to sink. The heartwood of most species is much more impermeable than the sapwood. Thin boards or short sections absorb water more rapidly and become waterlogged sooner than thick long pieces. Since water diffuses or moves most easily in the longitudinal direction, the wood close to the end surfaces of submerged lumber will normally have a much higher moisture content than wood 0.3 m or more from the ends.

For these reasons it is extremely difficult to calculate the weight of water that a given ship or boat will take on if continuously afloat over a period of years. The problem of doing so is further complicated by the fact that, in hull planking for example, drying continues from the inboard face even though the outboard face is in continuous contact with water. Under such conditions, solid planking never becomes saturated, particularly if it is made of an impermeable species such as cedar. On the other hand, keels are usually thick enough so that, if made of heartwood of a hardwood species, they can remain in water for years without becoming waterlogged. By contrast, sapwood may become waterlogged very quickly.

The specific gravity of a piece of wood is an excellent index of the amount of wood substance the piece contains and, therefore, is also an index of its strength properties. Average specific gravity-strength relationships based on a large number of species of both hardwood and softwood varieties show that some properties, such as maximum crushing strength parallel to grain, increase approximately in proportion to the increase in specific gravity, although in a few cases some species increase more rapidly. Again, although there appears to be a general correlation between specific gravity and strength, the rule does not hold true for all species, particularly those whose wood contain relatively large amounts of resin, gum, and other extractives, which add to the weight but do not contribute equally to the strength as would a like amount of wood substance. In addition, species vary in the structural arrangement of their fibres. For these reasons, two species that average the same specific gravity may exhibit different strength characteristics.

Conditions of growth often cause wide variations in weight of a species of wood; some trees grow faster than others of the same species; some are more vigorous than others. Trees with wide annual rings have grown more rapidly than trees with narrow rings. In softwoods, as a general rule, exceptionally slow-growth or fast-growth material is lighter in weight and weaker than normal-growth stock. In hardwood, fast growth is generally indicative of good strength properties, although slow growth does not necessarily mean weakness. Locality of growth, within the normal range of growth of a species, has relatively little effect on the quality of lumber.

All wood subjected to normal use conditions contains a degree of moisture. The exact quantity of moisture depends on the amount of seasoning the material has received and from the perspective of long-term usage, the atmospheric and related conditions to which it is exposed. The strength of wood is intimately related to the amount of moisture it contains. As wood dries, most strength properties are increased. On the other hand, there are great differences in the strength of different species of wood at the same moisture content. A relationship also exists between the strength of woods at the same moisture content and their weight, the drier woods being stronger.

A final important point is that wood shrinks most in the direction of the annual growth rings (tangentially), about half to two-thirds as much across these rings (radially), and very little, as a general rule, along the wood grain (longitudinally). Typically, the heavier species of wood shrink more across the grain than the lighter ones. When freedom from shrinkage is more important than hardness or strength, for example, in the planking of small boats, a light-weight species would ideally be chosen. When it is important to combine hardness with strength or low shrinkage, as in the case of tree-nails, some exceptional species such as the *Quercus* spp. and *Alnus* spp. would be selected.

Table 3 lists the mechanical properties of the most important commercial species within the Cypriot environment, and will aid in the demonstration of their cultural association through trade potential and desirability. The figures quoted in Table. 2 represent new and as yet unpublished research (Burnet). Previous data was gleaned from sources as diverse as the University of Thesalonika (Tsoumis 1991) and the South African Forestry Department in Pretoria (1977). However, both sets of figures appear to be derived from Kollmann (1951). The figures presented in Table 2 were prepared by Peter Chapman, Research Division, State Forests of New South Wales (1995), from small specimens of wood supplied by Mr. Alexandrou, Principal, Cyprus Forestry College, Prodromos. As noted, some difficulty was encountered due to the small size of some specimens. The analysis overall is the first to be conducted as a comparison of Cypriot species.

Timber utilisation

The wood of the *Pinus brutia* is harder than that of most other pines and after seasoning is more inclined to surface check or splitting (a form of drying degrade often related to excessive temperatures) than other species. For this reason slow seasoning is of great importance for maximum wood utilisation. Timber from relatively young trees, those of about 50 years growth, is hard and heavy, and consequently very strong. Timber from the *Pinus brutia* was used in construction as flooring, doors and ceilings, for ship building and musical instruments. Its bark was important for the tanning of hides and its resin for pitch and tar production.

The timber from the *Cupressus sempervirens* is fine, even and straight-grained. The relatively high figure quoted for hardness reflects these fibre qualities and its lack of retained sap and resin. Growth rings are distinct and young timber can be knotty. A fragrant scent repels insects. The timber dries rapidly with little degradation or splitting. It is easily worked and finishes well but it is difficult to make a clean bore unless the timber length is well supported. Likewise it is difficult to mortice. These factors reflect the density of the wood (Wenger 1984.585). The timber is liable to termite and marine-borer attack (Bolza and Keating 1972) if planking is subjected to a stagnant water anchorage. Timber from the *Cupressus sempervirens* was used in the ancient Near East for construction, shipbuilding, mummy cases, religious carvings, and the doors of sanctuaries and religious structures (Kings I 5:22; Chronicles II 3:5). The doors of St. Peters in Rome were hung for over 1100 years without showing signs of decay (Record and Hess 1943:7).

The *Platanus orientalis* (often confused with the sycamore by North American authors) was utilised for durability and strength although the wood weight was lighter than most *Quercus spp.* Principle uses were for chariot axles and hubs of wheels, including water wheels. It was used in the construction of Egyptian coffins and mummy cases. The wood grain is usually irregular, fairly easy to work, finishing smoothly but likely to warp unless carefully seasoned (Record and Hess 1943:428).

The wood from the *Juniperus spp.* was used in ship construction as ribbing because the natural curvature of branches from mature trees attained ideal proportions for this purpose (Manniche 1989).

Table 3: Sample mean test results

SAMPLE MEAN TEST RESULTS.						
	Compres-sion	**Bending**	**Bending**	**Shear**		**Moisture**
	Strength	**MOR**	**MOE**	**Strength**	**Density**	**Content**
Botanical Name	**(MPa)**	**(MPa)**	**(MPa)**	**(MPa)**	**(kg/m3)**	**0%**
Acer obtusifolium	68	154	13190	17.7	771	10
Arbutus andrachne	93	172	15700	22.0	934	10
Cedru brevifolia	56	86	6890	12.7	593	11
Cupressus sempervirens	50	94	7880	13.6	521	11
Crataegus azarolus	56	117	10440	13.8	699	11
Juniperus oxycedrus (c)	53	105	7190	15.5	572	11
Juniperus Oxycedrus (p)	65	109	8940	14.9	714	10
Laurus nobilis	67	141	8050	16.5	785	10
Myrtus communis	80	154	13910	19.7	911	9
Olea europea	65	51	7390	23.8	1053	6
Pinus brutia	50	82	6800	13.5	677	9
Pinus nigra	37	82	5710	10.8	513	11
Pistacia lentiscus	78	147	10400	15.4	953	9
Pistacia terebinthus	59	94	8620	21.9	971	9
Platanus orientalis	62	120	11180	15.8	699	9
Quercus coccifera	88	193	13260	22.6	1050	9
Quercus lusitanica	82	145	11900	22.0	1074	10
Ceratonia siliqua	63	90	8100	17.5	844	9

Table 4: Specimen Test Results (Compression and Shear)

SPECIMEN TEST RESULTS (Compression and Shear)							
		Compres-sion	Shear				Moisture
	Specimen/	Strength	Strength	Shear	Sloping	Density	Content
Botanical Name	Size (1)	(MPa)	(MPa)	Plane (2)	Grain (3)	(kg/m3)	%
Acer obtusifolium	A/i	71	20.1	R	>15	769	9
Acer obtusifolium	B/i	65	18.9	Q	>15	775	10
Acer obtusifolium	C/i	68	14.1	R	>15	771	10
Acer obtusifolium mean		68	17.7			771	10
Arbutus andrachne	A/i	96	22.1	Q	6	971	10
Arbutus andrachne	B/i	82	20.7	Q(4)	7	898	10
Arbutus andrachne	C/i	102	23.2	Q	7	934	9
Arbutus andrachne mean		93	22.0			934	10
Cedrus brevifolia	A/ii	53	13.5	T	>15	575	11
Cedrus brevifolia	B/ii	-	8.6	Q	>15	590	10
Cedrus brevifolia	C/ii	59	16.0	T	>15	613	12
Cedrus brevifolia mean		56	12.7			593	11
Cupressus sempervirens	A/ii	47	12.0	R	>15	484	11
Cupressus sempervirens	B/ii	51	12.8	Q	>15	487	11
Cupressus sempervirens	C/ii	65	17.6	Q	>15	624	10
Cupressus sempervirens	D/ii	36	12.0	Q	>15	489	11
Cupressus sempervirens mean		50	13.6			521	11
Crataegus azarolus	A/ii	55	13.9	R	>15	706	11
Crataegus azarolus	B/ii	54	15.0	Q	>15	677	11
Crataegus azarolus	C/ii	59	12.5	R	>15	714	10
Crataegus azarolus mean		56	13.8			699	11
Juniperus oxycedrus (c)	A/ii	48	15.5	T	>15	512	11
Juniperus oxycedrus (c)	B/ii	53	15.2	R	>15	559	10
Juniperus oxycedrus (c)	C/ii	59	15.7	Q	>15	644	1
Juniperus oxycedrus (c) mean		53	15.5			572	11
Juniperus Oxycedrus (p)	A/ii	-	-	-	10	669	9
Juniperus Oxycedrus (p)	B/ii	66	16.6	Q	10	712	11
Juniperus Oxycedrus (p)	C/ii	65	13.1	R	10	761	10
Juniperus Oxycedrus (p) mean		65	14.9			714	10
Laurus nobilis	A/ii	63	16.3	R	6	793	10
Laurus nobilis	B/ii	72	17.2	T	6	790	10
Laurus nobilis	C/ii	65	16.1	R	7	771	10
Laurus nobilis mean		67	16.5			785	10
Myrtus communis	A/ii	73	20.1	T	9	820	9
Myrtus communis	B/ii	81	21.3	T	8	904	10
Myrtus communis	C/ii	83	17.2	T(5)	9	971	9
Myrtus communis	D/ii	82	20.3	T(5)	9	950	9
Myrtus communis mean		80	19.7			911	9
Olea europea	A/iii	68	30.0	Q(6)	1	1099	5
Olea europea	B/iii	62	18.1	Q(6)	1	1006	6
Olea europea	C/iii	64	23.3	Q(6)	1	1054	5
Olea europea mean		65	23.8			1053	5
Pinus brutia	A/ii	44	11.5	Q	6	633	8
Pinus brutia	B/ii	56	14.9	Q	3	770	9

Pinus brutia	C/ii	52	14.2	Q	6	628	9
Pinus brutia mean		50	13.5			677	9
Pinus nigra	A/ii	40	9.6	Q	>15	555	11
Pinus nigra	B/ii	36	10.4	Q	>15	503	10
Pinus nigra	C/ii	34	12.3	R	>15	482	11
Pinus nigra mean		37	10.8			513	11
Pistacia lentiscus	A/ii	-	15.5	T(5)	6	960	8
Pistacia lentiscus	B/ii	78	15.2	T(5)	7	947	10
Pistacia lentiscus mean		78	15.4			953	9
Pistacia terebinthus	A/ii	75	27.1	Q(5)	3	1077	9
Pistacia terebinthus	B/ii	42	21.3	R(5)	2	996	10
Pistacia terebinthus	C/ii	60	17.4	T(5)	3	840	10
Pistacia terebinthus mean		59	21.9			971	9
Platanus orientalis	A/ii	56	15.4	R	10	686	8
Platanus orientalis	B/ii	61	16.1	R	10	699	8
Platanus orientalis	C/ii	68	15.8	Q	10	712	9
Platanus orientalis mean		62	15.8			699	9
Quercus coccifera	A/ii	91	22.9	Q	>15	1019	10
Quercus coccifera	B/ii	85	20.5	R	>15	1056	9
Quercus coccifera	C/ii	89	24.2	Q	>15	1077	9
Quercus coccifera mean		88	22.6			1050	9
Quercus lusitanica	A/ii	84	17.0	R	6	1036	10
Quercus lusitanica	B/ii	82	22.4	R	6	1088	10
Quercus lusitanica	C/ii	78	24.3	T	5	1096	10
Quercus lusitanica	D/ii	83	24.4	Q	9	1078	10
Quercus lusitanica mean		82	22.0			1074	10
Certonia siliqua	A/iii	61	16.9	R	4	844	9
Certonia siliqua	B/iii	66	18.1	R	4	844	9
Certonia siliqua mean		63	17.5			844	9

Notes on above:

(i) indicates dimensions of 6 x 20 x 20 mm (DxWxL)
(ii) indicates dimensions of 8.5 X 20 X 20 mm
(iii) indicates dimensions of 12.5 x 20 x 20 mm

(2) indicates plane of shear failure:
R = radial / longitudinal plane
T = tangential / longitudinal plane
Q = intermediate between R & T

(3) Numerical value indicates slope of grain.
e.g. 6 indicates a slope of 1 in 6.

(4) Some decay in specimen
(5) Near pith
(6) Severe wavy grain

Table 5: Specimen Test Results (Bending)

SPECIMEN TEST RESULTS (Bending)							
		Bending	**Bending**				**Moisture**
	Specimen/	**MOR**	**MOE**	**Loaded**	**Sloping**	**Density**	**Content**
Botanical Name	**Size (1)**	**(MPa)**	**(MPa)**	**Face (2)**	**Grain**	**(kg/m3)**	**%**
Acer obtusifolium	A/i	161	13700	T	>15	769	9
Acer obtusifolium	B/i	148	12350	Q	>15	775	10
Acer obtusifolium	C/i	152	13510	Q	>15	771	10
Acer obtusifolium mean		154	13190			771	10
Arbutus andrachne	A/i	190	17150	Q	6	971	10
Arbutus andrachne	B/i	137	14950	T(4)	7	898	10
Arbutus andrachne	C/i	190	15000	Q	7	934	9
Arbutus andrachne mean		172	15700			934	10
Cedrus brevifolia	A/ii	-	-	R	>15	575	11
Cedrus brevifolia	B/ii	73	5290	Q	>15	590	10
Cedrus brevifolia	C/ii	99	8490	R	>15	613	12
Cedrus brevifolia mean		86	6890			593	11
Cupressus sempervirens	A/ii	81	6950	T	>15	484	11
Cupressus sempervirens	B/ii	92	7660	Q	>15	487	11
Cupressus sempervirens	C/ii	123	10000	Q	>15	624	10
Cupressus sempervirens	D/ii	81	6910	Q	>15	489	11
Cupressus sempervirens mean		94	7880			521	11
Crataegus azarolus	A/ii	115	10090	T	>15	706	11
Crataegus azarolus	B/ii	112	9730	Q	>15	677	11
Crataegus azarolus	C/ii	124	11490	T	>15	714	10
Crataegus azarolus mean		117	10440			699	11
Juniperus oxycedrus (c)	A/ii	100	9030	R	>15	512	11
Juniperus oxycedrus (c)	B/ii	108	7310	R	>15	559	10
Juniperus oxycedrus (c)	C/ii	107	6230	R	>15	644	1
Juniperus oxycedrus (c) mean		105	7190			572	11
Juniperus Oxycedrus (p)	A/ii	135	9450	Q	10	669	9
Juniperus Oxycedrus (p)	B/ii	117	9570	Q	10	712	11
Juniperus Oxycedrus (p)	C/ii	76	7810	T(5)	10	761	10
Juniperus Oxycedrus (p) mean		109	8940			714	10
Laurus nobilis	A/ii	138	7770	T	6	793	10
Laurus nobilis	B/ii	142	8690	T	6	790	10
Laurus nobilis	C/ii	142	7710	T	7	771	10
Laurus nobilis mean		141	8050			785	10
Myrtus communis	A/ii	159	12860	R	9	820	9
Myrtus communis	B/ii	179	15650	R	8	904	10
Myrtus communis	C/ii	131	12430	R(6,7)	9	971	9
Myrtus communis	D/ii	149	14700	R(6)	9	950	9
Myrtus communis mean		154	13910			911	9
Olea europea	A/iii	53	7810	Q(8)	1	1099	5
Olea europea	B/iii	44	6670	Q(8)	1	1006	6
Olea europea	C/iii	56	7690	Q(8)	1	1054	5
Olea europea mean		51	7390			1053	5
Pinus brutia	A/ii	69	6660	Q	6	633	8
Pinus brutia	B/ii	83	6670	R	3	770	9
Pinus brutia	C/ii	96	7080	R	6	628	9
Pinus brutia mean		82	6800			677	9
Pinus nigra	A/ii	91	6380	Q	>15	555	11

Pinus nigra	B/ii	72	5250	T	>15	503	10
Pinus nigra	C/ii	83	5500	T(9)	>15	482	11
Pinus nigra mean		82	5710			513	11
Pistacia lentiscus	A/ii	163	11320	Q(6)	6	960	8
Pistacia lentiscus	B/ii	132	9490	Q(6)	7	947	10
Pistacia lentiscus mean		147	10400			953	9
Pistacia terebinthus	A/ii	81	9800	Q(6)	3	1077	9
Pistacia terebinthus	B/ii	126	8610	Q(6)	2	996	10
Pistacia terebinthus	C/ii	76	7460	Q(6)	3	840	10
Pistacia terebinthus mean		94	8620			971	9
Platanus orientalis	A/ii	102	10390	T	10	686	8
Platanus orientalis	B/ii	111	11040	T	10	699	8
Platanus orientalis	C/ii	146	12110	T	10	712	9
Platanus orientalis mean		120	11180			699	9
Quercus coccifera	A/ii	185	10670	T	>15	1019	10
Quercus coccifera	B/ii	178	12560	T	>15	1056	9
Quercus coccifera	C/ii	216	16550	Q	>15	1077	9
Quercus coccifera mean		193	13260			1050	9
Quercus lusitanica	A/ii	165	12300	T	6	1036	10
Quercus lusitanica	B/ii	137	11080	R	6	1088	10
Quercus lusitanica	C/ii	158	12360	R	5	1096	10
Quercus lusitanica	D/ii	119	11850	R	9	1078	10
Quercus lusitanica mean		145	11900			1074	10
Certonia siliqua	A/iii	84	7160	T	4	844	9
Certonia siliqua	B/iii	97	9040	T	4	844	9
Certonia siliqua mean		90	8100			844	9

Notes on above:
(i) indicates dimensions of 6 x 12 x 103mm (DxWxL)
(ii) indicates dimensions of 8.5 X 12 X 130mm
(iii) indicates dimensions of 12.5 x 22 x 200mm

(2) indicates face to which load is applied:
R = radial face
T = tangential face
Q = intermediate between R & T

(3) Numerical value indicates slope of grain.
e.g. 6 indicates a slope of 1 in 6.

(4) Some decay in specimen
(5) Large knot
(6) Near pith
(7) 2mm Knots
(8) Severe wavy grain
(9) Two 4mm borer holes

Chapter Six: Patterns of Ecological Complexity in the Makheras

The Nature of Ecological Complexity.

Ecology has been defined in numerous ways, but the definition most acceptable to this thesis is: "the study of the principles which govern temporal and spatial patterns for assemblages of organisms" (Fenchel 1987:12). The emphasis is on principles, but the definition does not restrict ecology to the theoretical. It includes field observation and experiment in the quest for general principles. The definition does not deny its origin in natural history or the ability of ecology to explain the findings of naturalists as these elements of investigation remain a test of ecological competence (Fenchel 1987:12).

The six principles of ecology outlined by Grubb have been proposed as "laws of ecology". They are:

- A population must meet limitation of resources if not limited by predation, disease or other deleterious environmental factors (Malthus 1798)
- One biotype will replace another under a given set of conditions if it has a greater fecundity or lesser mortality (Darwin 1859)
- Following from 2, two biotypes which are not permanently subject to rarefaction by some third party or other environmental factor can coexist indefinitely only if they have different niches or are subject to greater intraspectic competition than interspecific (Gause 1934; but see Krebs 1972:231)
- A prey species and a predator species cannot coexist indefinitely unless the environment is heterogenous (Huffaker 1958)
- Density-dependent effects may either stabilise population size or produce cycles or produce 'chaos', depending on the degree of non-linearity of the relation between N_{t-1} and Nt (May 1974; 1986)
- Available energy decreases at each stage of predation (Lindemann 1942, Grubb 1989)

The complexity of ecosystems is not only in the observation of nature - a function of discoverable laws and principles - it is also a function of singular historical events and chance. Another dimension is added with the inclusion of evolutionary events. Evolutionary history is ontological, but there are also elements of historically singular events and of stochastic events. Only in very few cases is it possible to reproduce experiments on evolutionary events (Fenchel 1987:14).

Perception of the geographical distribution of diversity is strongly scale-dependent. While the general trend of decreasing diversity from the equator to the poles is clear at one scale, at finer scales of resolution a more detailed pattern is revealed. The sample size for a standard quadrat during the survey was 30m^2 which means that the diversity ratio of species recorded was uniformly greater than that recorded by Ogden (1995) who used 0.5x0.5m quadrats. Whittaker (1972) presented a framework for the discussion of species diversity which, with minor modifications (eg. Soule 1986), has been widely adopted:

Alpha-diversity relates to the number of species coexisting within a uniform habitat. It is an inventory of the species present within a community, often referred to as "species richness" in the plant literature or as "species packing" by ornithologists. In practice the concept is inhibited by the difficulty at times in defining "the community". As a working model it has been assumed that the quadrat as an example of the homogeneous compartment was sufficient. The homogeneity was visually assessed and included uniformity in topography, aspect, soil type, altitude and climatic factors. In common with published analyses (Huston 1979; Peet 1974; Wilson and Keddy 1988; Ogden 1995) it was recognised that the relative proportions of species is a component of diversity, but information presented here is confined to species richness as this factor can be easily interpreted in statistical form.

Beta-diversity describes the cross community level of diversity. It provides an index of the diversity across a landscape within bio-geographical region. For example, as the community composition changes along altitudinal gradients or varies with aspect, new species are encountered and others drop out. This species variation is termed beta-diversity. Two forest stands with the same number of species present (alpha-diversity) can differ in composition, thus exhibiting a beta-diversity. Beta-diversity can also be measured by similarity coefficients (Sorensen's coefficient) or by distance between stands in ordination space (Druitt et al. 1990).

Gamma-diversity. Cody (1986) described the differences in bird communities between similar habitats in different regions, and referred to this variation as gamma-diversity. In the case of plants, similar "habitats" might be taken to mean similar locations on major environmental gradients (for example, altitude). In practice gamma-diversity can be best differentiated from the other measures by regarding it as the total species inventory for a large geographical region containing all the main environmental gradients present in the total area under discussion, that is, both defined forests of Makheras and

Adelphi. Gamma-diversity can therefore be partitioned into alpha-diversity and beta-diversity. However, unlike beta-diversity, which is usually measured as differences or rates of turnover, gamma diversity is obtained by summing the species present in all the communities in a given region (Ogden 1995) (See Figure 71: Map of Species Density for the Makheras and Adelphi Forests).

Many floristic species are endemic to Cyprus while others are cosmopolitan, that is, are a more generally wide spread species throughout the Mediterranean, southern Europe or North Africa. Species patterning tends to reflect the geological formation of Cyprus and the influence of disturbance. The broad species patterning across the island was first described by Holmboe (1914) and Meikle (1975; Meikle 1986).

	Makheras:	Adelphi:
Altitude Range Average	881.68	637.04
Std. Dev	**228.57**	**134.77**
Number of Quadrats	**98**	**75**
Average Faunal Spec / Qd	**6.54**	**4.45**
St. Dev	**2.16**	**2.15**
Number of Faunal Species	**58**	**56**
Flora + Fauna	**132**	**121**
Basal Area	**20.38**	**34.69**
Std. Dev	**16.92**	**11.40**
Density of Trees (with measurable DBH)	**5.08**	**6.55**
Std. Dev	**4.53**	**3.68**
Density of Saplings	**5.30**	**8.90**
Std. Dev	**13.24**	**6.65**
Total Flora species	**74**	**65**
Average Flora Species per Quadrat	**9.26**	**5.75**
Std. Dev	**3.48**	**2.62**

Table 6: Gamma-diversity Comparisons for Makheras and Adelphi Forests

High diversity could be a result of more species recorded per community sample in any given quadrat (high alpha-diversity), or a finer mosaic of community types (higher beta-diversity). For example, were more species growing in association along altitudinal gradients in the Makheras Forest in comparison with the Adelphi Forest? Are there distinctive latitudinal trends within the communities? These questions may appear simplistic but they require a considerable amount of field research in different geographical areas to answer them. Relevant data focused specifically on these questions have hitherto been unavailable, but from the observations and flora collections made during the field survey, instructive comparisons can be made.

Assessment and understanding of species variety.

The importance of taxonomic resolution in assessing patterns and processes within the forest ecosystem is twofold:

(i) to assess environmental impact. The main requirement was description, explanation and predication from field sampling and correlation with excavation material to determine human utilisation, impact effect and change.

(ii) conservation of biodiversity. The aim was to provide information for the implementation of genetic conservation. Information required was species distribution, status, biology, minimum population size, and area of reserve.

The use of coarse taxonomy (that is, at family level) or rapid assessment meant that most species listed in the inventory could be identified either in the field or with only minimal laboratory research. Fine taxonomy (at species level) or slow assessment was used whenever appropriate, that is, when species identification at a fine level was possible, affordable and fast.

Coarse taxonomy for assessing environmental impact, such as wild fire or extensive felling, is adequate as a baseline study or monitoring. It may also be sufficient for inventory and distributional studies aimed at establishing representative measurements (Jones 1993). i) Inventories. In order to proceed with biodiversity assessment inventories of forest standing stock with a GBH in excess of 56 cm were recorded. All sub-strata and ground flora species were listed. Fauna observed, spoor and bone material or seed extracted from castings were also listed. These lists were compared between forests and again with known species from archaeological deposits. A further comparison with published information on the distribution of endemic species was then made. Whenever possible Latin species name was given with common name in parenthesis. The Greek common name was not recorded. For some species, particularly ground macroflora, genus only was recorded (see Table 7: Flora Species Identified in Forest Environments and Table 8: Faunal Species Identified in Forest Environments).

Figure 71: Map of Species Density for the Makheras and Adelphi Forests.

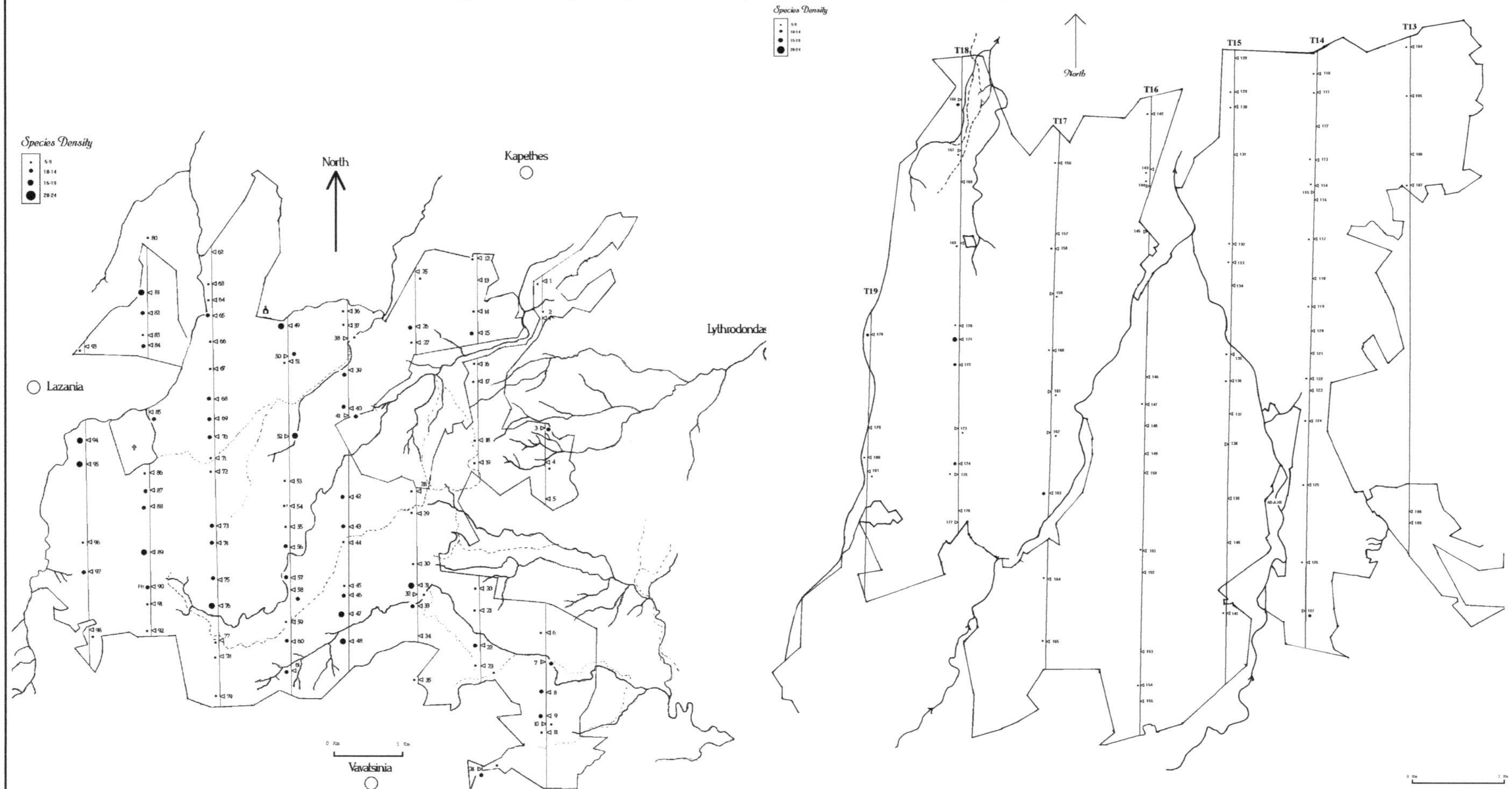

Makheras Forest, Cyprus

Adelphi Forest, Cyprus

Figure 72: Average biodiversity and tree cover per Altitude Band in the Makheras.

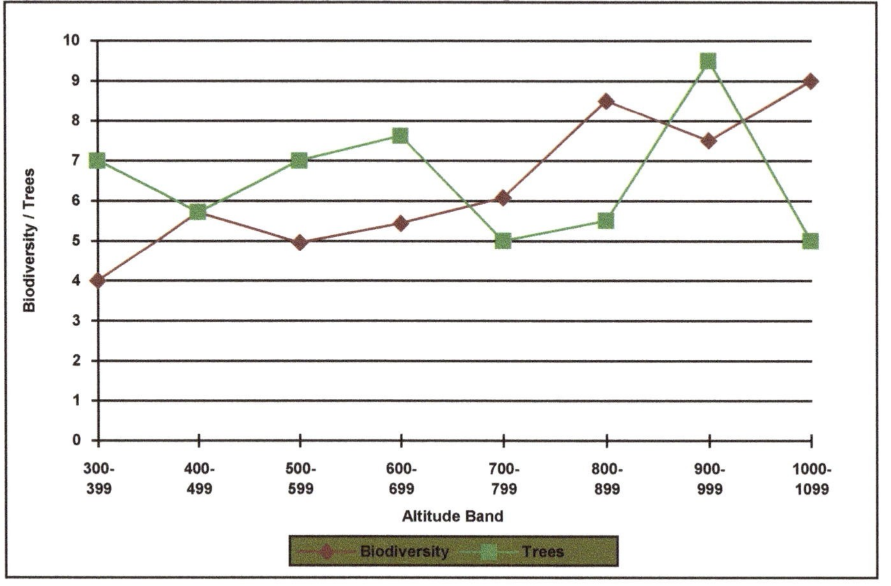

The steady increase in biodiversity as altitude increased reflects the different faunal species and habitat requirements.

Figure 73: Average biodiversity and tree cover by Altitude Bands in the Adelphi Forest.

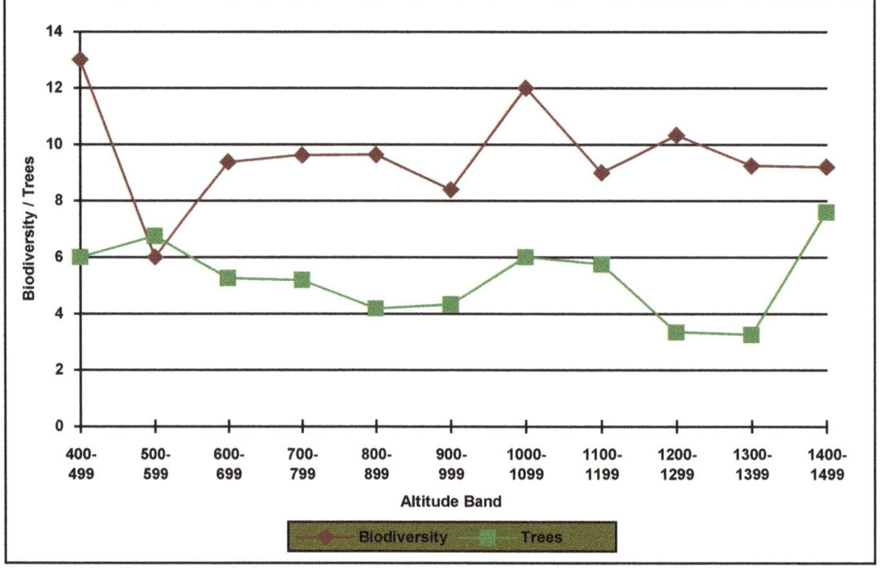

Acer obtrusifolium		Gagea juliae	*	Rhamnus palaestina	
Achillies sp.		Galium aparine		Rhus coriaria	
Agaricus campestris		Geastrum triplex		Romulea tempskayana	
Agropyron sp.		Geranium purpureum		Rubia tennifolia	
Allium ampeloprasum		Geranium sp.		Rubus sanctus	
Allium trifoliatum		Helichryysum italicum		Salvia cypria	
Alnus orientalis		Hordeum spontaneum		Scilla shyacinthoides	
cf. Andropogon		Inula viscosa		Scirpus lacustris	
Arabis purpurea		Lactarius delicious		Sedum cyprium	*
Arbutus andrachne		Juniperus oxycedrus		Sedum microstachyum	*
Arisarum vulgare		Lamium moschatum		Selaginella denticulata	
Artemisia sp.		Lathynis aphaca		Sinapis arvensis	
Asparagus acutifolius		Laurus nobilis		Smilax aspersa	
Asphodelus microcarpus		Lavandula stoechas		Smyrnium connatum	
Asplenium		Limodorum abortivrm		Sorghum	
Astralagus lusitanicus		Lithospenum diffusum		Spring ephemerals	
Avena barbata		Lonicera etrusca		Styrax officinalis	
Aymenocarpos circinnatus		Marrubium vulgare		Teucrium creticum	
Bamboo sp.		Mentha longifolia	*	Thymus integer	
Biscutella didyma		Micromeria chionistrae	*	Trifolium campestre	
Blackstonia cf. acuminata		Micromeria myrtifolia		Trifolium stellatum	
Bromus cf. madritensis		Morus alba		Umbilicus rupestris	
Calycotome villosa		Muscari comosum		Valeriana italica	
Capparis spinosa		Muscari incoustrictum		Veronica cymbalaria	
Carlina coryrubosa		Muscari parviflorum		Vicia cassia	
Cedrus brevifolia	*	Myrtus communis		Vicia sp.	
Cephalorrhynchus cyprius	*	Olea europa		Vitis vinifera	
Ceterach offiinarum		Orchis anatolica		Zizyphus spina-christi	
Cheilanthes pteridioides		Origanum majorana	*	Zosima absinthifoia	
Cistus salviaefolius		Orobanche cypria	*	Xanthoria parietina	
Cistus villosus creticus		Papaver somniferum			
Cladonia sp.		Parapholis sp.			
Cladonia cf. convoluta		Parmelia tiliacea			
Cladonia furcata		Peltigera cf. canina			
Clematis cirrhosa		Phagnalon graecum			
Crataegus azarolus		Phaonalon rupestre			
Crepis reuteriana		Pinus brutia			
Crocus hartmannianus	*	Pistacia lentiscus			
Crupina crupinastrum		Pistacia terebinthus			
Cupressus sempervirens		Plantanus orientalis			
Cyclamen cyprium		Poa bulbosa			
Dianthus strictus	*	Poterium spinostrum			
Equisetum ramosissimum		Prasium majus			
Evernia prunastri		Prunus dulcis			
Ficus sp.		Ptilostemon chamaepeuce	*	An * indicates that the	
Foeniculum vulgare		Quercus alnifolia	*	associated species is	
Fumana arabiea		Quercus coccifera		endemic to Cyprus.	

Table 7: Flora Species Identified in Forest Environments.

1	Parus ater cypriotes	coal tits	39	Cynipoidea	gall wasp
2	Erithacus rubecula	robin	40	Odonata	dragon fly
3	Regulus requlus	gold crest	41	Sus	pig
4	Alectoris chukar cypriotes	chukar	42	Caprine	goat
5	Turdus merula	blackbird	43	Oryctolagus cuniculus	rabbit
6	Carrulus glandarius glaszneri	jay	44	Equus asinus	donkey
7	Fringilla coelebs cypriotis	chaffinch	45	Vipera lebetina lebetina	blunt nosed viper
8	Lanius nubicus	mask shrike	46	Coluber cypriensis	Cyprus whip snake
9	Parus major aphrodite	great tit	47	Typhops vermicularis	Worm snake (asp)
10	Corvus cornix sardonius	hooded crow	49	Lacerta laevis troodica	lizard (stripped)
11	Falco tinnunculus	kestral	49	Agama stellio	lizard (geko)
12	Accipiter nisus	sparrowhawk	50	Chamaeleon chamaeleon	lizard (chameleon)
13	Anthus trivialis	pipit	51	Hyla arborea	frog
14	Corvus corax	raven	52		snake skin (unidentified)
15	Cuculus canorus	cuckoo	53		ants
16	Carduelis carduelis niediecki	gold finch	54	cf Helix aspersa	snail shell
17	Sylvia melanothorax	cistus hopper	55	Lithobius	centipede
18	Phoenicurus phoenicusus striata	red start	56	Triantelope	huntsman spider
19	Muscicapa	spotted flycatcher	57	Arachnid	spider (black)
20	Sylvia communis	white throat	58	Arachnid	spider (striped leg)
21	Hirundo rustica	swallow	59	Homoptera	cicada case
22	Oenanthe oenanthe	wheatear	60		scorpion
23	Columba palumbus	wood pigeons	61	Diptera	midges
24	Phylloscopus trochilus	willow warbler	62		grub
25	Prunella modularis	dunnock	63	Nematode	worm
26	Sylvia atricapilla	blackcap	64		bees
27	Aquila heliaca	eagle	65	Colias croceus	yellow butterfly
28	Budytes flava	yellow wagtail	66	Terenbrionid	beetle
29	Saxicola torquata	stone chat	67	Acrididae	short horned grasshopper
30	Rattus rattus	black rat	86	Diplopoda	flat backed millipide
31	Vulpes vulpes	fox	69	Thaonmetopoeoi pytiocampi	processionary moth
32	Lepus capensis	hare	70	Canis	dog
33	Hemiechinus auritus	hedgehog	71	Anguis fragilis	slow worm
34	Gallus gallus		72	Orthoptera	grasshopper
35	cf. Acomys mesiotis	Cyprian Spiny mouse	73		burrow
36	cf Heteroptera seutelleridae		74		midden
37	Isoptera	termites	75		egg membranes
38	Dermaptera	earwig	76		nests

Table 8: Faunal Species and Habitats Observed in the Makheras and Adelphi Forests.
Species identified with help from Dr. Michael Given, Cameron Loundes, Dr. Sebastian Payne, Ancient Monuments Laboratory, London, Pambos Christodoulou Cyprus Forestry Department and Dr. Dina Hales, School of Biological Science, Macquarie University.

Figure 74 shows a trend to a greater number of saplings occuring in quadrats that have a greater level of canopy cover, i.e. as canopy cover increases, so does the number of saplings. It is important, however, to note that there were no quadrats in the Adelphi forest that had canopy cover >70%, and thus there is no data for that quantity. Figure 74 was generated by selecting quadrats with canopy cover of 1, 2, or 3 and then finding the average of the number of saplings present in that quadrat.

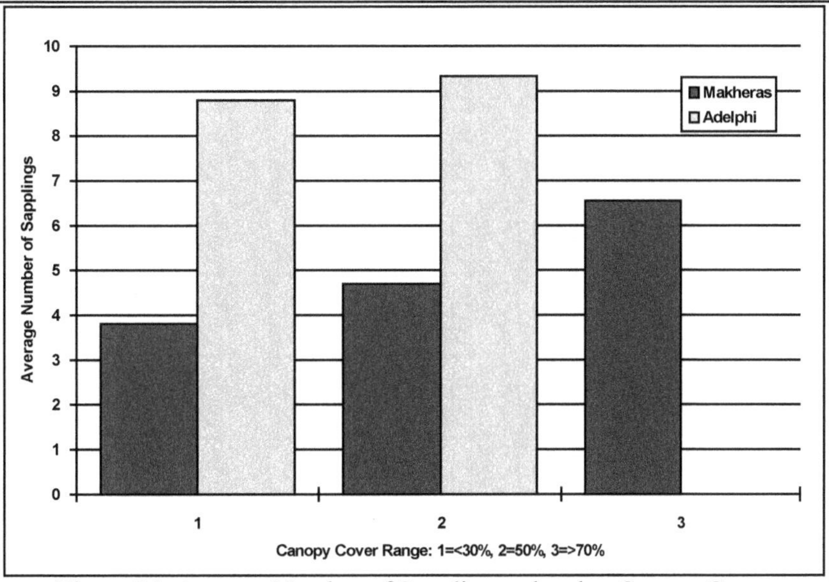

Figure 74: Average Number of Sapplings related to Canopy Cover

The correlation between canopy cover and the basal area of *Pinus brutia* indicates an increase in both forests when the canopy cover reaches 50%. A decline in basal area was also recorded when the canopy cover was greater than 70%. The decline in basal area representing *Pinus brutia* growing stock reflects the recorded density of *Quercus alnifolia* as a major sub-strata species. Based on this analysis future forest management will need to incorporate tolerant species as *Pinus brutia* replacements in order to sustain the forest structure and dynamic.

The depletion in sub-strata species in the Adelphi Forest is reflected in the lack of canopy cover in excess of 70%.

	Makheras	Adelphi
Burnt Average	1.167	1.364
Burnt Standard Deviation	0.471	0.481
Total Average	1.449	1.451
Total Standard Deviation	0.702	0.525

Table 9: Relative Canopy Cover over areas with a known Fire History.

The isolated presence of *Styrax officinalis* only in the mid-altitudinal band is the most significant feature of Figure 75. Clustering of *Cupresses sempervirens, Juniperus oxycedrus* and *Lauris nobilis* indicates modern reaforestation in areas adjacent to riverine associations. The presence of *Quercus coccifera, Platanus orientalis* and *Pistacia lentiscus* as a cluster at higher altitudes may be indicative of natural forest regeneration and forest dynamic systems outside sivicultural managent of the forest bioresource.

To facilitate the patterning of the sub-strata species illustrated in Figure 75 *Pinus brutia, Quercus alnifolia* and *Pistacia terebinthus* were excluded.

Canopy cover in relation to post-injury invasion was determined by correlating the species recorded at each quadrat described by clear felling; grazing and wild fire. In quadrats where clear felling was determind as the major disturbance secondary succession or invasion were calculated on a percentage basis. As many quadrats were reafforested mechanically with *Pinus brutia* this species was excluded from the results. Close to 100% of quadrats disturbed by fire were carrying *Cistus*, more than 50% of quadrats carried *Lavandula stoechas, Pistacia terebinthus* and *Quercus alnifolia.* Between 35% and 50% of quadrats carried *Asphodelus microcarpus, Crataegus azarolus, Galium aparine* and *Thymus sp.* Clear felling in the Adelphi Forest resulted in fewer species recorded overall with only *Cistus sp.* and *Asphodelus microcarpus* above the 35% range. Most species fell below 15% and included the major sub-strata associations.

Grazing damage was recorded at 10 quadrats. Species identified as dominants at a rate greater than 35% were *Cistus sp., Crataegus azarolus, Lavandula stoechas, Olea europea, Pistacia terebinthus, Quercus alnifolia* and *Thymus sp.* Only 25 species were recorded in total over these quadrats. Grazing damage in the Adelphi Forest was recorded only at a rate greater than 35% with *Asphodelus microcarpus* and *Cistus sp.* identified as dominant indicator species.

Post-fire vegetation in the Makheras was dominated at greater than 35% by *Asphodelus, Cistus sp., Crataegus azarolus, Graminae, Lavandula, Olea europea, Pistacia terebinthus* and *Quercus alnifoli*a. Only a limited number of species were recorded in the Adelphi Forest on areas with a known fire history. These species were a sub-set of those recorded for the Makheras but also included *Vicia cassia.* In both forests the greatest number of species were recorded between 10% and 20% with greater species diversity recorded in the Makheras.

(ii) **Rapid appraisals:** Rapid Biodiversity Assessment (RBA) is founded on the proposition that certain areas of ecological research, biodiversity assessment, conservation planning, resource management and environmental monitoring can be achieved without knowledge of species names (the Latin binomials). This proposition formulated by Beattie, Majer and Oliver (1993) reduces the characteristics of RBA to the "minimisation of the formal taxonomic content in the classification and identification of organisms". The taxonomic impediment encountered in Cyprus where many species were observed but not identified, severely inhibited the subsequent listing of taxa. It was important to understand the representation of the larger assemblage from the subsets examined. Distributional patterns and the way these related to other factors such as canopy cover, gradient, aspect and floristic associations were of importance (Jones 1993).

(iii) **Indicator groups**: Indicator species fall into four categories: faunal and macrofauna present after recent disturbance, flora colonising species present following disturbance, fauna indicative of undisturbed areas and old growth associations. All quadrats with disturbance recorded were grouped and the species analysed for comparisons.

Ants were recorded in many quadrants and appeared to be in association with the species association identified on disturbed areas. At greater than 35% the species were *Asphodelus microcarpus*; *Cistus sp.*; *Crataegus azarolus; Pinus brutia*; *Pistacia terebinthus* and *Quercus alnifolia*.

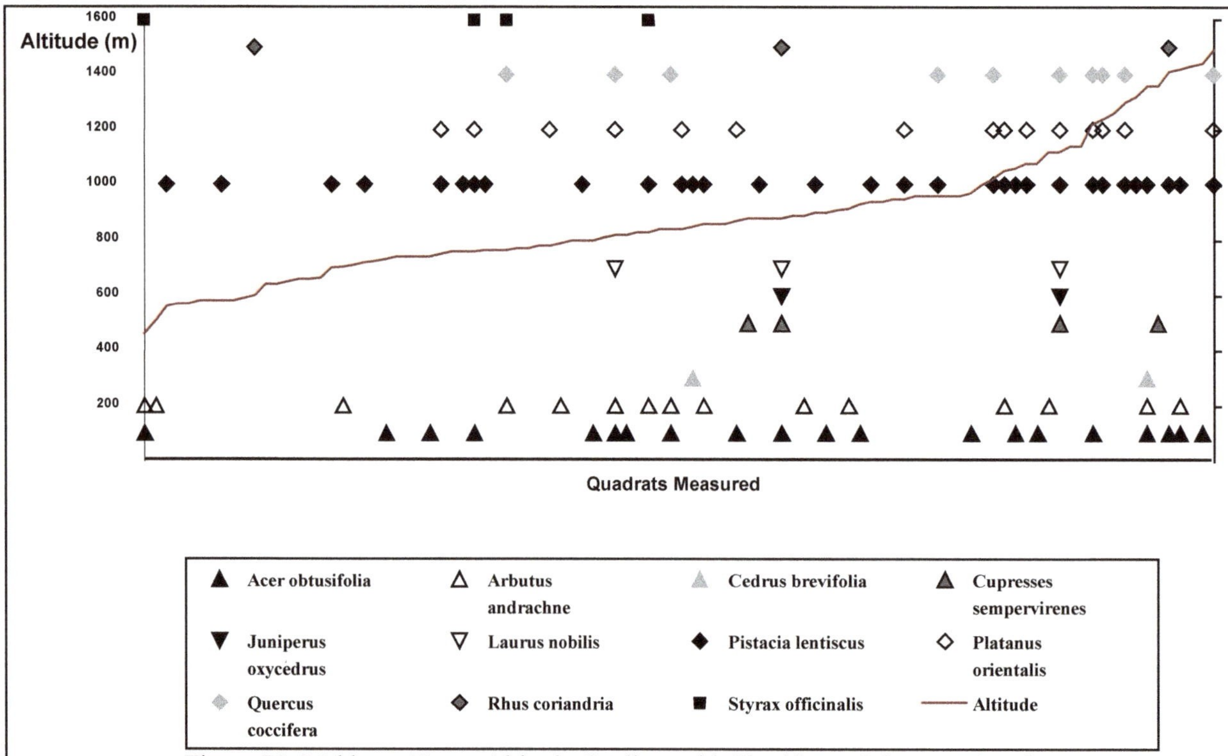

Figure 75: Makheras Forest Altitude Gradient with Significant Sub-strata Species.

The presence of ants may, therefore, act as an indicator of a disturbed area. Lichen was identified as a possible indicator for old growth stands. Taking species with an average presence greater than 50%, the species association included *Acer obtrusifolium*, *Arbutus andrachne*, *Asphodelus microcarpus*, *Cistus sp.*, *Lonicera etrusca*, *Lavandula stoechas*, *Pinus brutia*, *Quercus alnifolia*, *Salvia fruticosa* and *Vicia lunata*. Of this flora assemblage *Cistus sp.*, *Quercus alnifolia* and *Salvia fruticosa* were represented at 100%.

Understanding variability.

(i) **Assembly rules.** Altitudinal gradients are frequently used to classify species by association (Ogden 1995; Neiering and Lowe 1994; Kappelle, Kennis and de Vries 1995). The variation in altitude in the Makheras Forest is 1,000m. The pattern of sub-strata species appears almost consistent across this range. The most notable change occurs in a limited number of sub-strata species and ground cover flora along the altitudinal gradient. The following table represents the change in some sub-strata species along the altitude gradient. As this change was initially masked by the presence across the gradient of the *Pinus brutia*, *Quercus alnifolia*, *Pistacia terebinthus*, *Olea europea* and *Crataegus azarolus* these species were eliminated from the table.

Results for Makheras where bD >=10

	Makheras				Adelphi			
Number of Quadrats:	30	Match	99	All	18	Match	75	All
Diabase	30	100.00%	93	93.94%	n/a	0.00%	n/a	0.00%
Steep-Very steep	22	73.33%	79	79.80%	8	44.44%	50	66.67%
Northernish (N,NNE,NNW,NE,NW,N-S)	14	46.67%	39	39.39%	5	27.78%	25	33.33%
Southernish (S,SE,SW,SSE,SSW, N-S)	9	30.00%	38	38.38%	13	72.22%	31	41.33%
Alt-Avg	945.666		881.684		643.889		637.042	
Alt-St.D.	292.480		228.574		134.542		134.765	
Stage IV ecological Succession.	28	93.33%	78	78.79%	13	72.22%	46	61.33%
Storm Damage:	10	33.33%	25	25.25%	7	38.89%	21	28.00%
Complex Stands	27	90.00%	66	66.67%	9	50.00%	13	17.33%
Crown Condition:								
supressed	0	0.00%	2	2.02%	1	5.56%	2	2.67%
released	20	66.67%	70	70.71%	17	94.44%	54	72.00%
mature	9	30.00%	21	21.21%	0	0.00%	15	20.00%
Ratio of Tolerant Sp.								
avg	2.067		1.980		1		1.028	
std	0.814		0.845		0		0.165	
Site Class								
Avg(1=A, 4=D)	2.5		2.847		2.667		2.843	
Std	0.719		0.774		0.577		0.577	

Table 10: Comparative Table of bD with Geomorphological and Forest Dynamic Factors

The faunal species are related to various phases in secondary forest establishment. The emerging pattern of faunal species in association with habitat areas, corridors and food sources (See Figure : Map of faunal observations).

(ii) **Rarity.** Less than 15 percent of the listed flora were endemic to Cyprus. The frequency of these species however was important in understanding patterns within the forest canopy, particularly at high altitudes.

Establishing a Biodiversity Index (bD)

The dependant variables selected for inclusion in the biodiversity index were: canopy cover percentage over each quadrat; total number of flora and fauna species recorded; the presence of lichen and moss as old growth indicator species; number of saplings present and uneven age stand characteristics.

Canopy cover

The canopy cover of each quadrat was assessed subjectively during the survey and recorded as <30%; 50% or >70%. In quantifying this assessment, the canopy of quadrats was given a numeric value of 1 for canopy <30%; 3 for 50% cover and 5 for canopy cover >70%.

Light-requiring and shade-tolerant species respond to light in different ways. Light-tolerant or light-demanding species require sunlight for germination and their growth characteristics normally increase with light intensity. Shade-tolerant species respond differently, both in germination and growth under partial light without a decline in vigour. They may, however, require sunlight at canopy level to gain maximum growth and reproduction capacity. A rapid transition from partial shade to full sunlight can cause leaf loss through excessive moisture loss (Florence 1996:87). The difference in light requirements between the *Pinus brutia* which is light-demanding soft wood and the *Quercus alnifolia, Juniperus sp. Cedrus brevifolia* and *Cupressus sempervirens* that require partial shade for germination and establishment is significant. The importance of canopy depletion is an important factor in the maintence of forest structure and dynamics. It is also significant that the shade-tolerant species have a much longer maturation rate than the *Pinus brutia*.

Faunal density

Terrestial mammals, such as rodents, inhabit tree roots, stumps and rocky outcrops within the forest. Similar habitats are occupied by reptiles while the range of birds recorded nest below ground level (Coal tits), ground level (Chukars), in small bushes and shrubs (Cyprus warbler) or in tree crowns (Shrikes and Sparrowhawks). The distribution of faunal species was recorded in wide range of preferred habitats including broad forest classes such as open or closed forest, grassy slopes or dense understory. Attributes of each site type and vegetation cover determined the distribution of faunal species. Certain plant species (both above and below ground level) act as important seasonal food sources. The complexity of habitat requirements has not been investigated fully but there appears to be a definite correlation between forest complexity and animal density. Understorey vegetation seems to be an important factor for rodent and hedgehog presence. Sub-strata reduction favours the *Lepus* and *Vulpes spp.* The maintaince of faunal diversity and habitat quality requires continuity of shelter and food resources.

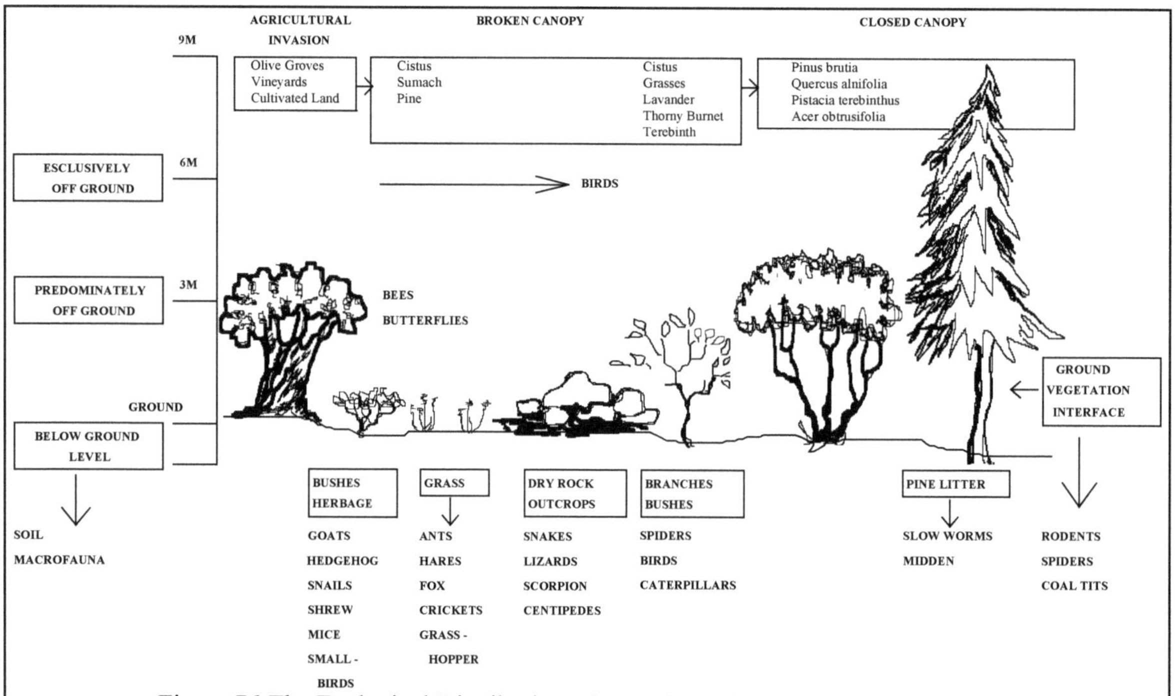

Figure 76:The Ecological Distribution of Faunal Species in the Makheras Forest

Faunal species were counted on observation of activity, scats and droppings, skeletal material and presence. Four gradings were established: <5 scored 1 point; <10 scored 2 points; <15 scored 3 points and <20 scored 4 points. Totals >20 scored 5 points but this figure was not reached in any single quadrat.

Floral diversity

Over one hundred different botanical specimens were recorded during the survey. Species were located in a wide range of forest environments from unstable scree to old growth refugia. In determining the weighting of species for the diversity index, the same scoring method was applied as to the faunal species (*i.e.* <5 scored 1 point; <10 scored 2 points; <15 scored 3 points and <20 scored 4 points. Totals >20 scored 5 points). Due to the timing of the survey not all species present at any quadrat were recorded.

Lichens and mosses

A presence/absence weighting was given for mosses and lichens recorded with a figure of 2 points scored for presence and 0 for absence. Lichen and moss were considered indicator species for old growth or refugia areas where other factors such as dominant species, may not indicate direct evidence of sustained growth.

Saplings

A count of all live *Pinus brutia* stems that fell below 56cm GBH[8] were recorded as saplings. Thie count was made to indicate regeneration potential and future growing stock. The saplings were given the following weighting: <5 scored 1 point; <10 scored 2 points; < 15 scored 3 points < 20 scored 4 points and < 25 scored 5 points.

Uneven aged stands

A weighting of 5 points was given to those quadrats exhibiting uneven age stand characteristics. Fell to waste

[8]This is a standard forestry practise to record growing stock and future forest potential.

of old mature and senescent trees under previous forestry practice has eliminated many trees that would act as habitat trees and provide animal refugues. The ecological importance of old mature and senescent trees in complex forest environments in the provision of canopy cover, as nutrient stores, habitat trees, seed supply and to maintain the floristic and structural characteristics of the forest also enriches its diversity.

Total scoring

A total score of 25 points for each quadrant with all the prescribed characteristics was possible.All scores above 10 points were mapped in each of the forests surveyed. Following Hawking's (1986:10) definition of a good theory two requirements were sought to accurately predict diversity in the eastern Mediterranean forest environments. Description of a large class of observation that contained only a few arbitary elements were obtained in order to make predictions about the result of future observations.

In mapping the data derived from the diversity index it was seen that isolation played a dominant role at quadrats with a high number of species recorded in the Makheras Forest. Other factors that appered to correlate directly with the quadrats of high bD were: geological parent material; steep to very steep slopes; northerly aspects; fourth stage of ecological development; natural disturbance (i.e. storm damage) and the complexity of stand composition (see Table 14). With the exception of stage of ecological development and natural complexity of stands, the same criteria did not characterise the high bD apparent in the Adelphi Forest.

The spatial arrangement of quadrats with a high ratio of bD in the Makheras at first appeared to be the result of isolation, that is, far from settlements, communication routes and with very little or no modern anthropogenic evidence (i.e. garbage or hunting debris). Almost all quadrats, however, were adjacent to abandoned agricultural terraces (see Appendix B). A well established sub-strata of *Quercus alnifolia* sometimes supplemented with *Arbutus andrachne* was characteristic of these quadrats. The presence of fox scats was also consistent with the level of bD. The correlation between the presence of foxes in quadrats and a high bD may be coincidental. The pattern of high bD quadrats in the Adelphi Forest in the area known as "Aloupophournos" (place of the foxes) may indicate former settlement in the area and not observed during the survey or ancient settlement areas adjacent to the forest perimeter.

An unconfirmed hypothesis that fox presence and high bD levels could reflect the impact on the forest resource by ancient cultures cannot be set aside. The presence of *graminae* species used as food resources in antiquity, and their proximity to quadrats with high bD levels may indicate a very complex set of relationships over time in forest response to utilisation and bioresource rehabilitation.

Figure 77: Map of the Makheras Forest showing the Biodiversity Levels as determined per quadrat.

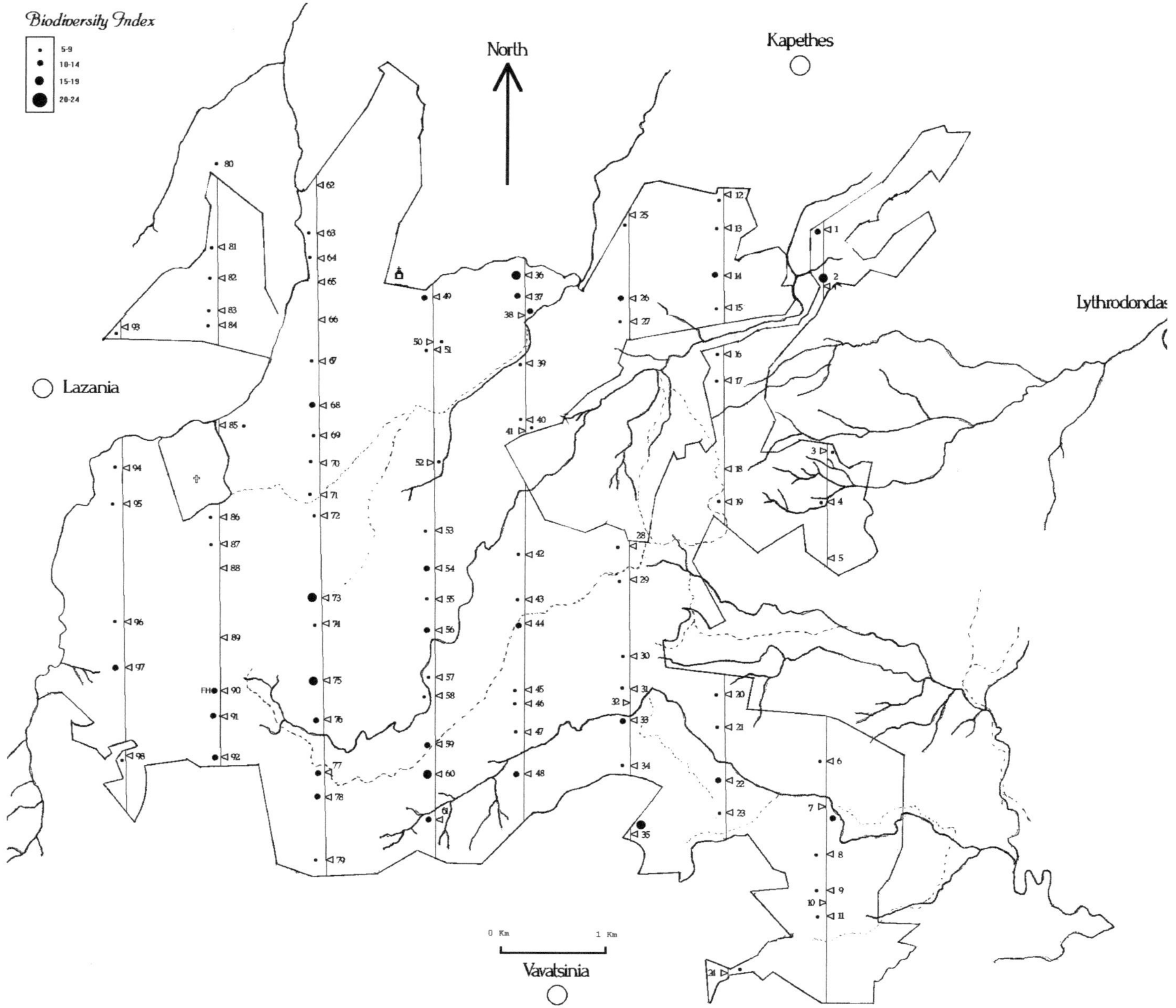

Figure 78: Map of the Adelphi Forest showing the Biodiversity Levels as determined per quadrat.

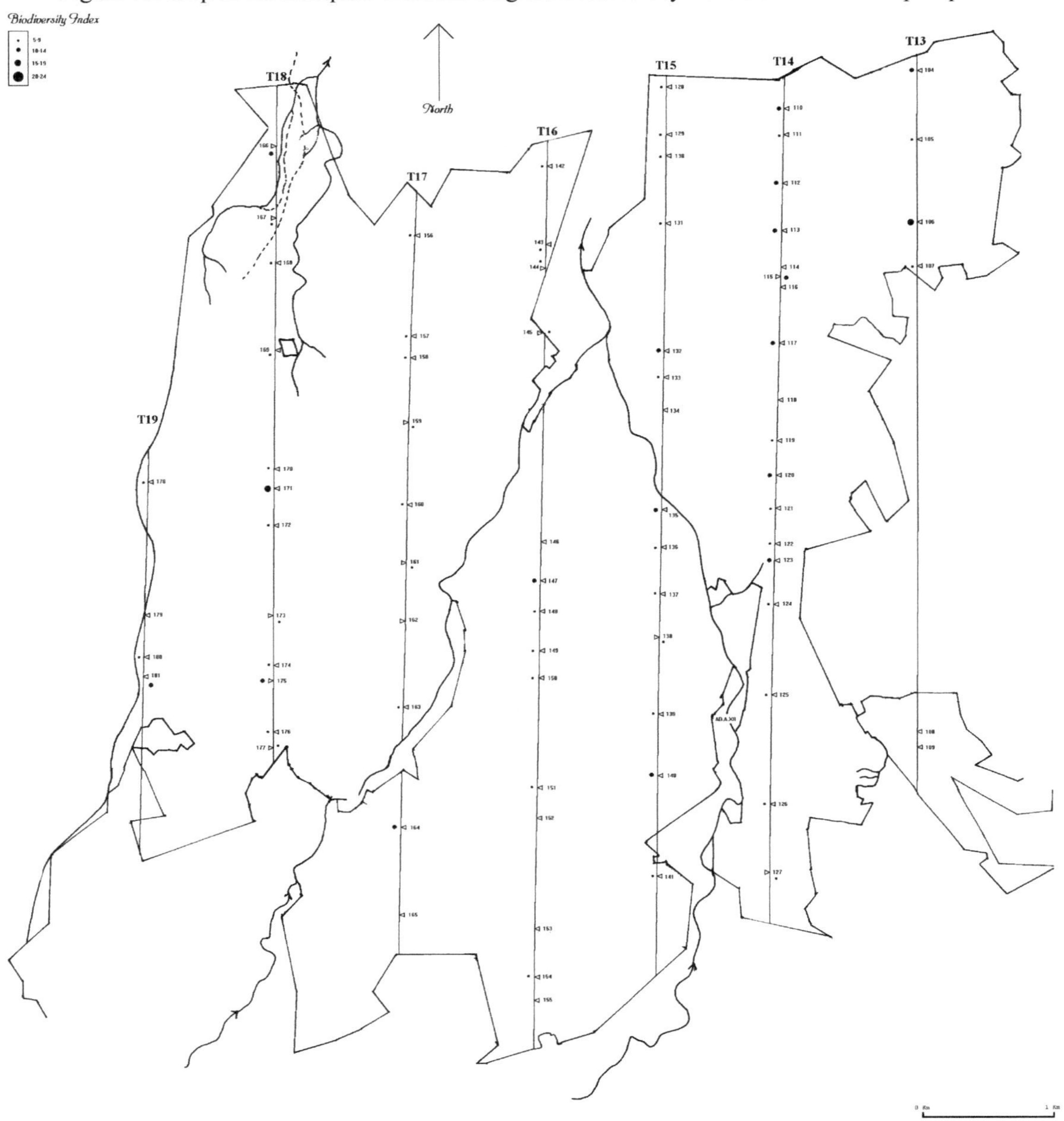

Chapter Seven: The Cultural Transformation of Landscapes.

The cultural choice to identify with a specific locality through habitat selection entails complex socio-economic factors. These are fequently related to specific bioresource availability and an intellectual or emotional ethos expressed through ritual or mythology. Observation and interpretation of environmental change that may have occurred over an extended time span, and as a direct result of human impact, involves becoming attuned to a locality through an increasing awareness of particular ecological relationships identified with a certain landscape. At an intellectual level this entails understanding activities and evolving social behaviour that enrich the life of that environment. In particular, aspects support life systems, and those that either establish an ecologically, and therefore socially, sustainable pattern of existence within that place (Berg and Dasmann 1978).

Living in a place so that one may understand its features in order to describe them accurately, requires learning the secrets of navigation - paying attention to landmarks. Landmarks represent a particular knowledge of a particular place. An understanding of a specific place provides insights into the many aspects of life in the diverse ecosystems pertaining to that landscape. Cultural adaptations to certain places are frequently reflected in mythological associations between people and place. Cultural perception of landscape does not represent a simplistic relationship between people and their natural environment (Meinig 1979:2), but rather the complexity of interaction between a "dynamic and historically contingent" landscape (Knapp 1994).

Places can be destroyed, that is, they can have their nature and meaning irrevocably changed. The manner in which a society describes its environment is often the result of an accidental separation of people from their landscape as the result of some catastrophe. The complex reuse of an inherited landscape can be determined through archaeological excavation. However, the utilisation of new environmental elements by successive generations, or element erasure, needs to be defined and understood at a landscape level (Roberts 1987).

Management and development of natural resources and their incorporation into a cultural environment also assumes a discontinuation of a natural equilibrium. For these reasons an environmental investigator must become a regional specialist, growing in awareness of the region as it appears in its contemporary context, so that in its physical geography, traces of the past may be observed. From the qualities the current landscape presents, the past conditions might be perceived. In being so attuned to the landscape one might learn the ability to see it as the former inhabitants would have in relation to their needs and capabilities (Sauer 1941).

Facts of geography, that is landscape and environment, are "place facts". Their association gives rise to the concept of landscape, whereas facts of history are "time facts". Their association gives rise to the concept of "period". By definition, the landscape has an identity based on a recognisable constitution. Its limits and generic relationship to other landscapes constitute a general world system. The structure and functions of landscape are determined by integrant, dependent forms such as mountains, valleys, and rivers. Landscape can also be considered as a dynamic organism. The concept of landscape is also distinctly anthropogenic in the sense that it has value or use for the people who are part of it, live in it, are limited by it, and who modify it. Thus the creation of a cultural landscape lies in the context of landscape found in the physical qualities of an area that is of human significance, and in the forms utilised within such an area (Sauer 1965: 325).

Landscape axes are part of human active and constructive perceptions. However, these axes do not move in the same direction, nor do they tend toward the same end. They cannot provide the same aspects of perceived reality. Tuan (1979) described landscape as "an ordering of reality from different angles" the result of which was an appearance of landscape achieved through "an effort of imagination exercised over a highly selected array of sense data". The conceptual depth of landscape is discovered through science in the systematic measuring and recording of physical factors, whereas the visual depth is revealed in the humanities through the study of artefacts, art and documentation. The first provides an understanding for the reason of things, the second is the ability to see their form. In scientific endeavour, the aim is to trace the phenomenon to its first cause, and to generate laws and principles. Through the humanities, one becomes absorbed in the immediate appearances of things enjoyable in their richness and complexity. On this axis it is not the uniformity of laws but the multiformity and diversity of interpretation which is encountered, a kind of "sympathetic vision" of things. These two views of truth are in contrast but not in conflict or contradiction, for each has its own dimension (Cassirer 1944: 169-170).

Space is fundamental in any form of communal life. Space is also fundamental in any exercise of power (Foucalt 1984: 252). Perceptional space is the space of psychology, the space of behaviour and the space of perception. Spatiality is perhaps the first principle of human life involving a primal setting at a distance from the world in order to enter into relationships. Thus spatiality is more than a necessary condition of human consciousness - it is its beginning. Spatiality is meaningful geographically because it constitutes a minimum definition of the personal with a capacity for detachment. Piget (1971) observed that space perception and representation followed a series of stages, each logically implying the preceding stages. The most fundamental spatial categories, both from a logical and genetic viewpoint, are topographical. They include segmentation by the means of notions of open/closed set, such as boundary, centre, order (*ie* subsequence) and continuity. Everything that is perceived is framed with the help of these notions - exclusively. A second level of abstraction is called progressive space: it is comprised of notions which all logically imply topographical constructions. At this level the category "straight line" is formed, as well as "perspective," section, surface and volume. Finally, the most sophisticated categories are elaborated each logically implying the foregoing: the development of Euclidean space occurs which is then used to perceive and represent the everyday environment. It seems probable that spatial phenomena which extend beyond the capabilities of direct perception, are framed with the help of topographical notions. For example, the social distances that are actually (and subconsciously) used in different communication situations can actually be described in topographical terms. These facts seem to point in a definite direction; whenever direct perception is impossible, topographical notions are used in order to segment the external world (Pintex 1974).

Merleau-Ponty (1964) emphasises that the world is inseparable from the subject and consequently we must say of time that it has meaning only because "we are" time. We can designate something by this word "time" only because we are at the conjunction of the past, present and future (Evernden 1992: 81). Our partial spaces are founded in and upon historical and sociological situations, that is, situations of reference as the net of relationships between people and the world or situations not solely dependent on people but with which human relationships are of concern.

Environmental Archaeology.

An archaeological approach to landscape systems examines three primary dimensions: space, time and form (Knapp 1994). Spatial archaeology deals with a set of elements and relationships representing human activities at every scale, the tracts and artefacts left by them, the physical infrastructure which accommodates them, the environment upon which they impinge and the interaction between all these aspects (Clarke 1977: 9).

Land use radically affects the survival or destruction of archaeological materials. This factor is central to any investigation into the environmental context of an archaeological site. An additional element in archaeological research is the awareness that each excavation investigated has its own relationship to assemblages of real world features that have been continually in a state of dynamic flux.

Land utilisation must also be considered in the contemporary regional context of a settlement site. Occupational studies which incorporate the historic structure of settlement need to appreciate the momentary equilibrium between habitat requirements and evaluate these in relation to resource supply. Environmental advantage or disadvantage may be relative to a particular stage of community development, land use accommodation, and the dynamic structure of the community. Change is often slow, as the rationalisation of land use may meet opposition from traditional customs and expectations.

The methods employed to determine the evolutionary ecology of the Cypriot landscape depend on a number of factors. A significant component of the field research undertaken was to establish a contextual relationship between archaeological sites and their contemporary environment and to define and describe the creation of a cultural landscape in relationship to the exploitation of natural resources over time. One method of determining the complexities of biodiversity and cultural interaction is examining the landscape to assess the unique qualities of tree species associated with that landscape, and make some correlations, albeit tentative, between these factors and the remains of wood from ancient artefacts.

Table 11 provides a brief overview of cultural periods on Cyprus, defined largely by pottery types. The abbreviated time blocks in the left hand column are those used to describe the known activity within or adjacent to, the forest boundaries. The main table is derived from Peltenberg (1982) and Knapp, Held and Manning (1994).

Factors relating to historical interpretation.

The critical role of field surveys in understanding the spatial relationships between Cypriot ancient cultures was established by Catling's map of then known archaeological sites on Cyprus (Catling 1962). The current policy held by the Cyprus Department of Antiquities is to promote field surveys, now a major component of archaeological field research (Adovasio et al. 1975; Frankel and Webb, 1995; Hadjisavvas 1983; Held 1988; Knapp et al. 1992, 1994, 1995, 1996; Manning and Conwell 1992; Sorensen and Rupp 1993; Todd 1989; Herscher 1995).

T*	Revised	Traditional	C-14 Dates
T1	Early Prehistoric	Neolithic-Chalcolithic	7900-2300
	Akrotiri Phase	"proto-Neolithic"?	7900-?
	Khirokitia Culture	Aceramic Neolithic	7500-5200/4900
			5200-4900
	Sotira Culture	Ceramic Neolithic	4900-3200
T2	Erimi Culture	Chalcolithic	3200-2300
T3	Pre Bronze Age	Philia-Early/Middle Cypriot	2300-1550
	Pre Bronze Age 1	Philia Culture (PCU)	2300-?
	Pre Bronze Age 2	Early/Middle Cypriot	2300-?
		(Early/Middle Cypriot)	2000-1700 BC*
	Pro Bronze Age	Middle-Late Cypriot	1700-1000 BC
	Pro Bronze Age 1	MC III- LC I	1700-1400 BC
	Pro Bronze Age 2	LC II	1400-1200 BC
	Pro Bronze Age 3	LC III	1200-1000 BC
T4		Geometric I	1050-950 BC
		Cypro-Geometric II	950-850 BC
		Cypro-Geometric III	850-750 BC
		Archaic I	750-600 BC
		Cypro-Archaic II	600-475 BC
		Classical I	475-400 BC
		Cypro-Classical II	400-325 BC
T5		Hellenistic I	325-150 BC
		Hellenistic II	150-50 BC
		Roman I	50 BC - AD 150
		Roman II	150-250 AD
		Roman III	250-356 AD
T6		Byzantine	330-1191 AD
T7		Medieval	1191-1489 AD
T8		Venetian	1489-1571 AD
T9		Ottoman	1571-1878 AD
T10		Colonial	1878-1960 AD

Table 11: Cyprus - General Chronological Scheme.

Texts and Inscriptions.

It has been customary in the past to rely heavily on ancient texts as an historical source, in order to determine environmental relationships, commodity production, distribution and utilisation. The use of textual material raises many fundamental problems, a principal issue being the range of ambiguities in the word lists naming trees and timber in ancient literature. For example, there is a difference in meaning between the Latin words *materia* and *lignum*, the former refers to wood used for building or craft making and the latter refers to wood used as fuel. In ancient Greek there was no such distinction, the single word *xylon* referred to all wood types. A further difficulty is encountered in identifying species. Theophrastus (*HP* 3.8.2) knew that the names attributed to the various species of oak growing in Macedon differed from those given to the same range of species growing on Mt. Ida in the Troad. Similarly, the Arcadian names for the main species of pine described by the Greeks were the opposite of those accepted in the rest of Greece. However, as some of the examples presented by Theophrastus were based on hearsay, these discrepancies between species might be explained by the collation of information from various sources (Meiggs 1982:18). Confusion between cedar and juniper can create additional problems. The Greek *kedros* and Latin *cedrus* do not always mean cedar. In Greek the word *kedros* also includes the *Juniperus spp*, the *Cedrus spp.* being relatively unknown to mainland Greeks.

Tree names in Akkadian, Hebrew and Egyptian textual sources are even more controversial. The only secure word known in Akkadian is the word for cedar, *erinu*. The identification of pine, juniper and other important commercial trees, however, remains obscure. Loret (1916) and Glanville (1931:16, 31) argued that the common term used by the Egyptians for imported wood (ash-wood), also pertained to fir, or possibly fir and pine, rather than cedar (Meiggs 1982:55). Standard text scholarship in botanical translation (Thompson 1948) has been expanded by more recent research (Knapp 1983; Hoffner 1995). However, the assignation of tree names to words in ancient textual material remains insecure (Mason 1996 pers. com).

A further difficulty is that most ancient writers appear to have borrowed freely from previous written works. Pliny the Elder, who is credited with the most accurate descriptions of ancient Cyprus (Meiggs 1982), relied on the works of Theophrastus (ca 370-288/5 BC), a pupil of Aristotle, and on the oral tradition ascribed to Homer. Pliny's plagiarism can be detected clearly in his description of the palm (*NH* 13.33) which is almost identical to the one by Theophrastus. Pliny's original contribution to the botanical corpus was thought to be his record of the largest cedar known to have been growing on Cyprus at that time. It was reported to have been harvested to provide the mast for a galley commanded by Demitrius Potioreates (ca 300 BC), and is reputed to have had a recorded length of 130 feet (*NH* 16:203-206). The same account, however, is related in Ezekiel (Ezekiel 27:7).

Considering the above, it appears that the relevant Mediterranean literature is inadequate for the identification of tree species pertaining to specific locations or the utilisation of timber resources. Botanical species appear to have been cited by writers who have little knowledge of botanical science, a fact deplored by Dioscorides during the first century AD (Riddle 1985). Furthermore, some authors, including Theophrastus, Eratosthenes and Strabo, relied on oral accounts as the source of their descriptions, or based their scholarship on documentary sources in the library at Alexandria. Nibbi (1994) not only reviews the error in species identification in ancient textual sources but cites the continuation of misinformation in current literature. Setting aside these traditional references, several sources are available and can be regarded as accurate, and therefore, of critical importance in understanding environmental factors related to timber utilisation and export. They are, firstly, the epigraphic evidence such as the association occurring in the third century BC between the Sacred Grove at Kourion and the god Apollo. An inscription indicates that the sanctuary at that time was dedicated to Apollo Hylates, Apollo of the Woodlands (Sinos 1990:20)[9]. No oaks grow in the vicinity of the sanctuary today, but in the shaded, moist areas of the aqueduct head that is adjacent to the temple, many ground cover species associated with a sheltered forest canopy can be found *in situ*.

Secondly, the artistic representations by indigenous craftsmen, such as those on cylinder seals, clearly depict tree forms which can sometimes be readily identified. Ohnefalsch-Richter (1893:101-141) identified eleven forms of stylised tree depicted on cylinder seals from Cyprus. The oldest depiction shows a realistic representation of a tree or branch which Richter attributed to one of the native conifers. The simplified form of the tree, however, precludes identification at species level. Other examples discussed by Ohnefalsch-Richter are of interest for they appear to suggest that the harvesting of lower tree limbs became progressively severe. Unlike the natural conifer, which is self-pruning in dense forest stands where lateral limbs are lost close to the bole, the artificial pruning of branches leaves a stump retained by the tree, which becomes fixed by resin and creates a very knotted structure in the timber. A final form in Ohnefalsch-Richter's list of types is undoubtedly a palm tree. With the further development of a distinctive glyptic craft during the Late Cypriot Bronze Age, another type of tree representation, described as the "bouquet tree" (Poroda 1948), became a unique addition to the collection.

The third source is comprised of dynastic lists, including accounting systems, maintained by strong centralised governments. Although this source does not often specify country of origin, or applies an archaic name form, (such as Alashiya to Cyprus (Petrie 1898:44-48; Moran 1992:104-113), through accurate identification of the species, the country of origin might be determined. These records include references to the importation of cedar from the temple list of Ramses III (c.1195-1164 BC). A reconstruction of a temple stele by Hayes (1949) reads: "(I) ap[proached (?)]...[ced]ar (of) the Plateaus of Cedar...which (I) made, cutting (down) the tree(s)" (Ward 1971:63). From the Palermo Stone (recto vi;2-4) we read: "Bring forty ships filled (with) cedar logs ... shipbuilding (of) cedar wood, one Praise-of-the-Two-Lands ship, 100 cubits (long), and (of) *meru* wood, two ships, 100 cubits (long)... making doors of the royal palace (of) cedar wood" (Pritchard 1950:227). Pritchard notes that the word translated as cedar probably applied to that species and other coniferous woods also, and may be applied to any timber transported by sea.

Environmental context of early settlement on Cyprus.

Successful sea crossings to Cyprus from mainland areas to the north or east of the island appear to have depended on the island's shape and the conterclockwise sea currents which flow around it (Held 1993). Palaeontological and archaeological evidence have shown that very few animals reached Cyprus on their own during and after the Pleistocene (Boekschoten and Sondaar 1972; Held 1992). Human crossings from the coast of Anatolia would also be more successful than crossings from Syria (Held 1992). However, Held's study is based on present sea currents and it is not known whether the same patterns would have been present at 10,000 BP.

The proto-Neolithic southern coastal environment has been reconstructed by the presence and skeletal adaptation of the extinct *Phanourios minutus* (pigmy hippopotami) and *Elephas cypriotes* (dwarf elephant) remains excavated at Akrotiri *Aetoremnos*. Parts of the Akrotiri Peninsula may have supported a marsh or lagoon environment suitable for the pigmy hippopotami. An open lagoon is also suggested by boat remains located in the present salt lake (Swiny pers. com.).

The earliest known human settlement on Cyprus is the collapsed proto-Neolithic rock shelter of Akrotiri-*Aetokremnos*. This site yielded quantities of Pleistocene fauna in association with stone tools and hearths

[9]The ephilet "Hylates" derives from the ancient Greek word YLN-Hye meaning woodland or forest. This has been ascribed to Stephanus Byzantius (5th centuary A.D.) who described an extensive forest located in the area of the sanctuary under this name.

(Simmons 1988; 1989, 1991; Swiny 1988; Simmons and Reese 1993; Simmons 1991). Excavation enabled the identification of two distinct layers of deposit. The lower deposit consisted mainly of disarticulated hippopotamus bones and a small amount of chipped stone. The upper layer contained some hippopotamus bones, and a greater quantity of chipped stone, shell and bird bone including great bustard, goose, dove, ducks. The inclusion of this material may indicate a change in subsistence. At least 220 individual hippos and three elephants were represented among the approximately 250,000 bones and fragments recovered. About 20% of the bone material was burned. Small amounts of fish, turtle, snake and land snail were also represented. Shell beads, ground stone ornaments and hearths were also recorded (Herscher 1995). Radiocarbon dates place the archaeological material in the late 11th or early 10th millennium B.C. (Manning 1991:875). Between the Akrotiri-*Aeokremnos* cultural phase and the Aceramic Neolithic occupation of Cyprus there appears to be a gap in the archaeological record, albeit that intermediary sites may not been located due to geomorphological factors.

The Aceramic Neolithic period is represented by more than twenty sites spanning a period of almost 2000 years. The two representative archaeological type sites are Khirokitia *Vouni* and Kalavassos *Tenta*. These sites may have been established by new settlers with a cereal based economy and a reliance on animal husbandry. Fine stone bowls and round house construction typify this period. Another gap in the archaeological record appears between the Aceramic Neolithic settlements and the Ceramic Neolithic or "Sotira-Culture" best exemplified by Sotira-*Teppes* (Keswani 1994) and Ayious Epiktitos-*Vrysi* from more than 60 known sites. This period may represent another wave of colonisation from overseas (Peltenberg 1982:111).

Geographical descriptions of an aceramic Neolithic site, Koletria *Ortos*, (Simmons and Corona 1993) and a later ceramic Neolithic site (Manzarani 1994) are important in placing early island settlement in an environmental context.[10]

The aceramic Neolithic site of Kholetria *Ortos* was located by the Canadian Palaipaphos Survey Project (CPSP) in the early 1980's (Fox 1988). It occupies an isolated hillock overlooking the Xeros Potamos River approximately seven kilometres from the southern Cypriot coastline. *Ortos* is steeply sloped and bounded on the northwest to south west by a bend in the Xero Potamos River. The hill top provides a commanding vista both down the valley towards the Mediterranean, and inland toward the Troodos massif (Fox 1988).

A rich assemblage of artefacts and economic materials has been recovered. Preliminary analysis of more than 35,000 pieces of chipped stone, from cores to microflakes, indicates that all stages of reduction were practised. With the exception of two obsidian blades, most are of local cherts. Several varieties of sickles, exhibiting a clear silica sheen, were recovered as well as an abundance of well-made blades (Kingsnorth 1993).

The site is on the igneous/calcareous interface with the immediate area of limestone composition with adjacent hills to the south east of pillow lava formation. Geologically Kholetria *Ortos* is raised terrace covered by an alluvial soil of varying depth. Landslide slumped Pakhna formation chalks cap the hill, while the saddle and lands to the east are Moni formation clays (Lapierre 1971). Numerous igneous cobbles from the fluvial deposit occur at surface level. Igneous rock types suitable for stone tool manufacture are to be found in the Xero Potamos river bed. Cobbles of vesicular andesitic basalt from the Lower Pillow Lavas and the microgabbro pebbles appear to have been utilised. Lefkara formation chalk and chert cobbles are also readily available. An outcrop of Lefkara chert-bearing formation is located 2.5 kilometres to the north of the site on the western side of the valley. Moni chert formation outcrops occur to the east of the site (Fox 1988).

On the hill adjacent to the north-east a range of plant species was noted in an abandoned vineyard which had suffered subsequent heavy erosion post-cultivation. In the summer of 1993 the dominant shrub storey was composed of *Quercus coccifera*, *Pistacia terebinthus*, *Olea europea* (both wild and cultivated forms), *Crataegus* sp., and *Amygdalus communis*. The ground cover was sparse and bordered on the "batha" type. However, *Capparis spinosa*, *Phagnalon graecum*, *Galium aparine*, *Asparagus acutifolius* and *Ammi visnaga* were present under shrub cover along the perimeter of the hill top plateau. The overall condition of the vegetation is reflected in the soil depth which had been severely depleted. Groves of almond trees, many of which had died, had been pit planted about forty years ago when cultivation of the hill top ceased (Burnet 1993b).

Further to the south east, another hill with a limestone cap outcrop sheltered vegetation in rock crevices. Here the range of species was rich and diverse. No evidence of occupation was noted. The area is currently utilised for goat grazing, thus reducing the capacity for species to regenerate in places other than rock cavities.

[10] Following the established archaeological tradition on Cyprus every site described uses two place name elements: the first, non-italicised, place name (Koletria) refers to the nearest town or village on a cadastral map to which the site is located. The second, italicised, name (*Ortos*) refers to the toponym, as known locally, or by tradition, nearest to the site itself (Knapp 1994).

To the south, an igneous outcrop carried a good soil depth on the northern face, though signs of erosion were noted particularly on animal paths. The hill slopes were steep to very steep. Species observed included *Olea europea* (wild form), *Crataegus* sp., *Poterium spinosum* and *Acer obtusifolium*. The gully incision was deep and may have remained a small perennial stream under extensive canopy cover. The outcrop provided good cover as a wild life habitat.

Two further limestone outcrops were investigated on the eastern side of the Xero Potomas valley and they produced further evidence for a diverse plant community. The area between the hills was under dry land cultivation or fallow. The steeper hillsides and hill tops were used for grazing. The hillside opposite the abandoned village of Phinikas carried the only standing tree, a well developed and mature *Crataegus* which was used as a shade tree by animals. Along the water-line of the dammed river wild *Origanum heracleotium* and *Mentha pulegium* were growing and appeared particularly prevalent in gully lines leading down to the water's edge.

In the nearby village of Phinikas, a number of mature trees were recorded including *Cupresses sempervirens* and *Pinus brutia*. A senescent tree which had been uprooted after death demonstrated the potential capacity for tree growth in the area. Although not measured the diameter at breast height was in excess of one metre.

Along the western side of the valley the steep limestone scarp contained numerous small springs which were flowing. Steep chimneys lead from the valley floor to the plateaux terraces above. The cultivation of these terraces is less extensive and coppices of *Quercus coccifera* have developed on abandoned fields. The fluvial terrace opposite the Kholetria *Ortos* site carries a heavy cobble layer on the surface. In comparison with *Ortos*, the density of the cobble stone appears greater in this field. There was no evidence of ground stone or chipped stone implements. The terrace is under dryland cultivation. Along the perimeter of the field similar plant species to those noted on the eastern side of the valley were observed.

Following the gully line up the northern face of Kholetria *Ortos*, a spring source was identified under the low spreading branches of a *Pistacia terebinthus*. The spring was dry although evidence for periodic flow was observed. Moss associated with moisture was noted and a natural basin with washed limestone gravels and alluvial deposits was located directly below the spring head, suggesting partial flow each year. Above the spring head, a mechanically formed terrace had been made. This terrace may be responsible for the decline in the spring's capacity. Another terrace below the spring head may have likewise affected the direction of the ground water supply. The vineyard planted on the second terrace has been abandoned. Plant species included moisture indicators such as *Galium aparine* and *Rubus sanctus*.

Below the spring head several old *Olea europea* trees were grouped together, surrounded by circular stone walls possibly for protection from grazing animals. The walling of ancient olive trees appears to be a traditional Mediterranean practice as similar structures can be observed on Malta, Sardinia and Corsica. The possibility of ancient building reuse cannot be upheld in this context for although the structures were circular, their construction of rubble stone does not correlate with the building techniques employed at other early Neolithic sites.

The vegetation on the Kholetria *Ortos* site is dominated by the vineyard established approximately 30 years ago when mechanical cultivation of the area became uneconomic. Previously the area had been under grain cultivation. Ploughing by oxen and hand harvesting were traditional techniques. Large stones from the field had been gathered and placed along a limestone ridge which transects the hill top. One mature *Olea europea* stands in the vineyard to the west of the dividing ridge. A mature *Amygdalus communis* has grown in the south of the vineyard since its establishment. Smaller specimens have seeded along the perimeter of the vineyard. Ground cover plants include *Anemone coronaria*, *Asparagus acutifolius*, *Asphodelus microcarpus*, *Capparis spinosa*, *Papaver rhoeas*, *Sinapis arvensis* and *Thymus capitatus*.

The geological formation of the site area would have supported three major tree associations: *Pinus brutia* on the Pillow lava formations, *Cupressus sempervirens* on the limestone formation and *Quercus lustitanica* as possibly the dominant tree overall. The *Pinus brutia-Olea europea* association occurs primarily on the pillow lavas in the mid-altitudinal range frequently with *Acer obtusifolium* as a shrub storey in the moister, more sheltered sites or on northerly aspects. *Cupressus sempervirens* is a species which forms dense, pure stands on limestone formations and flourishes in well watered aspects. The *Pinus brutia-Quercus lustitanica-Olea europea* association, which was probably the dominant forest cover at Kholetria *Ortos* during site occupation, is largely confined to the deeper soils and topographical aspect is of less critical importance.

These associations suggest an open canopy woodland with between 30 to 50 % overhead cover at the time of Neolithic occupation. The shrub understorey would have been comprised of *Quercus coccifera*, *Pistacia terebinthus*, *Crataegus azarolus* and *Olea europea*. These species are found on or near the site. They represent remnants of the former forest cover and are indicative of the tree associations. Single specimens of *Quercus lustitanica* are found a short distance to the north of the site. The range of ground cover species is indicative of a sheltered woodland environment with a dense shrub understorey which would have inhibited ground surface evaporation. The river floor would have carried an association of *Platanus orientalis*, *Alnus orientalis* and *Nerium oleander*. The reduction in the forest cover which was thus dominated by broad leaf species could be attributed to goat grazing. Ring barking of a *Cupressus sempervirens* specimen by goats in the village of Phinikas was photographed during the survey of this archaeological site.

The lack of evidence for additional occupation sites in the area suggests a forested region with a clearing made only for settlement at Kholetria *Ortos*. The faunal remains excavated included *Dama mesapotamia*, a species requiring a diverse browse habitat, met only within a forested area. The range of ground cover species which are generally considered moisture indicators could be seen as remnants of forest ground cover species which have survived in sheltered pockets, mainly between rocks in isolated outcrops where grazing has been restricted.

The forest cover thus observed would have provided a range of products for the Kholetria *Ortos* settlement through readily available fuel and workable timber for small household items, winter fodder and forage in the form of acorns for domesticated animals, in particularly pigs, and a wide range of browse species for deer, goats and sheep. Many of the plant species are known traditionally for their medicinal properties. Harvesting of the *Quercus* species, the fruits of the *Amygdalus communis*, *Crataegus azarolus* and *Olea europea* would have provided valuable nutritional additions to a subsistence diet. The forest habitat would also have sheltered smaller animals and a range of game fowl. During the survey evidence of fox, hare and game fowl were noted.

The Kandou *Koyphovounos* Neolithic site is being excavated by a team from Athens University led by Professor Eleni Mantzourani. The site is located above the confluence of two streams which feed into the Kandou Dam constructed during the 1950's. The altitude of the site is 243 metres above sea level. It is situated on a sandstone/chalk plateau with views towards the sea in the south, a typical attribute of most Neolithic settlement sites. From a small knoll at the highest point of the natural plateau the settlement under excavation lies on a gentle slope with a northerly aspect.

The vistas from the excavation area are far reaching. At 125 degrees south, the Salt Lake on the Akrotiri Peninsula is visible. To the north, views through the Kouris Valley to the Limassol Forest and beyond to the Troodos are possible. To the north west, there are valley views to the Sortira region approximately three kilometres to the north west. The Kandou *Koyphovounos* site has, therefore, a strategic command of both valleys near their confluence, and most probably the communication route from the inland regions to the sea, and to former settlements along the southern coastline.

The soil on the Chalk Plateau, represented by the Kandou *Koyphovounos* site area, is variable though mostly thin except in isolated depressions. The sandy facies give rise to light-coloured soils and on the chalks and limestones a red loam develops if protected from erosion (Christodoulou 1959).

The region in which the Kandou *Koyphovounos* excavation is situated is drained by two important creek lines, the minor lying to the east and a more perennial source to the west, and the Kouris River, one of the largest rivers on Cyprus, with its catchment in the Troodos Mountains to the north. The area is further served by an aquifer which is replenished directly from rainfall on the Kandou Sandstone outcrops, from streams in valleys cutting through the sandstone unit, and from the Kouris River below Kandou Village where the river gravel rests directly on the Kandou Sandstone unit. The water from the aquifer is of a quality suitable for domestic use (Hadjistavrinou & Constantinou 1977).

Human impact on the environment surrounding the Kandou *Koyphovounos* excavation site has been prolonged and severe. The clearing of vegetation to increase land under cultivation was commonly practised until the close of the nineteenth century. The landscape has therefore either prospered or suffered through improvement or neglect according to changes in the land use pattern. The region of the Chalk Plateau originally carried a forest cover of two distinct types, the *Certonia siliqua-Quercus coccifera* association (carob and oak) with *Pistacia lentiscus* as a possible understorey, and the *Cupressus sempervirens-Juniperus phoenicia* association (cyprus and juniper) of the coastal strip with an element of *Olea europea* (wild olive). It is the latter association which appears more prevalent in a remnant form on the slopes of Kandou *Koyphovounos*. In its original form this coastal forest would have produced a dense canopy and favourable wildlife habitats. The occurrence of *Pinus brutia* (calabrian pine) in the area is indicative of a secondary invasion after forest clearance, and represents a rehabilitation phase in the ecology and environment.

The aspect of the Kandou *Koyphovounos* excavation site is northerly. Aspect determines the microclimate of specific sites and the northerly aspect is most favourable as it is less affected by direct insolation and was considered generally cooler and more sheltered (Burnet 1993c).

A study of the present ecology of the Kandou *Koyphovounos* site commenced on the north eastern scarp of the site at an angle of 345 degrees north. At this point the sandstone cap, which formed the upper surface of the scarp, showed signs of prolonged weathering. The calcareous soil was shallow, though where caught between rocks it had developed deeper pockets.

The slope was steep and water-formed fissures between the rocks provided evidence of winter seepage. This was further supported by the density of moss growing in the area. A shrub strata consisted of *Pistacia lentiscus*, *Ceratonia siliqua* (carob), and *Calycotome villosa* (prickly broom). A rich ground cover with up to 12 species growing in close proximity was encountered. The species included *Galium aparine* (cleavers), *Thymus capitatus* (wild thyme), *Helichrysum stoechas* (everlasting sun gold), *Lithospernum hispidulum*, *Phagnalon graeca* (Grecian fleabane), two varieties of moss, graminae sp., *Papaver rhoeas* (field poppy), *Convolvulus arvensis* (field bindweed), and *Geranium tuberosa* (cranesbill). Seed from many species was collected.

Directly above the scarp a gentle slope with an aspect of 45 degrees east presented a shallow, stoney soil supporting a garigue vegetation type composed of *Pistacia lentiscus*, *Calycotome villosa*, *Cistus incanus* (Rock rose), stunted *Ceratonia siliqua*, and with a patchy ground cover of fine graminae species, samples of which were collected. A scatter of dead *Ceratonia siliqua* of stunted appearance was evident and indicative of previous fire damage to the site area. *Thymus capitatus* was prevalent in sheltered positions near the taller shrub growth.

On the plateau to 140 degrees south of the excavation area the soil was again shallow and stony. The slope was slight and the aspect open to the south. *Juniperus phoenicia* (common Cyprus juniper), small stunted *Ceratonia siliqua* regrowth, with a good ground cover of graminae species and *Thymus capitatus* grew well where the soil had been previously partially cultivated. On areas where no cultivation had occurred the soil surface was compact and stony. A regular scatter of anthropogenic material was observed over this area.

Where the topography levelled the graminae cover increased though there was little surface cover on the slopes, even those with a slight incline. Some moss cover was observed on areas between rock outcrops. *Olea europea* (wild olive) was noted in association with limestone outcrops and the redeposition of unworked limestone blocks.

Vegetation diminished markedly on the southern slope below the scarp edge. At the foot of the descent, some small *Pinus brutia* represented regrowth after wild fire. On the southern slope a shallow gully appeared to lead to a rock hewn cistern set to the west of the wash line (about 10 metres above the road alignment). The cistern measured 5m in length, 3.30m at its widest point, and 1.50m at it western end. The water holding capacity of the cistern appeared consistent with the needs of a small community. Feed into the cistern appeared to have originated at a spring head about 25 meters up slope. A narrow limestone ridge to the west of the road appeared to contain a cluster of tombs.

An isolated hill opposite the Kandou *Koyphovounos* excavation site to the east carried a heavier tree crop dominated by *Pinus brutia*. There was little evidence of past burning on the hill crown. A fairly dense scatter of sherds, mostly of Roman date, were noted and some diagnostic sherds were sampled. The tree species were comparable with those on the Kandou *Koyphovounos* site area and immediate vicinity. Accumulated pine needle litter appears to have aided the retention of ground moisture and reduced erosion while inhibiting ground cover growth under the *Pinus brutia* canopy.

This hill to the east of the archaeological site was transversed from south to north and the Kandou *Koyphovounos* excavation photographed from the western scarp. The vegetation on the western side had been damaged and the composition and density was similar to that carried opposite. A thick cover of *Cistus incanus* with scattered *Calycotome villosa* dominated the vegetation on the western scarp. *Origanum marjorana* was noted in sheltered positions. At the foot of the scarp in a shallow valley lying to the north east of Kandou *Koyphovounos* a gully had been cut through approximately 50 centimetres of alluvial deposit to the underlying bedrock. The valley carried some scattered *Pinus brutia* and plantation *Ceratonia siliqua* with cultivation.

The northern scarp of the Kandou *Koyphovounos* site hill was of eroded limestone and chalk with an isolated knoll to the north-north west. The vegetation cover was again dominated by *Thymus capitatus* and *Calycotome villosa*.

The range of vegetation, particularly the presence of *Cupressus sempervirens* and *Olea europea* with the quantity of *Pistacia lentiscus* on the hill top plateau, confirms the projected forest composition of a *Cupressus sempervirens-Juniperus phoenicia* association with an element of *Olea europea* and *Pistacia lentiscus*. These species were represented as remnants of the original forest cover. The presence of *Pinus brutia* as a secondary invader can be seen in the current context as a sign of gradual land rehabilitation. The wealth of small ground cover species was indicative of a formerly more sheltered and moister environment.

The area at the time of the Neolithic settlement might therefore be considered as one within an evergreen forest with access to a range of natural resources. Cultivation most probably occurred on the natural terraces carrying the terra rossa on the western scarp above the river gorge.

It is believed that under widespread forest canopy cover the rivers in the area would have maintained a perennial supply, thus providing access to additional food sources. The dense cover provided by the forest would have created suitable habitats for a range of species such as deer, goat, and game birds.

The transition from the Ceramic Neolithic to the Chalcolithic appears to have been peaceful and no subsequent break in occupational evidence for the island occurs (Keswani 1994). The number of settlements grew and changes in the architecture for site types suggests an increasing residential mobility (cf. Peltenberg 1982a:110; Todd 1989:12; Keswani 1994). Settlements in the western part of the island produced a dynamic and artistically prolific material culture (Given 1992). Erimi-*Pamboula* in the south represents the period site type although recent excavations at Lemba-*Lakkous* and Kissonerga-*Mosphilia* now rival it in cultural sophistication (Peltenberg 1985;1991). The Chalcolithic Period saw a continuation of the subsistence economies established during the Neolithic although on a more widespread scale (Catling.1966:551) with perennial springs, particularly on the slopes of the Kyrenia Range attracting settlements.

Settlement site position during the Bronze Age may owe more to factors such as water supply, workable soils and communication routes (Frankel and Webb 1992). Two principal site types can be identified, those that acted as port cities along the southern coastline and the inland settlements. Excavation at Marki *Alonia* (Frankel & Webb.1991:1992:1996) indicates substantial biophysical resource availability indicative of a local environment capable of supporting distinctive floras during the Early Bronze Age. The southern area of the Marki site has been dated to between 1835 and 1730 B.C. In the northern sector of the excavation Philia Phase ceramics in the fill for later construction suggests that the site must date to the beginning of the Bronze Age, thus the occupational time span is extensive.

Marki *Alonia* is situated at an altitude of 294 metres above sea level (asl). The annual rainfall is between 300-400mm. per annum. The principal soils are brown earths and calcarenite deposits. The geology is an interface of Lefkara formations and upper Pillow Lavas. Observation in the area of this geological interface noted the presence of vegetation growing most abundantly along the geological line where surface water flowing off the harder pillow lava soils penetrated the softer calcareous soils facilitating plant growth. Modern land use is a mixture of cereal dry farming and uncultivated land. The archaeological site of Marki *Alonia* is situated on the edge of the north-eastern foothills of the Troodos Range and overlooks the Alykos River which rises in the Makheras Forest.

The Marki landscape has the capacity for supporting a foothill forest association of *Pinus brutia*, *Quercus coccifera* and *Olea europea*. While there is no direct evidence for that association combination there are hints of its possibility. Single trees of *Morus alba* and *Ficus* occur along the Alykos alluvium and groves of ancient *Olea europea* and more recent plantings of *Amygdalus communis* occur. An area of cultivated alluvium opposite the archaeological site, when viewed from the ridge line adjacent to the site's cemetery, appears to have old irrigation channels running parallel across it at regular intervals. Marki *Alonia* is a typical example of sites that are located on the geological interface between the volcanic and calcareous formations. Many other major sites conform to this pattern including Tamassos, Alambra and Ayios Sozomenos. This site patterning may relate to the proximity of copper bearing geology to the east and north-east of the Troodos foothills (Catling 1966:816) and the trade routes along which copper ingots passed to the coastline. The importance of agriculture and the need to produce a regional surplus may also be an important factor in site location.

There appears to have been a smooth transition to the Middle Bronze Age culture on Cyprus with no major changes other than a gradual increase in wealth and foreign contact. In the Middle Bronze period the pattern of settlement sites in general move away from open terrain and may reflect periods of increased flooding due to wide spread canopy depletion and consequently a more rapid run off of surface water.

Archaeobotanical Analysis from Excavations.

Plant remains recovered from archaeological assemblages may be biased in favour of those species most able to survive extended periods. They may also represent a collection of species from a wider area concentrated at a specific location. Modern contamination through ant activity or rodent burrows may also confuse the information attained through excavation and flotation techniques.

Key: 1= Khirokita *Vouni (*Waines and Stanley Price 1977, Miller 1984, Hansen 1989); 2=Kalavassos *Tenta* (Hansen 1989); 3= Cape Andreas *Kastros* (Van Zeist 1981); 4= Ayios Epiktitos *Vrysi*; 5= Marki *Alonia (*Adams and Simmons 1993,1996); 6= Lemba *Lakkous*; 7= Maa-Palaeokastros (Milsicek 1988); 8= Nylouthkia; 9= Kalopsidha (Helbaek 1966); 10= Apliki (Du Plat Taylor Helbaek 1962); 11= Salamis (Renfrew 1970, Hjelmqvist 1970); 12= Kalavasos *Ay. Dhimitrios* (Hansen 1989); 13= Hala Sultan Tekke (Hjelmqvist 1977); 14=Tamassos (Schoch 1988); 15= Idalion (Stewart 1974).

Taxa	1	2	3	4	5	6	7	8	9	10	11	12	13	14	15	Total
Acer														1		1
Adonis dentata				1												1
Ajuga sp.												1				
Alnus orientalis														1		1
Amaranthus sp.												1				
Amaranthus retroflexus		1														1
Amygdalus communis				1												1
Anagallis arvensis										1						1
Anchusa sp.												1			1	1
Arbutus andrachne														1		1
Arctium cf. lappa				1												1
Asperula arvensis				1												1
Asphodelus sp.					1											1
Astragulus sp.		1														1
Avena sp.	1			1								1				2
Avena fatua													1			
Boraginaceae sp.				1	1											2
Bromus sp	1			1												2
Bryonia dioica													1			
Buglossos cf arvensis					1											1
Buxus sempervirens														1		1
Calendula sp.												1				
Calendula arvensis				1												1
Capparis spinosa					1								1			1
Carthamus lanatus													1			
Cedrus brevifolia											1					1
Celtis sp	1														1	2
Celtis australes				1												1
Centaurea spp.					1											1
Ceretonia siliquia							1									1
Chenopodiaceae					1											1
Chrozorophora verbascifolia																
Chrysanthemum coronarium													1			
Cicer arretinum				1	1			1			1				1	5
Cistus														1		1
Citrus medica													1			
Compositae					1											1
Coriaria myrtifolia														1		1
Coriandrum sativum										1						1
Coryllus avellana											1					1
Cruciferae				1								1				1
Cyperus sp.				1	1											2
Cynara cardunculus													1			
Echium sp.												1				
Echium italicum													1			
Echium sericeum													1			
Emex spinosa				1												1

Eragrostis barrclieri				1												1
cf.Euphorbia sp.												1				
Fagus											1					1
Ficus carica	1		1	1	1	1	1	1			1	1	1	1	1	10
Fumaria sp.				1	1							1				2
Galium sp.												1				
Galium spurium										1						1
Galium tricone		1	1		1					1						4
Genista spp.		1														1
Gladiolus segetum													1			
Graminae	1	1		1	1										1	5
Gymnosperm					1											1
Gypsophila cf. obionica				1												1
Heliotropium sp.					1											1
Hordeum sp.												1				
Hordeum sativum						1									1	2
Hordeum hexastichum										1						1
Hordeum vulgare	1	1	1	1	1			1	1	1	1					9
Juniperus														1		1
cf. Labiatae												1		1		1
Lathyrus blepharicarpus													1			
Lathyrus sativus	1			1		1					1					4
Leberinthus		1														1
Leguminosae		1												1	1	3
Lens culinaris	1	1	1	1		1		1	1				1		1	8
Lens esculenta	1									1	1					3
Lens sp					1							1				1
Linum usilatissimum			1	1												2
Lithospermum sp.	1															1
Lithospermum cf. apulum												1				
Lithosperum arvenese		1		1								1				2
Lithosperum officinale		1										1				1
Lolium sp	1			1												2
Lolium cf. rigdum				1				1		1			1			3
Lonicera														1		1
Lupinus alba													1			
Malva sp.				1	1							1				2
Medicago					1											1
Medicargo officinale	1	1													1	3
Melilotus sulcatus													1			
Morus sp.						1										1
Neslia apriculata										1						1
Neslia paniculata						1										1
Olea europea	1		1	1	1	1		1		1	1	1	1	1	1	10
Onobrychis sp.												1				
Onopordon cf. illyricum													1			
Oxalis sp.					1											1
Papaver sp.												1				
Papaver dubium				1												1
Phalaris sp.				1												1
Picris sp.					1											1
Pinus brutia										1				1		2
Pinus nigra										1						1
Pistacio sp.	1	1	1	1		1	1	1	1			1		1	1	10
Pistacia terebinthus					1								1			1

Pisum sp.	1		1	1							1					4
Polygonum sp.	1	1		1	1											4
Pomeideae														1		1
Platanus orientalis														1		1
Prunus dulcis					1											1
Prunus insititia	1			1							1		1			3
Prunus webbi														1		1
Punica granatum													1			
Pyrus malus				1								1				1
Quercus coccifera						1	1				1			1		4
Raphanus raphanistrum										1						1
Rubus sancta		1														1
Rosa sp.											1					1
Schoenus sp.												1				
Schoenus nigricans	1	1														2
Scorpiurus sulcatus													1			
Secale cereale				1												1
cf Setaria sp.	1			1												2
Solanum cf nigrum				1	1											2
Spergula arvensis		1														1
tetraclinis articulata														1		1
Trifolium resupinatum		1														1
Trigonella brevidentada										1						1
Triticum sp.												1				
Triticum aestivum				1		1							1			2
Triticum diococcum	1	1	1	1	1			1							1	7
Triticum mondcoccum	1	1	1	1			1									5
Triticum vulgare				1					1	1	1					4
Umbelliferae												1				
Veronica sp.				1												1
Vicia sp.					1							1				1
Vicia cf. cracca		1				1				1						3
Vicia ervilia		1									1					2
Vicia taba											1					1
Vicia lathyrus spp					1											1
Vicia narbonensis	1															1
Vitis vinifera				1	1	1	1		1	1	1	1	1		1	
Zizyphus spina-christi													1			

Table 12: Macrobotanical remains from archaeological sites on Cyprus.

Several problems in the interpretation of archaeobotanical material have been encountered.

Hansen (1989) cited the proximity of archaeological deposits to the ground surface (within one metre in many locations) as a primary cause of possible contamination. Bioturbation may introduce modern intrusive material and the preservation of ancient archaeobotanical remains can be extremely poor. Poor preservation and incomplete specimens has in some instances reduced the possibility of positive identification of species. Also, bias in the publication of results, that is, the inclusion of cultivativars and the exclusion of "weed" species is reflected in Table 12. Categorisation of the data by resource use and possible location has therefore been attempted.

Cereal cultivation appears to have been an important activity during the Neolithic Period with *Hordeum sativum* (hulled barley) the dominant crop, followed by wheat varieties *Triticum dioccum* (emmer wheat), and *Triticum monococcum* (einkorn wheat). Barley is an annual grass growing to a height of about 100 cm. and flowering in April. It occurs naturally in the dry, deciduous oak park belt east, north and west of the Syrian desert, and on slopes facing the Jordan rift valley. It has since adapted to the drier steppes and semi-desert areas of the Mediterranean growing from sea-level to about 1500m (Hansen 1991). *Triticum dicoccoides* is a wild form of wheat wide-spread in Israel, north Jordan and southern Syria where it grows as a common grass. It shows close affinity to rocky places and the relatively rich soils developed on basalts and hard limestone (Hansen 1991:90). *Triticum monococcum* likewise is an annual grass growing to a height of up to one metre and flowering in May and June. It originated in the hilly area bordering the Syrian desert and the Mesopotamian basin on the north. It is considered the progenitor of the cultivated einkorn. The wild species may reach an altitude of 2000m asl on basaltic and limestone slopes (Hansen 1991).

Legumes appear as part of the dietary pattern. *Lens culinaris* is a domesticated lentil common as an annual herb in the Mediterranean region. Its habitat depends on variety, for example, *L. nigricans* grows on rocky hill sides and in *Pinus brutia* woods from sea-level to circa 900 metres, *L. orientalis* is found in oak scrub and pine forests between circa 450 and 1300m asl (Hansen 1991:56).

Fruit varieties were found in isolated situations. *Vitis vinifera*, the wild grape, grows near rivers or streams in deciduous forests, where vines climb over trees. It is found between sea-level to circa 1600m asl (Meikle 1977). *Rubus* species, either raspberry or blackberry, require a cool, moist situation (Meikle 1977:611).

Tree crops are indicated either through fruit or charcoal specimens. *Pistacia* species, assumed to be terebinth, is a deciduous shrub or small spreading tree which grows to a

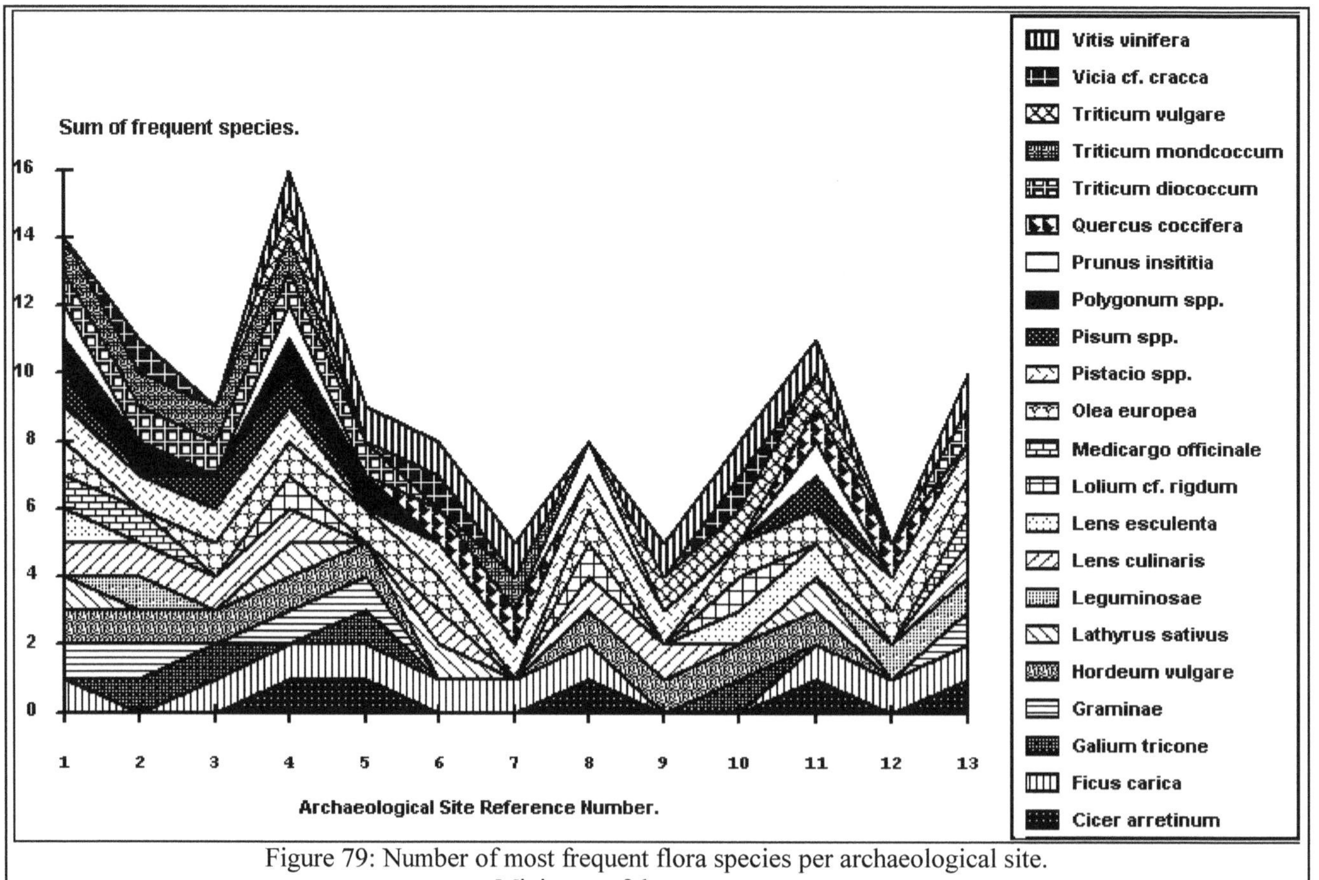

Figure 79: Number of most frequent flora species per archaeological site.
Minimum of three occurences

height of about six metres. Its habitat is on dry, rocky slopes, hill sides and pine forests ranging in altitude from sea-level to about 1200m (Chapman 1967:48; Meikle 1977:369). Bullgloss, *Prunus insitilia* represents a wild prunus species which still exists as a rare relic in the remoter forests (Waines and Stanley-Price 1977).

Pasture grasses identified from Neolithic excavation fall into three main categories: marshland species (*Spergulia bocconii*; *S. marina*; *Trifolium resupinatum*, and bog rush); dry limestone associates (*Genista sphacelata*; *Astragalus sp.*; *Vica ervilia*, *Gallium sp.* and *Lithospermum sp.*); and those growing as ground cover in woodland (*Genista spahacelata* var. *crudelis*, *Astragalus laniger*; *Gallium divaricatula*). The accessibility of these types indicates the possibility of extensive marshland, as projected by Myres (1914) and the utilisation of dry limestone hill sides for crop cultivation. The presence of woodland ground cover species enforces the close association between village settlement near suitable grazing for the range of browse and foraging animals identified from the faunal remains.

During the Chalcolithic period, the cultivation of cereal crops saw the maintenance of barley as the dominant grain and the introduction of *Triticum aestivum* (bread wheat) which may indicate an improvement in agricultural techniques. Lentils (*Lens culinaris*) remained a staple food source while the reliance on a wider range of tree crops developed. *Ficus carica*; and olive (*Olea europea*) appear as dietary additions. The utilisation of *Morus* sp. and *Quercus* sp. identified from charcoal specimens, may relate to the development of pottery firing techniques and the utilisation of these species in early pyrotechnology.

Although it is not possible to determine the relative quantity of arboreal and non-arboreal vegetation in the region based on the quantity of macroscopic plant remains recovered from excavation, the increased variety of species indicates that new resources were being exploited (Hansen 1991:121).

Cherry and Halstead have argued against long-distance transhumant or nomadic pastoralism prior to historic times believing that farmers in fertile lowland areas were primarily dependent on crops with livestock playing a vital but subsidiary role in subsistence (Halstead 1996).

The range of botanical species recovered in the archaeological record increased substantially during the Bronze Age to included *Prunus dulcis*, *Olea europea*, *Figus carica*, *Citrus medica* (citron) and *Punica granatum* (pomegranate), a species list indicative of some external influence, most probably Phoenician. The introduction of medicinal species such as *Bryonia dioca* (bryony), *Echium italicum* and *Carthamus lanatus* (woolly safflower) may also be indicative of external trade and cultural exchange (Hjelmqvist 1977).

Weed species of cultivated fields appear in the archaeological record as distinct from pasture species which may have grown in association with early cropping techniques. The areas under cultivation appear to be

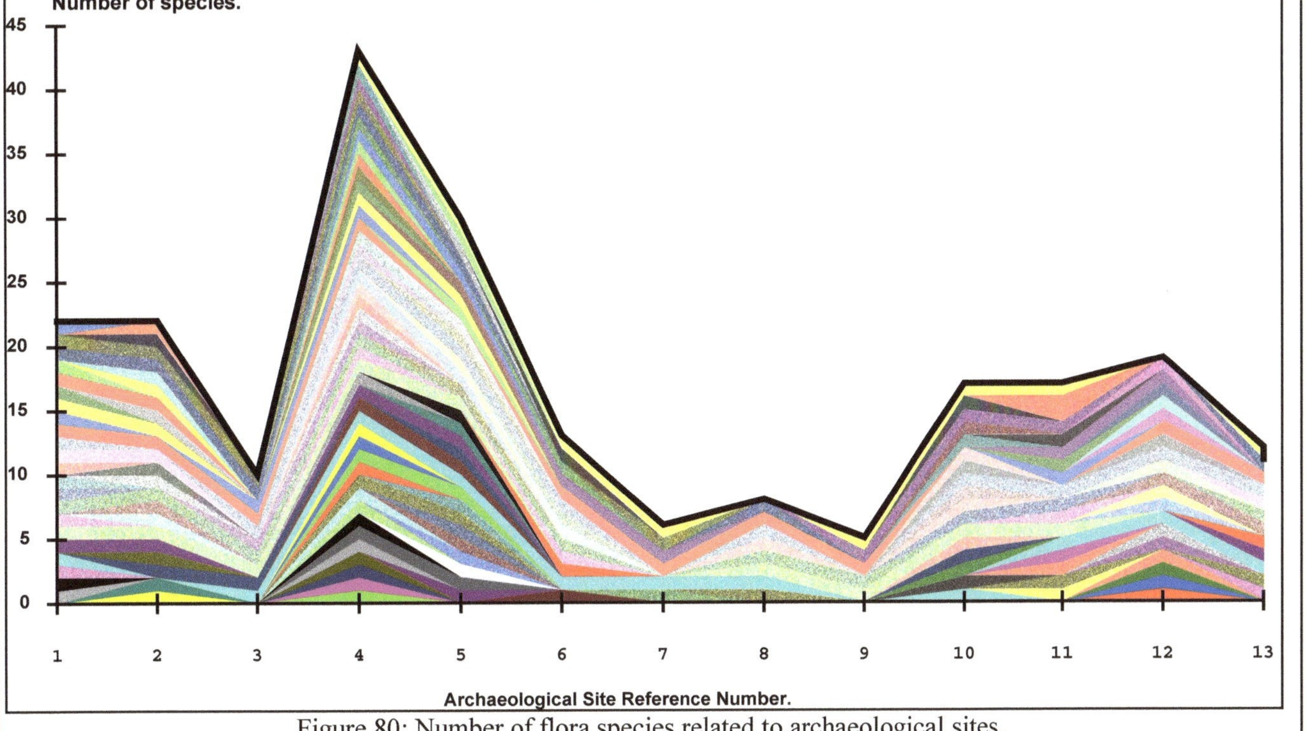

Figure 80: Number of flora species related to archaeological sites.

upland dry slopes where the igneous/calcareous interface provided optimum conditions for a "double-based economy combining herding with farming in the richer soils" (Frankel 1973). Barley continues as the most common and widely spread cereal under cultivation though an increase in wheat types indicate cultural connections with Greece (Hjelmquist 1979).

The transition from a hunting-based reliance on deer to the utilisation of oxen and more intensive sheep, goat and pig husbandry is a significant trend over the period of the Bronze Age. The ratio of deer to other species remains high in the southern districts where associated pottery finds depict motifs of hunting scenes. A decline in deer faunal material in the assemblages in the northern locations appears consistent overall with an increase in sheep numbers in settlements on or near the newly exploited Mesaoria Plain. The decline in deer numbers in the north is the clearest indicator of a changing environmental pattern, perhaps the result of higher population density, more intensive agriculture, and a retraction of the forest cover. Equids, either horses or donkeys, appear in the archaeological deposit and may have been associated with either hunting or chariot warfare. Cattle appear to have been raised for secondary products, milk or traction, with the latter probably holding the higher priority (Halstead 1977).

The Iron Age was a period of intensified arable agriculture which probably influenced by the wide spread introduction of the iron plough during the period of Roman expansion (Dimbleby 1984:69).

Chapter Eight: Exploitation of Bioresources.

Early human activity during the Neolithic period included the introduction of ungulates. All populations of feral domestic species or continental game extant on Cyprus or appearing in the archaeological record were introduced during the Neolithic period or later (Schule 1993). The original mammalian fauna of Cyprus is believed to have been composed of many herbivorous species with different habits and ecological needs (Schule 1993). The presence of large carnivores is unlikely and the native genet appears to have become extinct (Bate 1903).

The impact of Neolithic grazing on the mountain forest area of the Makheras can only be assumed. Croft (1991) suggested a population spread of fallow deer from feral domesticates could occupy the island within 150 years. Ungulate browse preference for broadleaf species such as *Acer obtrusifoila*, *Alnus orientalis*, *Platanus orientalis*, *Olea europea*, shrubby ground cover and riverine species such as *Rubus sanctus* would have a slow but accumulating impact. In areas of unevenaged mixed forest, initially only lower branches of mature trees would be eaten and saplings destroyed. Long term impact on sapling regrowth would inhibit broadleaf species regeneration and open areas for invading species such as *Pinus brutia* or *Quercus alnifolia*. Large concentrations of animals either herded or freely dispersed would gradually transform the fluvial terrace areas adjacent to water courses. If observations of feral herds currently occupying highland forest areas can be applied to range patterns of earlier herds it could be suggested that the natural corridors used were along ridgelines. Concentration of droppings were recorded along spurs and ridgelines. Small shrubs were ringbarked by browse patterns most notably those growing on gentle slopes above creek lines. The regular movement of herds across the very friable volcanic soils of the Makheras induce rapid erosion patterns. Large trees can be debarked by regular rubbing. The amount of time involved in inducing permanent change patterns would be in excess of 300 years, based on the natural maturation rates of *Acer obtrusifolia*, *Alnus orientalis*, *Olea europea* and *Pistacia terebinthus*. *Platanus orientalis* coppices well and would resist incursion for extended periods. As the ungulates were introduced as part of the subsistence farming techniques employed by the early setters of Cyprus, hunting would have curbed over population and would have restricted much of the described damage to forest successional patterns. Most introduced species, *Dama dama mespotamia* (fallow deer), *Ovis* (sheep), *Caprines* (goat) and *Sus* (pig) as free range game animals were either extinct or severely reduced in herd numbers by the late Byzantine Period (Croft 1991).

Chalcolithic extension of grazing and agricultural impacts on vegetation.

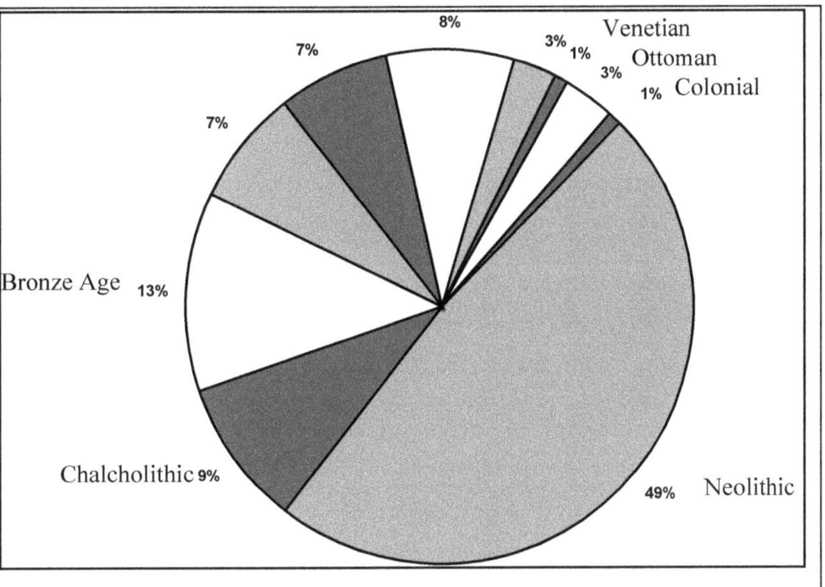

Figure 81: Time period percentages.

It is probable that the demand for agricultural land encroached more on the forest reserve than indiscriminate harvesting of the resource itself. Moreover the exploitation of the forest sub-strata may have created an internal imbalance in the forest dynamic. Sub-strata species are mostly hardwoods and would provide premium fuel for ceramics, lime plaster and copper production. The landscape cannot be viewed as a simplified pattern of degradation sequence since the former mosaic and the present vegetative cover continue the evolutionary sifting of species in a form of dynamic flux constantly responding to the demands placed on the ecological resource of specific localities. Local knowledge of habitat requirements and the need to sustain the forest resource has been recorded by Theophrastus (*HP* v.8.1) who cited the command of Cypriot kings to protect trees and not cut them down. This edict may possibly represent the earliest record of forest management, but it is also characteristic of a cultural system in which trees signified wealth and stability.

Cultivation areas along alluvial terraces partly cleared by

grazing may have been expanded by fire. Fire in the broadleaf forest is different in behaviour and effect than that occurring in a coniferous forest. Forests are generally very sensitive to fire and many ecosystems deteriote if plant cover is removed. Fires frequently are the principal cause of deforestation. The frequency of fire in pine-dominated stands determines the composition and structure of many representative species. The disapearance of some species creates dramatic shifts in associations and their function within the forest environment (Council of Europe 1976). Some Cypriot species such as *Quercus alnifolia*, *Olea europea* and *Arbutus andrachne* coppice after low intensity fires while other species are completely destroyed. *Rhus coriaria* (sumach) is an important post-fire invader and will also occupy abandoned agricultural fields and terraces at high altitudes.

Bronze Age developments and forest utilisation

The early Bronze Age (*circa* 2300-1700 BC) introduced a range of technology not merely confined to the pyrotechnology from which its name is derived. Agricultural innovations such as the plough, the introduction of cattle and equids dramatically changed the cultural orientation of society which may have instigated a dramatic increase in population (Swiny 1989b:187). Settlements typified by Sotira-*Khaminoudhia* were undefended. Houses were of rectangular design clustered together and approached through narrow corridors. The dead were buried in rock-cut chamber tombs in cemeteries located outside the settlements (Swiny 1989).

Knapp (1994) argued for an early Bronze Age economy stimulated by environmental and technical modifications to the environment, such as forest clearance with bronze axes, associated with new farming techniques in an open environment. There is no archaeological evidence to support bronze axe heads and controlled use of fire is the most plausable forest inhibiting strategy.

The impact of pyrotechnology on the forest resource would have effected only those sub-strata trees that are of hardwood types, that coppice well and would be found growing in the vicinity of the smelting site. Field experimentation has demonstrated that a mature, healthy donkey can carry a load of 200 pounds a distance of 27 kilometres in a ten hour day. However, it is more likely that the harnessing of animal power would be for the transportation of the smelted product rather than the conveying of fuel wood. Most settlement sites within the Makheras lie within two hours walking distance of each other. Transportation of fuel wood or charcoal to an industrial site is, therefore, feasible. However, the smelting areas located within the forest boundary do not appear in size to have demanded more fuel than could be procured locally. Very few kilns were located in the Makheras in contrast to those located in the Panagia Bridge Division of the Adelphi forest. This may indicate a variation in fuel recruitments and supply.

Timber utilisation.

The demand for construction timber for major building works or ship building has been in the past regarded as a major factor in species shifts. Madon (1881) contributed an insightful account on forest resource depletion in observing that construction of ships masts and spars demanded the selection of timber from mature trees. Likewise, the planking required in vessel construction requires strong lengths which can only be supplied by mature trees from which the sapwood must be discarded. Timber for construction also demands well-grown trees. Fellings to produce export lumber by necessity would be confined to fully mature specimens, because young trees are unsuitable. In an uneven-aged forest stand, the felling of mature trees would release the younger pole and sapling age group and provide them with a wider resource pool, that is, access to soil nutrients and available moisture, enabling them to attain strong onward growth.

The Late Bronze Age culture on Cyprus was comparable with other great civilisations of the period in the Mediterranean. Sophisticated urban centres and port cities such as Enkomi, Kition and Hala Sultan Tekke provided extensive foreign trade links. Large concentrated urban populations caused considerable pressure on agriculture, and this, together with the extensive copper and shipbuilding industries, impacted upon the Cypriot environment. Land degradation was further exacerbated by unsettled conditions towards the end of the period, with foreign raiding of coastal settlements and probably civil strife between communities (Given pers. com. 1994).

The pollen taxa recovered from the level of early Kition occupation in the Late Bronze Age indicates a mixed forest component dominated by *Pinus* sp. but with *Quercus*, *Cedrus* and *Fagus* together with *graminae* of a cereal variety and Ambrosia (Gifford 1978:63). The extent of the forest cover on the coastal plain and hinterland may be thus considered as still extensive, though the increase in the *Pinus* ratio may indicate a high level of human impact and extensive forest disturbance. *Pinus* regenerates and colonises deforested areas more readily than broad-leafed species.

Forest products as trade commodities.

The exploitation of the Cypriot forest resource that began during the Bronze Age was the result of internal and external economic and political strategies. The trade system of the eastern Mediterranean moved towards the development of urban nodes in strategic locations (Kreutz 1976). The dominance of water transport in the expansion of metropolitan nodes in the eastern Mediterranean periphery meant that Cyprus needed to maintain a dual economy, one based on the seaboard and thus within the trading system, and the other on a self-sufficient agricultural economy. Knapp (1992) has cited the increase in size and number of settlements throughout the island as indicative of successful adaptation to "environmental constraints". The economy developed more complex attributes possibly as a direct result of agricultural innovations from external sources. Specialised products such as woollens, textiles, stone carvings, beads and gaming stones were produced (Swiny 1986, 1989; Knapp:1990a, 1992).

The whole Eastern Mediterranean from the 14th to 12th centuries B.C. underwent a period of unprecedented international contact and stimulation in trade (Smith 1965). Evidence from the widespread distribution of trade items of the adoption and adaptation of foreign goods has been observed throughout the entire region (Knapp 1992). I would, therefore, suggest that while the production and export of copper as a major trading commodity is undeniable, trade in timber as a second major resource is also of critical importance when assessing the complexities of foreign economic relationships, some of which relied more heavily on the Cypriot timber supply than on its copper production.

During the latter half of the 18th Dynasty (1567-1304 BC), Egyptian control over the Levant was weakened by the rise of the Hittite empire, with the eventual loss of the Syrian ports and suzerainty over Lebanon and Palestine. By the end of the 19th Dynasty (1304-1195 BC) Egypt had lost control over all the major areas that had previously supplied ship-building timber. Alternatively, Cyprus had plentiful supplies of timber within easy access of natural harbours and sites. Hala Sultan Tekke and Kition with such harbours, show clear signs of contact with Egypt during the Late Bronze Age (Peltenberg 1985; Astrom 1996). It appears logical on the grounds of accessibility, supply and demand, together with established trading contacts, that by the time of Ramses III (1198-1167 BC), Egypt was obtaining foreign timbers from Cyprus.

A rich forest resource can be utilised in many different ways, four of which appear particularly pertinent to the trade and tribute relationships connected with Cyprus. In selecting these four aspects of forest utilisation, a range of forest bioresources have been listed as exploited or possibly exploited. The four areas of utilisation are: heavy construction, domestic requirements, funerary and ritualistic use, essential oil and medicinal preparations.

Construction timber.

Timbers from the Ma'agan Mikha'el (ca. 430-390 BC) shipwreck in Israel indicate a light structure mostly of *Pinus brutia* (Parker 1992:248). The Kyrenia wreck off the north coast of Cyprus was likewise principally of *Pinus brutia* with a pulley-block of *Morus* (mulberry) and tenons of *Quercus* (Steffy 1985; Parker 1992:232). Lying in the Sea of Galilee to the south of Kibbutz Ginosar, the Ginosar wreck's keel was of cedar and jujube, with cedar planking and oak tenons (Parker 1992:194). The diverse species used in the Ginosar wreck suggests timber reuse.

Domestic wood use.

Combs, utensil boxes, spindles and chairs are all items of domestic use made from timber. Moderately hard, fine and homogeneous woods are suitable for this purpose, in which neither the annual growth rings or the medullary rays are too prominent. *Quercus* spp., *Juniperus* spp. and *Olea europea* are primary woods utilised for these objects.

Wood used in musical instruments has special qualities and characteristics: the higher the key, the finer the woody zones should be. Such timber is often isolated within a forest and it is rare that an entire tree specimen could be used because only the fine-zoned parts tend to be suitable. All wood used for such purposes must be thoroughly seasoned.

The timber utilised by wheelwrights in the construction of chariots needed to be even-grained, long-fibred, tough and dense, and free from knots. Wheels, axles and shafts are the principal components of the wheelwrights craft. The wheels consist of the nave, spokes, felloes and tires. The nave or hub was made generally from the timber of the *Quercus* spp. or *Platanus orientalis* as the wood needed to be hard and dense so as to prevent the loosening of the spokes, which were morticed into the nave. The felloes, morticed together in a circle, were generally of split wood. Whenever possible the wood was curved naturally, and as the pieces are only short there may have been an ample supply of timber with the suitable curve. Sawn felloes are much weaker than those split from wood with a natural curve. A bent rim used in light chariots may have been made from one piece of steamed, split wood, *Quercus* spp. or *Acacia raddiana* (Acacia) being the species most often employed. Spokes were made from a thoroughly tough and strong timber, not subject to shrinkage in hot, dry weather. Poles required strong, straight timber lengths, and the axles were usually fashioned from *Quercus* spp. or *Platanus* spp.

Funerary applications of timber.

Sarcophagi and coffins of cedar have been identified with the funerary practices of Egypt (Lucas 1948:488-490; Meiggs 1982:409) and appear closely associated with the 18th Dynasty. Merrilles (1968) noted the presence of Cypriot Base Ring pottery found in the burials of the professional or middle classes from sites throughout Egypt. Finds excavated by Kanwanti (1984) from the Saqqara tombs north-west of Teti's Pyramid also include Cypriot Base Ring ware. Forty burials were recorded from this site, over half of them in wooden coffins (Kanawati 1984:59). The juxtaposition of wooden coffins and Cypriot pottery may suggest a correlation between two imported commodities: the Base Ring ware, and the single-width timber used in the coffin construction, a feature indicative of imported timbers.

Burial number 9, located above the NE corner of the Mehi tomb at Saqqara, in Egypt, consisted of a well preserved wooden coffin with only a few planks displaced from the lid which had been skilfully joined. The planks extended from the raised foot to the shoulders onto which the separately made head fitted. Likewise the beard, wig and arms were fixed into place with wooden tenons. A nearby wooden anthropoid coffin had a lid carved from a single length of timber (Kanawati 1984:67-69). A quantity of Cypriot pottery was recovered from about one third of the burials excavated at this site.

To date, very little work on the identification of timber species associated with archaeological deposits has been undertaken. The importance of species identification through polarisation microscopy, electron microscopy and other techniques as an archaeological tool has far reaching applications which provide information regarding the structure of cell walls and other diagnostic features hidden to light microscopes. Cell walls are composed of a thin primary wall and a much thicker secondary wall, the latter made of three layers. The smallest visible building units of cell wall are the fibrils which appear string-like under the electron microscope. The orientation and weaving of fibrils vary between species which can thus be identified.

Medicinal products from the forest sub-strata.

In the fourth century BC, Aristotle wrote: "... it was found on the island of Cyprus that there was a mountain larger and higher than all others, which was called Troodos, where there were many different kinds of plants useful for the art of medicine, and if I attempt to talk about each one individually, time will not be sufficient to tell everything ..." The role of Cyprus as a supplier of medicinal herbs was well known and most probably a trading system in this commodity was already established before Aristotle's time. Dioscorides visited Cyprus in the first century AD to collect herbs and mineral extracts (Riddle 1985) and Galen also comments on the medicinal herbs of the island (*De Antidotis* XIV 7). The export of herbs from Cyprus is tentatively suggested by extracts from the Linear B texts and from cargoes excavated from ancient shipwrecks. Coriander found on the Ulu Burun wreck seems certain to have been of Cypriot origin (Bass:1991) and terebinth resin used for embalming, incense and perfume was also carried on the Ulu Burun wreck in large quantities (Ward Haldane 1990:57).

Decoration on the Mastaba of 'Ankhm'ahor' at Saqqara, Egypt, includes a processional scene of gift bearers. The third bearer has a basket with two looped handles attached to the top edge which the hieroglyphics beside identify as "best cedar oil". Another adjacent sequence also depicts a bearer with an enormous, elegantly shaped jar with a small lug and curved, slender handle reaching from the vessel shoulder to the broad rim crowning a slender neck (Badawry 1978:32). This vessel appears to be of Cypriot origin.

The demand for volatile oil used in the embalming process may have opened a market for the range of plant extracts available from species growing in the forest sub-strata not available in Egypt. The mention of oil in the Amarna correspondence (Petrie 1898:45-47) emphasises the Egyptian need for oil imports. The paucity of the forest resource in Egypt was recorded by Theophrastus (*HP* IV:3-5) and it is reflected in the lists of imports recorded.

Impacts during the Classical Period.

During the Classical period, Cyprus was an important source of grain for Athens (Christodoulou 1959:123). In 1926 ancient African and Syrian forms of wheat were recorded as growing on the island (Flaksberger 1927). Olive oil and wine were also notable exports for this period.

The copper mining and smelting site of Ayia Varvara *Almyras* has produced the most complete chain of metallurgical production in ancient Cyprus (Herscher 1995). Finds and radiocarbon determination date periods of the site to Cyprio-archaic, Cyprio-classical, and late Hellenistic. Chalcopyrite and pyrite ores were mined by the open cast method from a pit approximately three metres square and two metres in depth. An associated gallery-like entrance into a hill has also been excavated. Chunks of ore appear to have been broken out of the hard host rock and carried to the nearby smelting site for further treatment. Mortars and pounders were used to crush the ores in several specific locations. Furnace installations have been excavated at a distance of approximately thirty metres east of the mine. Two furnaces, one Classical and one Late Hellenistic, have been interpreted as roasting installations. Free-standing smelting furnaces have been dated to the Late Cyprio-archaic, the Classical and the Late Hellenistic periods. Preliminary spectrographic analysis and metallography show that two kinds of copper metal occur on the site; one a raw copper product with considerable impurities of iron and sulphur, and the other a refined copper with few impurities .

The first recorded description of naval activities on Cyprus was given by Herodotus in the fifth century. Warships from Cyprus under the command of Gorgus of Salamis, his brother Philaon, Timonax of Amathus and Penthylus of Paphos contributed to the fleet of Xerxes formed to invade Greece in 480 BC. According to Herodotus Cyprus provided one-eighth of the naval force by supplying 150 vessels (*Histories* VII.90).

The capacity of Cyprus to produce a large number of ships was again demonstrated in the early fourth century when Evagoras of Salamis in revolt of against Artaxerxes II, launched a fleet of 200 ships. Of these vessels, 130 were provided from his own and allied Cypriot kingdoms. The revolt was defeated and it is possible that the Phoenician cities of Cyprus, especially Kiton, were critical in providing opposing war ships.

The revolt of Evagorus may be linked with recent archaeological discoveries including ship building ramps that have been dated to the end of the fifth and early fourth centuries B.C. at Larnaca *Bamboula*. This area is less than a kilometre south-west of the ancient site of Kition *Kathari* where a temple of Astarte was excavated in 1976. Outlines of ancient vessels can be discerned on the external ashlar blocks of the temple. The *Bamboula* hill was first excavated in 1929 and had been severely damaged by British in-filling in 1879 as part of the malaria eradication program. The area North and East of *Bamboula* appears to have been a sheltered cove that may have served for the beaching of boats during the Bronze age. However, towards the end of the fifth, or to the beginning of the fourth century, the South bank was totally rebuilt. At least six ship sheds were constructed consisting of stone ramps approximately 2 metres wide with a downward slope towards the harbour at about thirteen degrees. Stone paving appears to have extended more than ten metres. Beyond the paved ramp an additional sloping track of packed earth has been excavated continuing at least nineteen metres toward the northern sector of the site.

A stone pedestal excavated in 1990 carries an inscription in Phoenician referring to a trophy erected by King Milkyaton of Kition following a victory over his enemy, presumed to be Evagoras of Salamis in 392 B.C. The dating of the ship sheds is therefore linked with the period of Evagoras revolt.

Hellenistic demands on the forest resource.

The late fourth century B.C. campaigns of Alexander the Great involved intense naval activity to which Cyprus contributed. These campaigns also marked the transition from Persian to Macedonian dominance of the region, first under Alexander, and then under his Suleucid or Ptolemaic successors. Initially, the Cypriots were allied to their Persian overlords, and contributed significant naval forces in opposition to Alexander's *Anabasis*. Cypriot and Phoenician vessels were critical elements in the fleet of over 400 war ships that opposed Alexander's allied fleet at Miletus and Lade in 334 B.C. (Arrian, *Anabasis Alexandri* 1.18.5-6). Following the battle of Issus both Phoenician and Cypriot naval forces deserted the Persian camp and joined forces with Alexander. The Cypriot fleet was later heavily engaged on the Macedonian side at the siege of Tyre, and contributed to the eventual success of the campaign.

Arrian recorded the inclusion of quinqueremes, the most powerful and advanced military vessels of the time, captained by Cypriots representing King Pyntagoras and Androcles of Amathus and Pasicrates of Courium. (Arrian, *Anabasis Alexandri* 11.21.8-22.2) It might be inferred that Cypriot shipwrights were capable of constructing the most powerful and advanced military vessels then afloat (Creevey 1995).

Following the death of Alexander in 323 B.C. the Eastern Mediterranean was the centre of intense conflict among his successors. The strategic position of Cyprus and the wealth of its shipbuilding timber became of critical importance, particularly to the Ptolemies of Egypt. The paucity of Egyptian timber resources required a secure external source, without which the Pharos and Ptolemies of Egypt were unable to defend their realm.

Ptolemy I (Soter) engaged in diplomatic relations with Cyprus in 321 B.C. establishing alliances with four kings: Nicocreon Salamis, Pasicrates Soli, Nicocles Paphos and Androcles of Amathus. The alliance was forged as a punitive measure against the city of Marion. The campaign marked the beginning of Ptolemaic suzerainty over Cyprus that endured almost 300 years. The basis for Ptolemaic control was the strategic importance of the island, and the Egyptian need to secure a regular supply of ship building material.

Theophrastus refers to the cutting of shipbuilding material on Cyprus on two specific occasions. First when describing the timbers preferred by shipwrights, Theophrastus notes the preference for *Pinus brutia*. Second, he remarks on the ship timbers cut for Demetrius Poliorketes eleven-banked ship. These timbers were recorded as thirteen fathoms, or about 78 feet in length, smooth and without knots. This reference stems from a discussion of the height and breadth attained by Cedar trees. This reference may allude to Demetrius' ship being constructed from *Cedrus brevifolia*. Demetrius also appears to have been responsible for the export of timber for the construction of 100 triremes (Meiggs 1985; 145).

Analysis of pollen cores from southern Argolid in Greece (Van Andel and Runnels 1981) indicates an increase in shrub pollen during the later Hellenistic and early Roman period with an associated high incidence in gully erosion during the former period (Alcock 1994). While there is no direct correlation between agro-pasturalism in Cyprus and Greece, an awareness of possible similarities in production capacity and economic demand should be considered. A greater density in settlement sites would be expected on better quality soils and settlement extension beyond optimal ranges would indicate agricultural expansion in response to population growth or increased external pressure (Alcock 1994).

Roman occupation and forest exploitation.

The most significant Roman source detailing ship building practices on Cyprus was compiled by Ammianus Marcellinus who described a rich timber resource on the island after centuries of intense harvesting. "Cyprus is so fertile, and so abounds in products of every kind that without the need of any help from without, by its native resources alone, it builds cargo ships (*omerarium navem*) from the very keel to the topmost sails, and equipping them completely entrusts them to the deep" (Ammianus Marcellinus, XIV.8.14).

The Roman exploitation of medicinal plants was recorded by Diorscorides in the first century AD. Pedanios Dioscorides of Anzarba, Cilicia, lived during the first century AD, an era known as the Greek Renaissance. He studied at Alexandria and Tarsus before becoming attached to the Asian-based Roman army of the Emperor Nero as a physician (Sadek 1983:1n). In his letter to Areuis, preserved as a preface to the Arabic editions of the *Materia Medica*, Dioscorides said of himself:

> ... we have travelled in many countries throughout many decades and we did not live in one place for more than a single decade. We have written this book in five treatises according to your request ... Plants should be collected when they are fully grown and I am not ignorant of the fact that maturity varies according to the character of each country. Some blossom and shoot their leaves in winter, while others flower twice a year.
>
> Anyone who wishes to know about drugs must observe the plants when they first shoot out of the earth and also when they are fully grown and complete their cycle and then when they begin to fade. Those who could observe the plant only at the beginning of its cycle will never know its power in maturity. Likewise, those that are familiar only with fully grown plants will never know the nature of the young plants as the herbs vary throughout the different stages of their growth concerning their leaves, stems, flowers and fruits. For this reason many people who did not observe the complete cycle made a lot of mistakes saying that some drugs were made from plants having neither stems, nor flowers, nor fruit, citing Gramen, Tussilago and Cinquefolium as examples. But those people who have observed the plants at several different stages can easily recognise them. (Dioscorides, trans. Sadek 1983:5)

The *Materia Medica* of Dioscorides described approximately five hundred botanical species of which one hundred and thirty were included in the Hippocratic collection (Hippocrates vol.VI). This reflects the tradition of Greek pharmacy upon which the work of Dioscorides is founded. The importance of the *Materia Medica*, however, is in the profound influence it had on later works and the introduction of plant species from the eastern Mediterranean to Roman colonies in southern Europe (Singer 1927:22).

Species included in the *Materia Medica* were common sub-strata plants on Cyprus and Dioscorides was known to have visited Cyprus to collect specimens. Some plants recorded by Dioscorides were *Crataegus azarolus* which produces medicines from both flowers and fruit predominantly used in the treatment of heart complaints. *Cupresses sempervirens* is rich in resinous compounds and the fruit or cones have been used medicinally since antiquity in the treatment of anaemia and glandular fever. *Juniperus* spp. has been used medicinally and as an aromatic for both cosmetic and culinary purposes. *Lauris noblis* produces fragrant oil used in the treatment of rheumatism. *Lavandula stoechas* (Mediterranean lavender) of the Makheras Forest is well known for its aromatic, antiseptic, and medicinal properties and is still harvested locally. *Myrtus communis* is also very rich in fragrant oil and has been widely used in perfumery and as an antiseptic. *Papaver* spp. (poppy) was, as noted above (Merrillees 1968), a primary export believed to be so valuable that a distinctive pottery style was developed to contain its precious contents. *Pistacia spp*, both *P. terebinthus* and *P. lentiscus*, and *Salvia willeann* (Cypriot Sage), also produced elements utilised medicinally (Georgiades 1992).

The effect of Roman occupation on the landscape may have been limited to an intensification of arable production within the existing pattern (Dimbleby 1984:70) whilst the establishment of many villa farmsteads and opening of communication routes between previously isolated areas enabled a spread of population numbers. The stress on the known woodland as a result of the Roman bath-house culture is, however, an unknown factor though it is believed that about twenty hectares of extended woodland was required in permanent production to service each bath-house (Dimbleby 1984:71).

The importance of sheep grazing in Roman occupied

areas may be significant in the forest and forest peripheries. Hodkinson (1988) suggests that plots of land cultivated intensively without the benefit of draft animals represent an intensification of land-use with increased productivity. Animals domesticated under this type of economy would contribute directly to the production of wool or food and their manure would contribute an important source of crop fertiliser. This system of agriculture could incorporate marginal land areas with some rotational cropping and fooder cultivation being required. This system of agricultural production was described by Theophratus but it is not known how widely the system operated.

A supplementary aspect of livestock husbandry in antiquity was the role of sacrifical animals. Animal sacrifice may provide details on herd-strategy and the role of meat in the ancient diet. Keswani (1994) has argued that the archaeological assemblage of faunal material suggests a variety of socioideological and ritual requirements rather than utilitarian optimisation. The economic importance of goats from the Chalcolithic period is reflected in maintance of large numbers of mature males, a factor that cannot be reconciled with practical economics. Lewthwaite (1981:61) has argued that cereal production on poor soils required quantities of dung and therefore the principal reason for maintaining ovicaprine flocks. During the Roman period the increased costs of animal husbandry contingent on a nuclear settlement pattern promoted the establishment of larger, consolidated flocks in addition to local herding by individuals (Gamble 1982; Halstead 1987; Cherry 1988).

Byzantine occupation of the forest.

The political and socio-economic status of Cyprus between 688 and 965 A.D. is not easily determined. In refuting Hill's hypothesis that Cyprus was in possession of frequently changing foreign dominance (Hill 1940:290), Jenkins suggests that the island was demilitarized and neutral with both Christian and Moslem communities in peaceful coexistence (Jenkins). Both communities had free access to port facilities and paid taxes to both Byzantine and Saracen representatives. The island does not appear to have been geographically divided from the account recorded by St. Willibald, a pilgrim who stayed at Paphos and at Constantia (Salamis) in 723 A.D. There was a regular passenger service to the island from Attaleia and Seleucia with traffic also flowing from Tarsus, Syria and Egypt (Theophanes).

The critical importance of Cyprus to the Arab forces were the naval ports used as a rendezvous for Arab fleets (Abulfaraj). Apart from using Cyprus as a naval base, the Arabs raided and plundered the island on at least three occasions. In 806 A.D. Harun laid waste to the island in retaliation to a breach in treaty agreements.. A further well-documented raid in August 911 A.D. by the Saracen admiral Dimyana (Damian of Tarsus) was brutal and destructive and was a breach in the treaty concluded in 688 A.D. by the emperor Justinian II and the caliph Abd-al-Malik (Hill 1940:286,287). A considerable proportion of the Christian population was enslaved and taken to Syria. Diplomacy later redressed the situation which repatriated the Christian community of Cyprus and gave an economic advantage to the Saracens. It is probable that the death at sea of Dimyana and the destruction of his fleet facilited the diplomatic moves (Jenkins ?).

An important trade in organic goods between Cyprus and the Islamic world is hidden behind the political adgendas documented historically. The importance of written documents in the Islamic culture demanded a supply of inks obtained from vitriols and plant materials readily available from Cyprus. Chemical perscriptions obtained from manuscript sources have included galls from *Quercus* sp., *Pistacia terebinthus*, leaves from *Myrtus communis* and *Rhus coriaria* (Shopin pers. com.).

Medieval documents compiled at the end of the fifteenth century record a range of products grown for export. These include olive oil, sugar, grain, cotton, wine, and carobs. Under the Lusignan dynasty (1191-1489 AD) Cyprus became an important centre for cane sugar production. Plantations were established on the rich coastal plains between Larnaca and Neo-Paphos. Historical sources named the royal estates at Kouklia as a prominent source of sugar. Excavations by Swiss-German expedition have located sugar refining installations adjacent to the Paphian Sanctuary of Aphrodite (von Wartburg and Maier 1989;1990). Although the structures have been largely destroyed part of the water system has been located. Additional mill installations and a refining complex situated on the coastal plain below at Kouklia *Stavros* have been excavated. This industrial site dates from the late thirteenth century and appears to have been in use through the Venetian period (1571). Another sugar refining complex has been located and excavated at Episkopi *Serayia*. The complex includes several milling arrangements preserved *in situ* and provides critical evidence for the mechanical processes driven by water power. A subterranean vaulted chamber below the grinding room housed the wheel emplacement that turned the millstone above. Well preserved double hearths were located above furnaces and provided the heating areas for large copper cauldrons of cane juice. Other important features include a rectangular cistern for washing the sugar mould pots and parts of an extensive water system including aqueducts, water channels and conduits. It is not known what quantity of fuel was required to fire the sugar refineries of the Medieval period.

The importance of the mountain areas as a retreat from the summer heat of the Mesoria surrounding Nicosia has been documented. The removal of the court to Makheras during the plague of 1393 has likewise been recored in early chronicles. The introduction of the *Lepus europeaus* (hare) for hunting may have been the most significant Lusignan impact on the forest bioresource.

Cyprus as a Venetian colony.

Grain was the chief export commodity during the Venetian occupation of Cyprus (Morris 1980:97). However shortages were experienced when land use was converted to cotton and sugar production after a locus plague (Chambers and Pullan 1992:108). Salt, presumably harvested from the salt lakes at Larnaca, was controlled by a Venetian government monopoly. Wine from the island became popular in Venice with three million gallons exported annually to Venice alone. *Sylvia atricapilla* (black-caps), a small migrant bird, were caught in vast quantities (over 10,000 per annum) and pickled as "Beccaficoes" for the Italian market (Bannerman and Bannerman 1971:173). Sugar and cotton continued to expand as valuable export crops (Morris 1980:97). The Venetian trading fleet in 1414, at the pinnacle of her naval supremacy, possessed 3000 small merchant ships, 300 large carriers and 45 war galleys. These vessels were manned by approximately 36,000 seamen (Morris 1980:117). Forty grain ships supplied Venice from Cyprus (Chambers and Pullan 1992:108). In the 15th and 16th centuries, Venice was unable to supply her shipwrights with local timber. However the control of the east coast of the Adriatic and island colonies provided timber to maintain the fleet (Meiggs 1982 :384). The main timber used in the construction of Venetian merchant vessels was *Quercus* sp. (oak) which produced timber lengths with characteristic weight and strength (Meiggs 1982:118,384).

The economic importance of the goat and its impact on the forest resource has been documented from this period. Condemnation of forest grazing obscured the pressure applied to local agriculturalist by food shortages and famine in Venice. The goat produced milk which was converted to cheese, hair for weaving and multiple litters which provided meat as well as herd replacement (Meiggs 1982:385). Christodoulou (1959:190) has described the goat as a measure of forest over-grazing rather than a prime cause. This argument is based on the browse preference of the goat and its ability to subsist in areas no longer viable for other animals . The pressure on agricultural production may have induced a pattern of forest felling and burning "to improve the soil quality" (Paulini 1934) similar to that practised in Venetian Italy.

Cyprus under Ottoman suzerainty.

There are very few records available from the period of Ottoman occupation period of Cyprus. Ottoman legislation governing the use of forest land and forest products has been described as "permissive" (Christodoulou 1957:109). The Medjelle (Article 1243) considered natural forests as common property. The Ottoman Land Code (Article 91) recognised *ab antiquo* rights to particular forest areas and their bioresources by villages. The traditional forest-based resource exploitation by the end of the Ottoman period included free range grazing, resin-tapping, furniture and domestic utensil manufacture and charcoal burning. Charcoal was exported to Syria and Egypt. In addition forest clearing for vineyard expansion was increased during the latter period of Ottoman rule. Briogne (1877), employed by the Turkish administration to report on the island's forest resource, estimated that in the previous twenty years the forest area had decreased by one third in area and by one half in standing timber volume (Thirgood 1981:125).

In 1874 the forests supplied fuel for the production of 400,000 kg. of charcoal; 22,000 kg of pitch which caused the death of between 8,000-10,000 trees, contruction wood and wood for export to Egypt (Montrichard 1874). *Gallium tinctoria* (madder), an important forest ground cover species, was collected and cultivated for export principally to Turkey and Persia via Allepo. In 1738 the French purchased 500 quintals reaching 100,000 kilograms which were exported to France in 1842. Export declined after 1860 and abandoned in favour of cotton by 1879. Ladanum derived from the *Cistus* sp. was collected for medicinal preparations and exported.

Colonial reafforestation under British rule.

Cypriot forests under British Colonial administration were delineated by Wild (1879) on the basis of appearance. Wild was employed by the Colonial Government to advise on the forest condition and means to improve productivity. He wrote:

> "... from Larnaca towards Santa Croce or Stravro Vuni, we meet no forests whatever, only bare hills, fallow land, and cultivation. Ascending Stravro Vuni from the Ayia Varvarna Monastery, we come to the plutonic greenstone. Near the monastery at the latter place we find a few self-sown Pinus brutia. The Stavro Vuni hill is covered with a low growth of myrtle (Myrtus communis), Pistacia lentiscus, thorny and aromatic plants with a few Pinus brutia scattered here and there. Going from there to Korno and Dalijibo we see nothing but barren hills and prickly shrubs. Soon after leaving Korno we get into the green stone again, and a little beyond the village of Dalijibo, on the road to Lithrodonda, a few scattered Pinus brutia of about 8-16 years old, and visible among the small scrub.
>
> Leaving Lithrodonda on the way to the Macheria Monastery somewhat older but still immature growth of Pinus brutia intermixed with Quercus

> alnifolia was encountered. From there higher and older growth was entered though it was very scattered and open canopy of Pinus brutia of 20-80 years mixed on northerly and north-westerly aspects with Quercus alnifolia. This pattern continued up to and around Machera and Chawni Hill and further with variation in either quality, age or composition until the heights of Adelphi and Troodos were reached."

In contrast to the view of Wild, Jeffery the Colonial Secretary wrote in 1916 that:

> "As one approaches the eastern ridge of the Troodos hills dominated by the central peak of Kionia the scenery becomes gloriously wooded. Within these narrow valleys the planes and walnuts, and the stately Pinus halepensis grow to a great size, and form shady glades covering the winding mule paths and the occasional perennial streams" (Jeffery 1918:214).

The re-afforestation policies established late last century under British Colonial rule changed the internal structure and complexity of the Cypriot forests. Described by the British foresters as "bare, denuded hills" symbolising over-exploitation (Baker 1869), the landscape perceived by the British was both foreign and oriental (Severis 1995 pers.com.). Efforts to introduce English landscape concepts and transform the Cypriot environment were an early mandate of British Colonial rule. The British perception of landscape may have provided an unfavourable bias towards the Cypriot environment, particularly when compounded with the Romantic idealism presented by Victorian painters depicting scenes from Greek classics (Snodgrass 1987:70).

In 1899/1900 the mountain slopes of the Makheras Forest were cultivated and sown to *Pinus brutia* and *Cupressus sempervirens*. The initial germination of seedlings was successful but seedling survival failed. In 1901 the same area was planted with 7,730 sapling *Cupressus sempervirens* with a survival rate of over half. In 1902 a further 8,000 seedling trees were planted over the same area again with a survival rate of about half (Thirgood 1987:264).

In an effort to overcome the slow growth of the indigenous tree species British foresters clear felled stands of *Quercus alnifolia* and *Arbutus andrachne* (Christodoulou 1959:113). This action is in direct contrast to the observation of Humboldt (cited Thirgood 1981:150) who said: "By felling trees, which are adapted to the slopes and summits of mountains, men of every climate prepare for future ages at once two calamities: want of wood and scarcity of water". The pine plantation stands, still with discernible rows of trees, remain after a series of heavy fellings, particularly during the First and Second World Wars. The ANZAC forces landing at Gallipoli carried with them timber from the forests of Cyprus and the Allied troops serving in North Africa and Palestine used charcoal from Cyprus as cooking fuel.

A transition from a conservation management practice advocated by Unwin to a production management of the forests was initiated by Chapman. Arguments for wider canopy gaps to induce natural regeneration were suggested (Waterer 1949) although Madon had argued against further canopy opening in 1881. In 1954, trees with a GBH of two metres were considered a "wasting asset" and undesirable. Large numbers of these old mature trees were scared or deformed by resin tapping and it was considered sound economic practice to replace them with new growing stock (Thirgood 1987:259). Clear felling under the old forest policy has created wide canopy gaps and although the *cistus* flourishes, pine regeneration has largely failed. Thirgood, employed as a forester in Cyprus by the British Colonial administration, described the British administration as "paternalistic" (Thirgood 1981:147). The reality of British forestry practice, however, might be better assessed through the forest production during the 81 years of British rule. During the First World War (1914-18) nearly one-sixth of the entire standing stock was felled yeilding approximately 400,000 cubic feet of timber. Fellings were concentrated and not subject to silvicultural control (Unwin 1931). The 1939-45 war period created further demands on the forest resource, with fellings in excess of natural maturation rates again occuring. An extensive road program to facilitate felling was also implemented (Thirgood 1981:136). At the beginning of the Republic in 1960 the total standing volume of the forests was recorded as: productive, exploitable pine forest 99,000 acres; poorly stocked and burnt pine forest 159,000 acres; riverain forest 12,000 acres (Thirgood 1987:242).

The extent of a colonial mentality is reflected in embedded attitudes, such as "Mediterranean woods are unsuited to modern requirements" (Thirgood 1981:151). Rackham in reply suggested that Thirgood had not investigated "ecological realities" (Rackham pers. com.).

The impact of the old mature tree canopy and sub-strata clearance is reflected in the recent rapid increase in alluvial deposits on lowland areas outside the forest. In 1995 a two metre sediment core extracted by Wells and Gabbet was examined at 70cms intervals along its length. Between 0 and 2 pollen grains per cm^3 were counted and considered statistically insignificant. The carbon-14 dates indicated that the sediments from the core were less than 50 years old. No charcoal was observed (Gabbet pers. com.). The rapid accumulation of sediments without indication of fire induced erosion suggests severe environmental impact through clear felling practices within the adjoining forest uplands.

Figure 82a: Map of Cyprus drawn by Bradley Creevey, 1994.

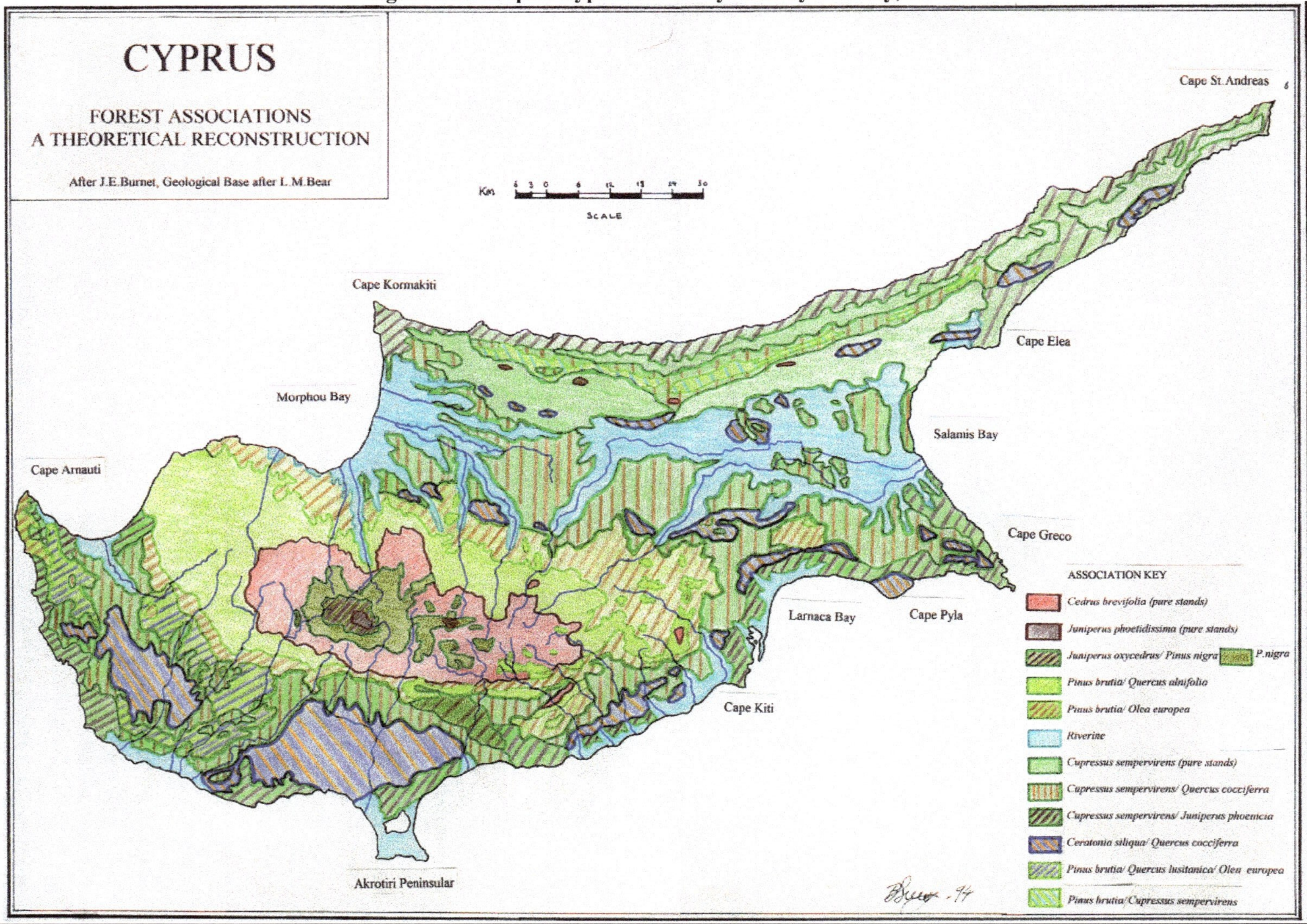

Chapter Nine: Anthropogenic Impact on the Forest Resource.

An understanding of the social complexity required in order to pursue forest bioresource utilisation can be approached through historical and archaeological research. The most important factor in comprehending the inter-relationship between landscape and cultural habitation within the forest environment is an assessment of the interplay between the two. It is, the cultural perception of place and its incorporation into the social expression of a community, and the impact this in turn has on the modification of the landscape or its resources.

The choice of the term "place" rather than the more conventional word "site" in discussing landscape utilisation follows Chang (1992). Sites in an archaeological sense are defined as places where material residues are located and identified. In a forest environment ancient people may have used places within the environment without leaving anthropogenic material. To describe such places as "sites" would lead to a confusion between spatial organisation and the landscape itself. Places used as grazing territories, paths, tracks, springs or watering areas along streams or rivers are interspersed among other areas of human activity such as seasonal or permanent occupation areas, field cultivation or woodcutting places. If a distinction is made through terminology of use areas and facilities then the conceptualisation of the total forest landscape as a system, rather than an array of site locations or settlement patterns, is facilitated (Chang 1992).

Held's Predictions for Archaeological Locations

The known spatial patterning of archaeological sites on Cyprus led Held (1992) to model a set of predictions as principle guidelines for the Sydney Cyprus Survey Project (Knapp, Held, Zangger and Johnson 1992). The first of these predictions was that no major settlement or site would, as a rule, be located above the 400m. asl contour.

Of the 49 sites recorded in the Makheras and Adelphi Forests 5% lay between 400m. and 600m. asl in the Makheras Forest and 59% between the same altitudes in the Adelphi Forest. In the Makheras, 85% were located above 600m asl compared with 41% in the Adelphi Forest. Sherd scatters occurring above 600m. were located in both forests.

Held also suggested that the combination of pine forest ground litter cover and steep topography would obscure anthropogenic evidence should any remain *in situ*. Furthermore, as a corollary prediction, he thought the occurrence of archaeological sites in the forest areas would be restricted to narrow river terraces where major

```
                 N      MEAN    MEDIAN    TRMEAN     STDEV    SEMEAN
Adelphi         29     603.4     580.0     600.4     107.8      20.0

               MIN       MAX        Q1        Q3
Adelphi      400.0     890.0     525.0     650.0

                 N      MEAN    MEDIAN    TRMEAN     STDEV    SEMEAN
Makheras        19     819.5     830.0     816.5     256.5      58.8

               MIN       MAX        Q1        Q3
Makheras     270.0    1420.0     680.0     910.0

Adelphi

                             -------------
               -------------I     +       I---------------        *
                             -------------
           ----+---------+---------+---------+---------+---------+--Adelphi
             400       500       600       700       800       900

Makheras

                                  ---------
                *       ----------I     +  I--            *    *
                                  ---------
           ----+---------+---------+---------+---------+---------+--Makheras
             250       500       750      1000      1250      1500
-----------------------------------------------------------------------------
```

Figure 82: Distribution of Altitudes for Archaeological Sites.

settlements would be unlikely.

The anticipated distance between settlement sites and a perennial water supply was believed to be no greater than 200m unless an easily identified source, such as a river, lay at a greater distance but less than one kilometre.

Should sites be located as settlement clusters, their component loci were expected to belong to consecutive periods, following a pattern suggested by Peltenberg (1985) for the early prehistoric period.

If the concept of axis is applied to the construction of an anthropogenic environment in the Makheras Forest, a reflection of inter-relationship between people and place is observed. The spatial configuration of social life within such a pre-determined environment can be expressed as an understanding of the bonds between life, time and space - an example of "situational ecology" (McNeely 1994), and as such acts as a framework that captures the constant reappraisal of resources by societies associated with a forest environment. The forest is not, nor has it been, a stop over place, but rather a site-specific combination of presences and absences, a particular interaction of physical resources, a specific conjunction of human artefacts and elements of the natural world that facilitates and focuses the interactions or activities which have occurred within its boundaries.

An important aspect of the forest as a social or cultural landscape is the artefact assemblages, both modern and ancient, that present a range of anthropogenic evidence on which an interpretative archaeology can be based. A total of 20 archaeological sites were located in the Makheras Forest and an additional 30 in the Panagia Bridge Division of the Adelphi Forest. Many of the sites recorded appeared from the surface material collected to be multi-period in occupation. The variability of site size over time was difficult to determine from the surface scatters. The majority of archaeological sites in the Makheras Forest as ascertained by the presence of coarse ware and tile appear to be independent mandras or agricultural centres. Judging by the range of pottery dating in some cases from the Hellenistic period, these cover many centuries of site occupation. In contrast, some sites appeared to be small single occupation establishments or the central estate that administered outlying mandras. Quantities of fine quality pottery were recovered from these sites with only a small proportion of coarse ware or pithoi.

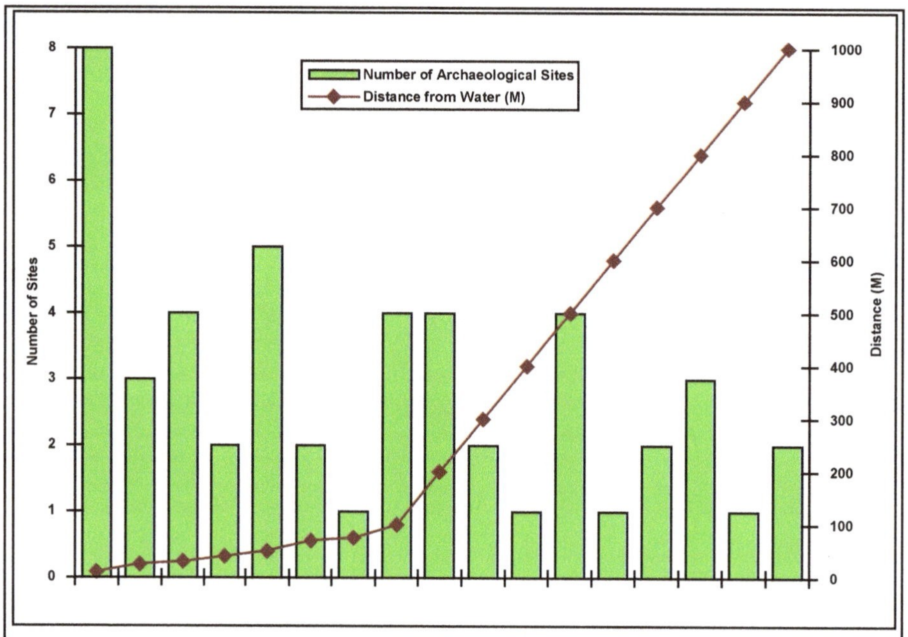

Figure 84: Number of Archaeological Sites related to distance from Water Supply

Sites recorded in the Panagia Bridge area were different in many respects and possibly represent Roman villas or agricultural centres located on the river alluvium. In both forests a series of watch or signal towers were recorded on spur lines.

Number of sites.	Topography
16	ridge
8	valleys
20	hill side
4	hill top
1	spur

Figure 83: Archaeological Sites by Topography.

Figure 85: Map of the Makheras Forest indicating Archaeological Sites

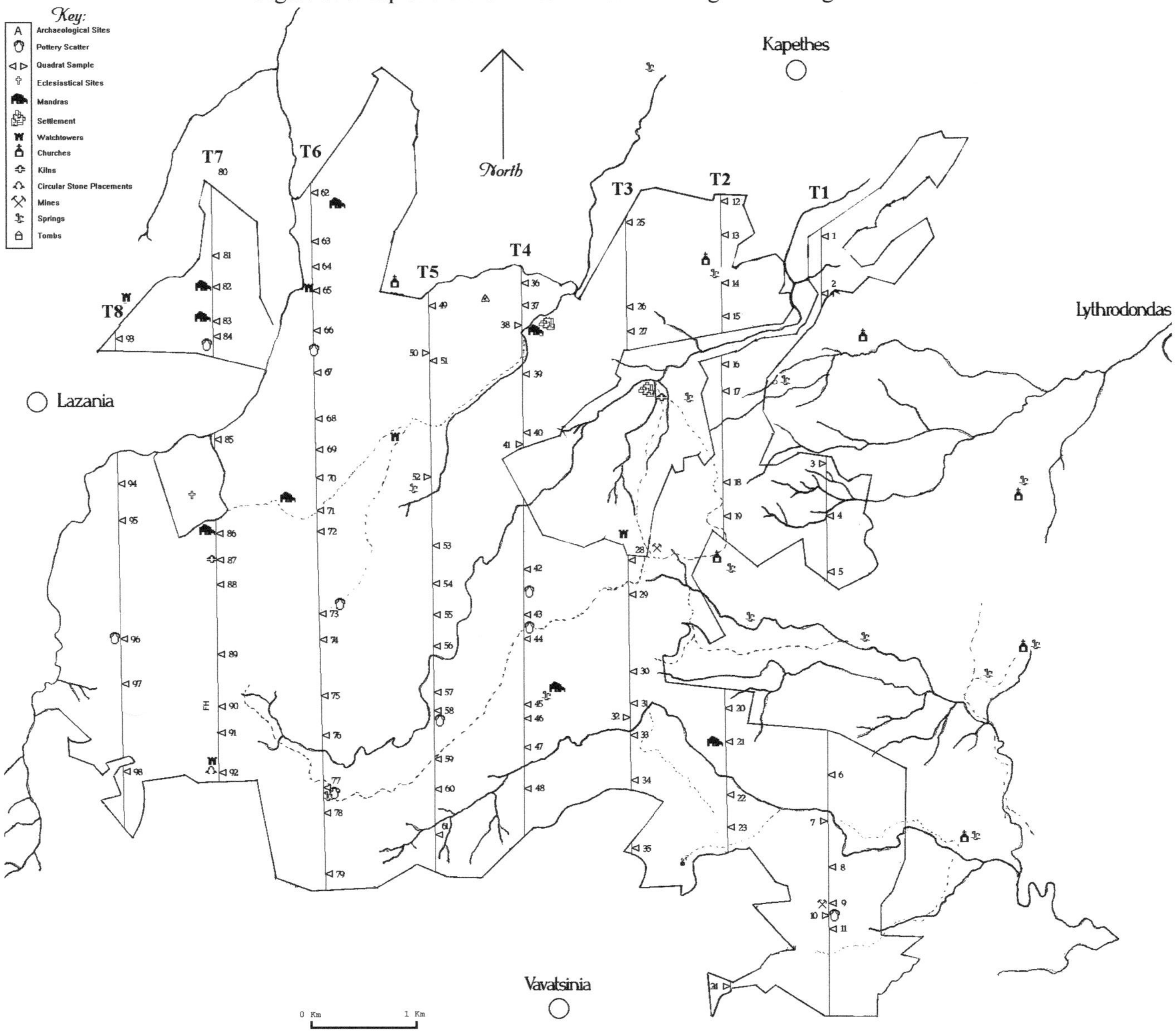

Figure 86: Map of the Adelphi Forest indicating Archaeological Sites

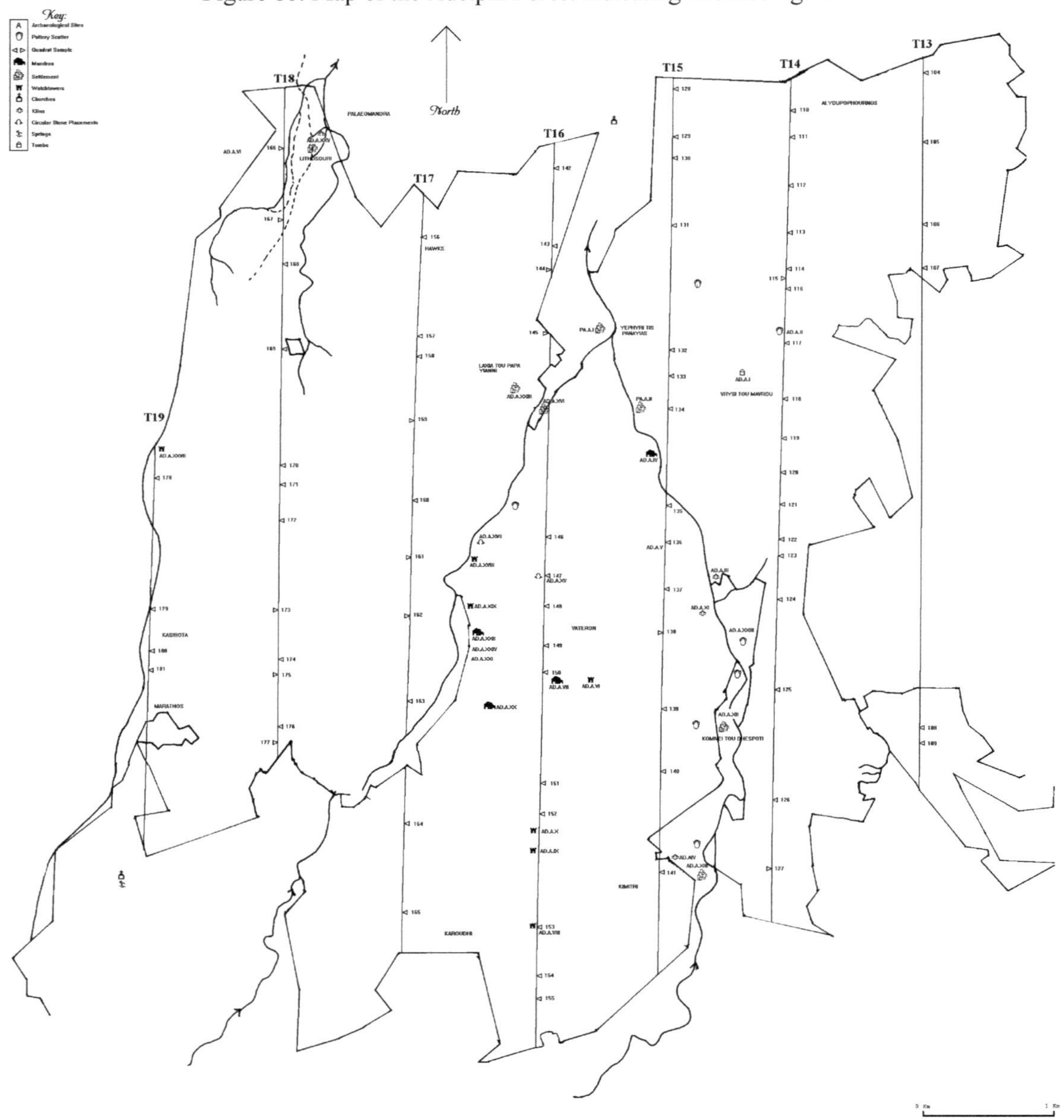

Analysis of Site Type and Bioresources.

Specific site exploitation has been determined by the availability of certain commodities. The resource potential of the forest cannot be limited to the traditional perception of utilisation: timber, mineral, geological, or agricultural exploitation. Site patterning suggests a complex series of economies or settlement processes (Barker 1991) which developed at different periods in response to shifting relationships between mining and agriculture dominated cultures. The importance of foreign trade relationships specifically linked to the timber industry, and olive oil and export wine production also require assessment. The loss of economic resources through external political control needs to be investigated in the light of power bases established during antiquity and carried through into the modern day.

The topography of the Makheras has determined its utilisation. Within the forest there are points that have significance as landmarks and places which have become landmarks. The natural features of the geography have provided visual signals in sighting paths across the landscape in poor visibility. Kionia, the highest point, is distinctive and can be seen from most vantage points within the forest. Other notable peaks include Petromouti, Tyrannos, Mouti tou Skasmenou, Dhyo Mouti and Mouti tous Troullous. The ridgeline linking these peaks affords excellent corridors for travellers. Therefore, natural landform features provide a focus for many purposes. Kionia appears as the central feature of the forest. It is visible from most places in the forest and adjacent areas. All paths lead to Kionia where they converge below the old forest station. There are possibly two considerations which arise from the visible perspective of such an important topographical feature: the significance of its height which lends itself to a hierarchical ordering of space, and the vista afforded by that space.

The river catchment for the Pedhieos and Yialias Rivers reach to high altitudes. The Pedhieos rises in the range west of Kionia and the Yialias immediately below Kionia's peak. The Agakis tou Mandra tou Klosmatou rises between the twin peaks of Dhyo Mouttes and flows south after its confluence with the Argaki tou Marathefti and Argaki tou Mandron near Kyprovasa Chiflik. The Mandron and Marathefti rise in the wide basin below Prophetis Elias Monastry. Many deep natural pools have developed in river lines which, for the most part, have cut channels through solid rock. These pools retain water during prolonged summer periods. All rivers flow through narrow gorges with few natural crossing points.

In contrast, the Panagia Bridge Division of the Adelphi Forest is divided into three main areas by the river gorges of the Phterirouni Potamas and the Platanistassa Potomas which have their confluence just above Panagia Bridge.

Figure 87: Spring of the Black Man

Natural springs are located throughout both forests. In the Makheras all but one spring was capped during the colonial period to provide piped water for forest stations and holding tanks for wild fire control. The uncapped spring located lies below the high ridgeline between Kionia and Dhyo Mouttes. The area is on the northern slope of the mountain and several mature *Lauris nobilis* and *Cupresses sempervirens* have grown around it. The pool below the spring has been roughly built with large boulders. Three pottery sherds were found in the spring basin. An old path from Kionia to Mandra tou Petri passes beside the spring. Upslope from this area there appears to be the remains of an old agricultural terrace now covered by mobile scree with an invasion of sumach. The sheltered nature of the area, the quantity of moss and lichen on rocks and trees below the spring suggest a long period of relative stability. All *Quercus alnifolia* thickets, however, bear coppice scars. The capped springs are sealed with cement and have galvanised pipes leading from them. In some cases a rough track has been bulldozed to these sites. Many springs in the Adelphi have also been capped to provide a water supply to forestry stations, for recreational areas and fire prevention. In contrast to the springs in the Makheras Forest those occuring in the Adelphi appear to have received little maintenance in recent years to their colonial capping structures as many forest stations have become connected to a wider water articulation supply. As a result the capping controlling the flow of many springs has disintegrated and the flow is returning to a natural water course. The environmental impact of extensive spring capping has not yet been assessed.

	Makheras	Adelphi
Church	1	0
Forest Hut	2	0
Industrial	1	1
Kiln	0	3
Mandra	6	7
Mine Shafts	1	0
Monastery	2	0
Pitch	0	2
Settlement	0	7
Stone Arrangement	0	4
Stone Structure	7	5
Tomb	0	1

Table 13: Comparative Table of Archaeological Sites in the Makheras and Adelphi Forests.

Land utilisation within the forest reserve needs be considered also in the contemporary regional context of a settlement site. Occupational studies which incorporate the historic structure of settlement need to appreciate the equilibrium between habitat requirements and evaluate these in relation to resource supply. The advantage or disadvantage of forest disturbance by bioresource utilisation may be relative to a particular stage of community development, land use accommodation, and the dynamic structure of the community. An assessment of environmental change brought about by human impact on a specific resource base, and the resulting transformation of the society as it develops in response to environmental factors, is frequently underestimated by regional archaeologists. Four elements need to be addressed in order to understand the complexities of a developing society within the defined boundaries of a location. They are: (i) Exploitation of a local area centered on a single natural resource, such as a permanent water supply or a deep loam suitable for cultivation. Local area exploitation can also include the harvesting of trees for specific purposes, the extraction of minerals or the security provided by an isolated refuge. Local area exploitation is often restricted to a limited set of site specific qualities. (ii) Catchment and range exploitation is a situation where a permanent locality is used as a base for agro-pastoral activities that may follow a seasonal pattern. This situation might also be applied to a centre which draws resources from the upper reaches of the valley in which it is located. (iii) Movement across the landscape involving natural corridors and specific site exploitation is applicable particularly when timber reserves or mineral deposits where exported out of the resource zone. (iv) Energy and resource loss from the resource base through external control occurs when an area or region is under foreign suzerainty, that is either economic or political control. The over exploitation of a natural resource to a point of resource exhaustion and non-renewal is classified under resource loss through external pressure.

Tracks and Paths.

Figure 88: A Typical Path in the Makheras Forest.

Human impact on the forest landscape is in some respect transient. Old paths linking the northern plain with the southern coastline, the villages and agricultural centres within the forest and natural springs are narrow footways built along the leeward side of the ridges. They are of stone construction, and although overgrown, they can still be discerned. The width of these paths varies from the substantial road link between Mandra tou Kambiou and Khelidonia with an average width of 2m, to the narrow mountain passes with an average width of 80cm.

Similar ancient roads occur in Israel, Turkey and southern Arabia. Avner (1990) dated roads in the Uvda Valley, Israel, to the fourth millennium B.C. and he believed that the entire system of roads in the Negev was already established by this date. Later improvements to the system such as the building of stone retaining walls to support the road on steep gradients, the levelling of the road surface and the marking of roads probably dates to the Roman period (Isaac 1980).

A possible alignment of the Roman road between the ancient Cypriot city kingdoms Amathus and Tamassos is marked on the Peutinger Map (of which a 13th century copy is extant) as XXIX Roman miles (Stlianou 1980). The distance between the two city kingdoms, as the crow flies, is about twenty-three English miles (Hill 1940). Mitford has drawn the Roman territorial boundary from a point between Amathus and Maroni in a slight curve north to Tamassos and it is reasonable to assume that the paths frequently recorded in the Makheras were part of the north-south communication route between the cities.

Hellenistic and Roman watchtowers.

Figure 89: MK .A. XIX The remains of a watch tower.

Hundreds of individual towers are scattered across the Greek mainland and also occur on many islands. They appear to occur in areas adjacent to commom city-state warfare . As the location of the Greek towers are in uninhabitable terrain,they appear to belong to defence systems. A few may have been attached to farming communities (Lawrence 1979). Ancient Greek text sources include references from Aristotle who wrote in 329 A.D. that young conscripts from Athens "patrol the countryside and are quartered in guard-posts (phylakteria)". Forest wardens or land superintendents also had guard-posts and mess rooms for keeping watch.

The sites most favoured for these towers were hilltops with extensive views preferably with at least one other tower in sight for communicating signals. Many of the smaller towers were located by roadsides and a chain of such were located along a mule path roughly one hours journey apart. Rectangular towers were most commonly built, their internal structure being hollow from ground level with the single doorway also at ground level. The height of the tower was governed by the wall thickness, one tower beside the ancient road in the Megarid measured 5.25m square and about 10m in height (Lawrence 1979).

Roman *Limes* and associated watchtowers followed main roads within the Roman provincial regions, particularly roads controlling main water sources, or traditional routes of travel in the Middle East (Gichon 1980). Some *limes* in Israel have evidence of later Byzantine occupation. However, to date excavations, have not been published. Watch or signal towers pertaining to *limes* in the context of Roman Transylvania were situated in front of forts on hills varying from three to ten kilometres apart. The towers were not placed in a straight line but in depth according to the requirements imposed by topographical features (Gudea 1980). Watchtowers, built in southern Egypt to protect the main commercial route between Alexandra and India, were of rough split stone construction with a single doorway. The remains of these towers in Egypt are in good preservation and appear very similar to the watchtower depicted on the Tarjan Column. Two stockades, a pile of logs to fire the beacon, straw piles for smoke signals and three galleried signal or watch towers are depicted in the scene on Tajan's Column known as "The Watch on the Danube". The watchtowers carved in relief on the column appear reminiscent of the system established under Domitian along the Gask ridge in Scotland and in Upper Germany on the Taunus and Odenwald *limites*. Archaeological evidence in support of watchtowers along the Danube is currently lacking.

Isolated towers in Greece and on islands in the Aegean may have specific non-agricultural use such as the protection of mines and mineral wealth, beacon towers, border defences and either coastal watchtowers or lighthouses for those located along the coasts (Cherry, Davis and Mantzouri 1991:292-293). Those described on the island of Keos (ibid) appear to date from the Hellenistic period and appear to dominate communication routes rather than state borders.

A large number of structures described as "cult sites" were located along ancient road ways in the Middle East. *Crenellations* or stone built cairns were situated on hills or cliff tops in direct relationship to roads. They were also found on elevated locations above water sources or paths leading to springs (Avner 1990).

Administrative boundaries, road alignments, military boundaries pertaining to the city kingdoms or cult sites may therefore be represented by a series of square stone structures recorded in the Makheras and Adelphi Forest. These structures are all located midway down a spur running from a ridgeline and they have strategic views over the surrounding terrain. The altitude at which these square stone structures were located ranged from 540m asl to 950m asl.

Figure 90: The remains of the first watch tower recorded in the Makheras Forest

The first square stone structure recorded was located on a spur line above Makheras to Philani roadway at an altitude of 950m asl. The area was originally described during the survey as a rubble pile from small building approximately 4m in length by 3m. Spoil from the

Figure 92: The remains of a central Helenistic tower surrounded by later domestic additions at Kyprovasia Chiftlik. Ottoman regulations prohibited the use of a second storey in housing by Greek Cypriots after 1823

structure had collapsed down slope in a southerly direction. The site is strategically placed with excellent views north down the valley towards Tamassos and to the Makheras Monastery to the south. A quantity of pithos, tile and ceramics were associated with the site. Its distance from a water source appeared to be about 500m.

About 200m further up the same spur another similar, but smaller, rubble pile was located with some bedrock exposed. This area may be entirely natural with the bedrock used for building blocks. A fairly distinct rectangle was discerned on the top of the rock pile, approximately 3m by 2m, with bigger stones around edge and fill of smaller rubble. No ceramics were visible at this second site.

Another square stone structure was located on a spur line descending from forest boundary cairn 27A, situated on track from Prophetis Elias Monastery to Kionia. Here the ruins of a small stone building made of rubble masonry approximately 5m by 3.5m was recorded. There were no tiles or pottery visible, however, the site is heavily overgrown. The location of this site may be significant due to its relationship to a possible mine site observed in an area of rich gossan just below the ridgeline. A path leads from the junction of the link between Prophetis Elias and Khelidonia at a point adjacent to the boundary cairn.

The stone structure in best repair, and therefore suggesting a building type comparable with those depicted on Tarjan's Column, was located above the path linking Mandra tou Kambiou and Mandra tou Jeramaias. It lies adjacent to the path from Mandra tou Kambiou to Kionia. The structure is an unroofed stone building of squared rubble construction situated on a hilltop with commanding views up the adjacent valleys on either side and to the Messoria. Walls are standing on four sides from a height of 30 cm. to 1.5 cm. One doorway facing south was noted but no windows appeared visible. A quantity of broken tile was apparent but no diagnostic pottery. Several old olive trees grew nearby.

Similar square stone structures were recorded in the Adelphi Forest. A small square stone structure was located in the corner of quadrat AD59 on a spur line

Figure 93: An example of broken roof tile and pithoi in wall construction signifying reuse of an occupation site within the forest.

descending from Muti tou Voudhokreanou. From a distance the area appears to be a slightly raised knoll. A pit-like depression occurred within the rubble filled structure. One corner and the outline of the western wall

Figure 91: An example of an Hellenistic tower foundation in the Adelphi Forest

was visible. One pottery sherd, a jar handle, was recovered near this corner area. A well defined path crossed the quadrat and lead directly to the structure. Two rows of rough terraces were located above the path and the soil appeared to be a deeper red loam with a higher moisture content. A series of similar stone structures were located on spur lines located above a path from Voudhokreanou to Vateron. The aspect of each structure was west-south-west and they ranged from 200m to about 50m from the river line. Pottery, particularly pithoi, was recorded at almost every site and several had adjacent flat platform areas similar to the level beacon platform depicted on Tarjan's Column.

A stone structure consisting of rough rubble walls, cut mid way into a spur line adjacent to a deep gully was located on a spur parallel to the first. Water access to the Piataostasa Potamos lay about 200m distant. The area appeared sheltered and the structure which was mostly collapsed was approximately 5m^2.

Three other stone structures of similar size were located on spur lines descending from high ridgelines. On transect line T16 AD64/51 the site recorded as AD.A.VI Kamitirkou was located. It was described as square stone structure with two standing walls of a width of approximately 55 cm. The northern and western wall foundations were well constructed and preserved measuring 4.5m and 4m respectively. A large amount of rubble obscured the remaining walls. A small level platform behind a terraced retaining wall was situated just above the structure. Two pieces of pithoi were recorded but not collected.

The foundations of a small stone structure in the Dhiplaera area at an elevation of 897 m asl were located beside an old path linking Ayia Marina Xyliatos with Stavros Dhiosmatou. No pottery was observed. The location afforded excellent views to Muti Zizitias and the valleys trending north.

Another small stone structure appeared to consist of two rooms. It was located on a ridgeline at an elevation of 540m asl in an apparent alignment (north-south) with a previously recorded circular stone arrangement (AD.A.XV). The building had foundation walls only in part with three large rubble heaps. Visible walls were 80cm thick. The structure may have consisted of one room and an enclosure.

Figure 94: An example of two distinctively different wall structures at the "Fort Site". On the right, a modified longand short wall of a later, possibly Medieval, period and on the left, a wall of dry stone construction designed to "self-enforce" due to the compound stone structure used originally as an outer curtain wall.

Two buildings suspected as being *limes* were also recorded. The "Fort Site" at an elevation of 940m asl is a large four roomed complex on a ridge line below the peak of Muti tou Knizou between Lazania and Phikardou. It commanded strategic views over the Pedhieos River valley and towards Nicosia on the plain. A steep slope fell sharply below the site and old paths ran on either side. A concentration of pottery occurred on the eastern slope below the site. Fragments of a cup or beaker were recovered and when joined gave a full profile. Two other diagnostic sherds were also collected from this area. However the sherds observed were predominately non-diagnostic body sherds. Above the structure, an area of bedrock was located with cutting lines distinctly marked. The innermost room of the complex was cut into the bedrock with its end wall standing to a height of 1.93m. Walls ranged in thickness from 70 cm and 120 cm, the average being usually between 70 to 80 cm. The floor plan of the site possibly suggested three rooms and an enclosed forecourt. No immediate water source was located, although later map investigation marked a spring about 50m distant.

Figure 95: A Hellenistic beaker from the "Fort Site"

Due to their similarity in construction, size and preservation as those located in southern Egypt (Sidebotham 1986:59), Greece (Lawrence 1979), Keos (Cherry, Davis and Mantzourani 1991) and parts of Israel, these square stone structures are believed to be remnants of Hellenistic or Roman watchtowers The role of watchtowers in securing district perimeters, communication lines, water resources and providing a bureaucratic infrastructure is an important factor in understanding the significance of rural and industrial areas during the period of the city kingdoms. Dated to the same period are the smelting centres at Khelidonia and the villa/settlement at Mandra tou Kambiou.

Tombs.

An area thought to be the location of a Hellenistic tomb or hilltop sanctuary, Vrysi tou Marrou, was recorded near Panagia Bridge in the Adelphi Forest. A dense concentration of small fine-ware fragments was located in a small saddle along the ridgeline dividing forest compartments 93 and 97 in the Panagia Bridge Division of the Adelphi Forest. No similar situation was observed in the Makheras. The site aspect is westerly with excellent views to the south through a gap in the ranges. The area is approximately two kilometres by foot from the densely settled sites along the river alluvium east of Panagia Bridge. The site is also located approximately 100m from the spring Vrysi tou Marrou (Spring of the Black Man) from which the toponym portion of the given name was derived.

The pottery scatter which included rims of ten different vessels was concentrated in an area of 1.30 m in diameter. Fragments about 1 cm in diameter were spread 15.4 meters uphill in a NE direction. This is assumed to be the action of the bulldozer during the construction of the fire break. A total of 290 fragments were collected for preservation as the fire break was due to be regraded during the winter of 1996. The presence of tree roots may indicate a deeper soil horizon suggesting an excavated for a tomb. A small scatter (three sherds) of similar manufacture were located at Ypsitas about one kilometre away.

Churches.

Figure 96: AD.A.II. A former Byzantine Church site and later an area used as a forest hut, now demolished.

Several Byzantine churches were known to have existed within the forests. Unfortunately, information on their location, date of establishment and dissolution is difficult to obtain. Hearsay from villages described the suppression of religious centres in the mountain areas following the Ottoman victory. An area recently bulldozed in the Makheras Forest may represent the site of a Byzantine church, as a quantity of Medieval tile was present together with native rounded tile. The site was recorded on early forest maps as being the location of a forest hut. The presence of some solid 20th century brick and a little 4cm Medieval or Ottoman brick probably relateed to the forest structure. The site was situated on ridge line with a north-west aspect. Water was located on site possibly from a spring head.

Forest Huts.

The ruin of Kionia Forest Rest House located at an altitude of 1300m asl, beneath the peak of Kionia, was the largest modern building recorded during the survey. The original structure appeared to have been two square rooms divided by a stone wall bonded into the western wall. The two doors, set adjacent to each other, opened to the east. A window opened to the north and another, later made into a door opening, to the south. The roof top was of moulded tile and the floor of local brick tile. The structure was later extended to include an additional room to the south with two southern-opening windows and a door onto one of the original rooms. Another door opened to the east. A larger room was constructed at the northern end and the original northern window closed and plastered as shelf space. A fire place was constructed on the northern wall with a drainage pit adjacent. The original roofing and flooring was re-used in the construction of the additional room, which used much less stone of a smaller size. The original building still has a heavy white wash coating in parts, whereas the later additions show little white work, paint or plaster. The entire building was shuttered with only the shutter bolts and burnt timber framing around the windows remaining. The walls for both building periods were 60cm (two feet) thick. Two internal roof beams provided a false ceiling in the kitchen. The internal wall between the two original rooms was 30cm thick. Second period wall thickness was 40cm. The new extension also included a covered verandah. Capping on the original building and continuing around the new extensions was of reinforced concrete, largely collapsed on the western side. Aggregate for the concrete appeared to be too large and not well pressed into the scaffolding forms. The masonry above appeared contemporary with the extension.

Five *Cupresses sempervirens* planted along the terrace edge showed burn marks similar to lighting strike burns. These extended about 3m above ground level to just above ground level for the southern most trees, 1.25 meters to above ground level for the third. The tree situated closest to the kitchen doorway was almost completely destroyed from burn de-barking, and the tree opposite the kitchen window showed about 60cm of burn scarring.

Kilns and Industrial sites.

At least two forms of industrial activity were recorded within the forests, with the possibility of a third also under consideration. Industrial activity was most notable for the residues produced which were of little economic importance and usually non-biodegradable. Metallurgical slag is a waste product of the smelting process through which metal is extracted from ore. Slag can act as a process indicator; a chemical analysis can reveal important information on the metallurgical processes involved at that site (Bachmann 1982:3) and it can also provide a reliable date of production. Slag heaps can be stratified, with each stratum corresponding to a specific production period, thus providing a relative chronological sequence of the various technological processes (Buchmann 1982:6). In addition, slag heaps may contain fragments of furnace lining, ore, flux, charcoal or other artefacts.

The distribution of slag in the Makheras Forest was concentrated at two sites: Khelodonia where a substantial slag heap was recorded, and at Mandra tou Kambiou where large quantities of slag was distributed across all fields. In the Adelphi small concentrations of slag were recorded at several sites, although only one area could be definitely described as a smelting site. Other industrial activities include pitch extraction using a double kiln process and possibly the production of pottery or glass.

Figure 97: The section face of AD.A.XI exposed by road works in the forest and suggesting an ancient kiln site

A fire pit or kiln was located approximately 500m from Mandra tou Kimitirin. The kiln had been exposed by road works and partially destroyed. It measured 2.35 metres in diameter and appeared to be surrounded by stones. (see plan) Samples of ash, charcoal and vitrified material were collected for analysis. One pottery sherd was collected from the road verge above the kiln and fragments of a crucible-like material were collected from a road gutter below the site.

On a slope above a very steep gully line trending north in the Sylitis area, a quantity of pithoi fragments and coarseware pottery were located. The area appeared to have a small platform or terrace adjacent to the pottery scatter. No structural remains were otherwise visible. An unusual rock formation was nearby and easily seen from a forest road above the site.

A double kiln similar to those traditionally described for pitch extraction was located on a steep ridge in the Vateron area. The kiln walls were carefully made of split stone 1m in thickness. The diameter of the larger pit was 1.7m. A smaller pit below had been largely infilled.

Figure 98: A double kiln site now overgrown and in an inaccessible forest region. Its original purpose appears to have been for pitch production.

A small stone building with agricultural terraces, an adjacent footpath leading to a natural spring and a double pitch kiln emplacement comprised this industrial site of Sykioes. A small pit for charcoal, ash dump and vitrified earth was also recorded. The building had collapsed although the foundations were partially visible. The kilns were intact and represented a typical example of the pitch extraction process. The site was situated within 25m of water, located on the southern side of the Phterirouni Potamos above the confluence of a creek line from the spring and the river.

Broken pithos sherds, some with a pitch-like substance adhering to their sides were collected from a bulldozed access road near Panagia Bridge. In winter, a natural seep to the north-north-west of the site was covered with a thick grass growth. Many hare droppings were observed in the area. The pithos fragments were located near a gossan seam exposed by gully erosion. A vegetation transect was drawn across the site from the south-south-east to north-north-west to include the gully line and embankment below the fire trail and the slope above the site. However, no vegetation was recorded that corresponded with the geobotanical species list published by Brooks and Johannes (1990).

Marrou (AD.A.III) was a small disturbed industrial area that included eight features. The former slag heap has been removed during the creation of a roadway linking Panagia Bridge with Askas. Slag extended along the roadway for 31 metres. No slag deposit was recorded on the cliff face below the road line. Area or locos 1 was a small slag and rubble heap with pithos fragments. Locos 2 included three rubble dumps; locos 3 contained gossans and ore (sample collected); locos 4 was an ash lode suspected to be modern (charcoal collected); locos 5 appeared to be a circular stone placement; locos 6 was a spoil heap containing ore, glass fragment, ash and charcoal; locos 7 was another rubble pile with broken rock, pottery and slag (samples collected); locos 8 appeared to be a preparation area with crushed ore, charcoal, heat fractured rocks and crucible fragments.

Figure 99: An artifical mound at Ad.A.XXVb partially excavated in 1997 to reveal a limestone paved kiln.

Industrial areas appeared to be concentrated by streams, where a significant confluence occurred or supplementary springs were located to provide a source of permanent water. Large slag deposits and extensive pottery scatters were concentrated in these areas. At Khelidonia, modification of the landscape to facilitate ore carrying sleds from the ridge about 7 km away are still apparent. These were cut to a depths of 3m in some places and angled to slow the gradational pull of load.

The impact of pyrotechnology on the forest resource would have affected only those sub-strata trees that are of hardwood types, that coppice well and would be found growing in the vicinity of the smelting site. Most settlement sites within the Makheras lie within two hours walking distance of each other. Transportation of fuel wood or charcoal to an industrial site is, therefore, feasible, but the smelting areas located within the forest boundary do not appear in size to have demanded more fuel than could be procurred locally. Very few kilns were located in the Makheras in contrast to those located in the Panagia Bridge Division of the Adelphi forest. This may indicate a variation in fuel recruitments and supply.

Mandras or farmstead sites.

Figure 100: Mandra tou Jeremias

Rural settlement densities, settlement visibility and chronology are difficult issues to address, as surveys are liable to produce unequal quantities of diagnostic pottery for different time periods (Mattingly 1993). There is little evidence to support intensive rural settlement by the pre-Roman societies of Cyprus in the forests. There are few Roman sites of the late Republic in the Makheras, although dense concentrations of late Roman pottery occur in the river valleys of the Adelphi. Issues that arise from the interpretation of the rural settlement pattern include the local organisation of land, labour and surplus production and the exploitation of regional resources by the state. The economic viability of the forest areas appear to have been continuous with greater occupation in the valley floors during Roman occupation and high altitude pastoralism occurring during the early Medieval period. The role of monastic establishments in providing capital investment to leasehold farms may be reflected in the size and quantity of pithoi sherds recorded.

Figure 101: Mandra tou Klosmatou

The production of fleece and yarn in the Makheras can be assumed from the evidence recorded at the "Loom Weight Site". The artificially flattened area on a spur opposite the Makheras Monastery seems typical of other mandras located in the forest. Concentrations of pithos and tile were mapped and three loom weights were recovered. The loom weights were in size and shape similar to those recorded by Stewart from the early Cypriot Bronze Age and those excavated at Marki *Alonia* by Frankel and Webb (1996:197-1999). Loom weights are poorly represented in archaeological excavations of tomb sites and their presence in the forest environment is of considerable importance. The finding of terracotta troughs also indicates either wool processing or dying being carried out *in situ*.

Immediately below the Loom Weight Site a long, stone built enclosure or structure 19.8m in length and 6m in width was recorded. The walls averaged 60cm in width. The bearing of long wall was 100 degrees. One large piece of pithos was observed and a rubber from a saddle quern appeared to be reused as fabric built into the wall.

Another farmstead known as Mandra tou Klosmatou appeared to be a typical example of long settlement within the forest. Access to the site was along a narrow dyke from the Kyprovasa Chiftlik to Prophetis Elias road. It appeared to have been a single-roomed structure with three stone wall enclosures. A considerable quantity of broken pithoi, tile and pottery had been reused in the construction of the walls currently standing. This reuse of materials seems indicative of long site occupation and similar wall fabric construction was observed at several locations. Access to a series of deep water holes in the creek line into which a permanent spring flows provided a ready water source for the site. It also appeared to be a good defensive position. A quantity of medieval pottery was located on area above creek line and below the house where the ground was slightly eroded. The building roof had collapsed raising the floor level over a metre. No windows were observed in the structure although remnant wooden fabric, presumably from the door, was noted. A sheep or goat fold appeared to be located between two walls. A range of domestic and semi-domestic trees and herbs were located in the area.

A larger mandra, Mandra tou Anayiou, may have been rebuilt on the site of a Roman *lime* as its internal room arrangement and enclosed forecourt were unique in comparison with other structures. The mandra was situated on a narrow ridge between two gully lines. Tracks converge at the site and the major link to Philaini and subsequently, Tamassos passed the building. The building had a westerly aspect with the main entrance orientated toward the south. Walls were rubble built with a flat, open platform type yard on the southern exterior. The entire complex was roughly rectangular. Some walls included reused pottery or tile material. Five "rooms" were mapped: Room 1 being called the large southern enclosure. The wall height for room 1 on the south, east, and westerly aspects was averaged at 40 cm. The northern wall of Room 2 was constructed of small blocks filled with stones and round tile. It survived to a height of 140 cm. Other walls appearred similar, though with less tile filling. The doorway was well constructed. Room 3 appeared to be another enclosure with roughly built walls of round river cobbles and some squared blocks. A shallow pit was located inside. The walls of this room extend to a height of 80 cm. Room 4 also contained a shallow pit of square form in the north-west corner. Traces only of Room 5 were visible, but a curious, elongated semi-circle was located similar in form to one observed at Mandra tou Jeramais.

Mandra tou Jeramais lies along a flattened spur trending south and overlooking the headwaters of the Argaki tou Ayiou Onouphriou. The site was situated in a high and isolated location with excellent views up and down the valley. An old path crosses the mountain opposite. The remaining standing structure consists of a two roomed enclosure with walls about 70 cm thick. Old pottery sherds have been reused in the wall construction. Two large tree stumps have been incorporated into the walls along the eastern and western lengths. Four fields were defined. Field 2 was almost entirely paved while field 3 is partially paved. Here the paving was much better laid with river cobbles carried from the valley below. Similar paving was locally associated with threshing floors used for the separation of millet and barley grains from their husks. Another structure, believed to be the dwelling, was completely razed and the area planted to seeding pine and cedar. Two areas yielded diagnostic pottery: Field 4 to the west and fields 1 and 3 in the vicinity of the former dwelling. Several large pithoi ruins were located in Field 4 on the eastern side. Part of a stone bowl was recovered near a semi-circular feature in Field 2. The site and immediate area carried a quantity of medicinal herbs.

"Bishop's Retreat", situated down the spur line running from the Makheras Monastery to Kionia road, was originally recorded as a low density scatter of pottery recorded over a quadrant as the result of mechanical terracing for reforestation. No pottery occurred in the area west of quadrant above the road. However, a concentration of early Medieval pottery was located in a cutting exposed by forestry terracing. The site therefore, appeared to extend 20m north of the quadrant with a distinct layer of ashy deposit above a level of pottery. A quantity of glazed and decorated pottery was collected for analysis. Portion of a free standing wall on a bedrock foundation was exposed by the road cutting. The site appeared to be a single occupation early Medieval dwelling. The pottery was overall very fine. The destruction of the dwelling by fire may be related to control measures exercised to contain an outbreak of the plague within the household.

Figure 102: Mandra tou Petri

Mandra tou Petri is located beside a forest road leading to the adjacent capped spring. It now consists of a bulldozed platform above spring on a ridge with a southerly aspect. A dense concentration of Byzantine tile or plain ware pottery was apparent with about 10 pieces recorded per square metre.

The richest sequence of pottery was recovered from the area known as Mandra tou Kambiou. The original mandra building was completely destroyed by construction of the present Forest Station structure, with the exception of an oven or

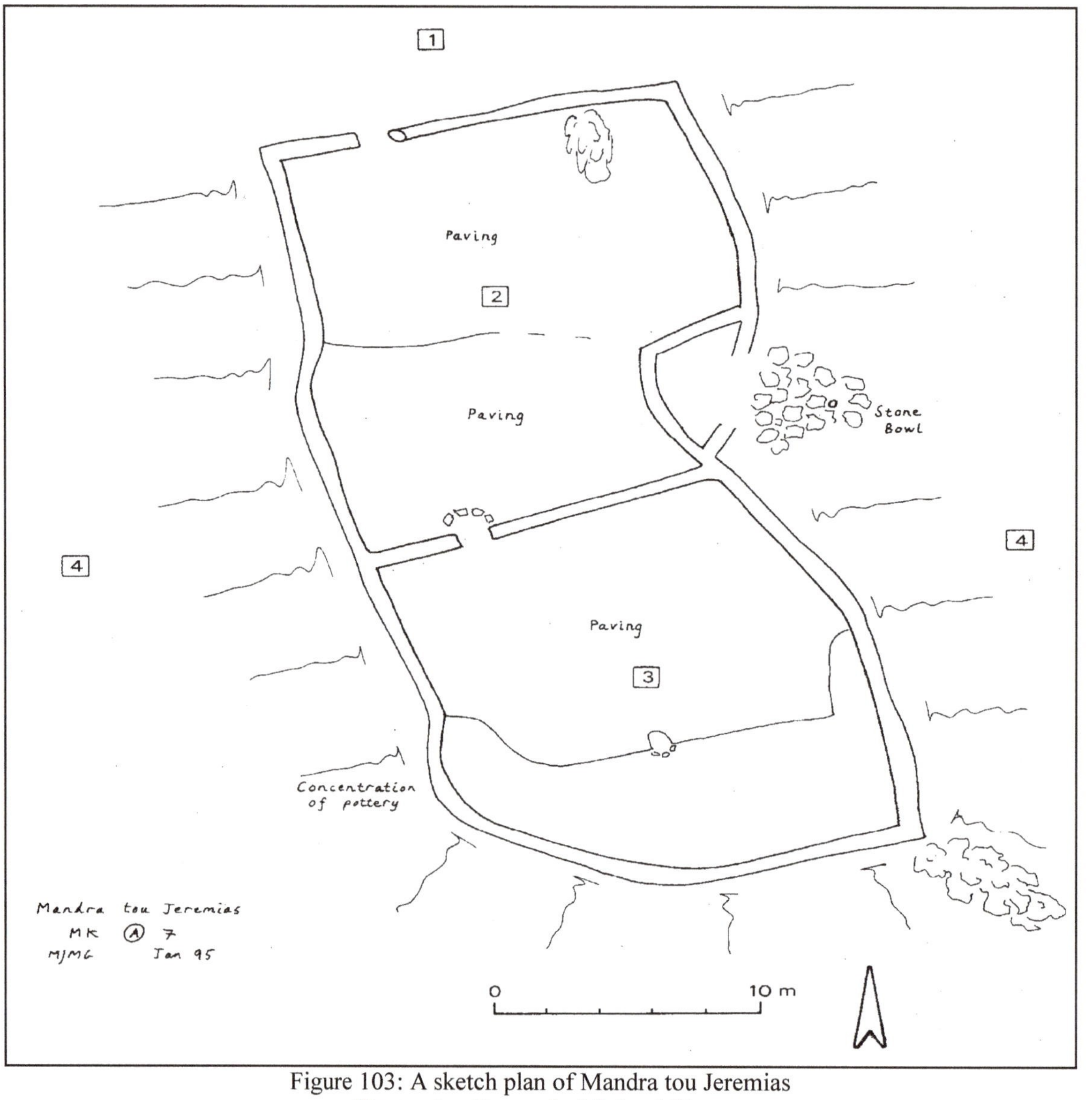

Figure 103: A sketch plan of Mandra tou Jeremias
Illustration Drawn by Michael Given

similar installation. Scatters of sherds and tile were very dense to north and east of Forest Station. Four fields were defined for the area:

Figure 104: Mandra tou Kamitirkou

Field 1: Lay below the road line and trended toward the eastern side of the Forest Station. To the south-west, a greater concentration of thicker pithos type sherds were located together with some large tap slag pieces. These decreased toward the Forest Station. The field was systematically walked and the recently ploughed surface afforded recognition of a dense sherd scatter. Finer sherds and those of a diagnostic nature appeared to be concentrated in the northern sector, that is, closer to the Forest Station.

Field 2: Tiles and pithos were concentrated in the north-east corner of this field, due perhaps, to bulldozer activity. Sherds were more concentrated in the south-eastern area. The north-western sector of the field is a recently constructed terrace but remains of a small square

Figure 105: AD.A. XX Situated near the river.

structure with a bottom opening remains. It was built of plastered masonry. The original pre-Forest Station mandra was probably situated on the platform directly north of the present building.

Field 3: Following the path from behind the forest station to the creek below where one large rock pool lay with a wider shallow wade below. Sherds from plain ware vessels (non-diagnostic body sherds) were observed, although many may have fallen from the platform above. Diagnostic sherds from two vessels were collected - one green glazed, the other yellow glazed. These sherds were scattered across the rocky gorge face above the rock pool and two were recovered from the pool.

Field 4: Lay adjacent to a capped spring north and west of a line of cedars. A concentration of fine ware (Roman and Hellenistic) was located within 10m of the spring. The area had been heavily ripped by a bulldozer in recent times. The scatter of sherds extended north to the water stand pipe and south into an olive grove. Numerous pithos sherds were observed but not sampled.

The mandras recorded in the Adelphi Forest were in a better state of preservation and no site had suffered mechanical devastation. Mandra tou Kamitirkou was recorded on a spur line about 100m below the watch tower or lookout recorded as AD.A.VI. The structure consisted of one large room with a narrow smaller, or secondary room built in one corner. The inner room was built against an earth bank cut into the spur line. Both rooms had doorways facing west down the spur line. A large court yard extended 18m to the west and the stone foundations were visible along its entire length. Evidence of wall robbing was apparent and rubble tumble extended down the scarps on the northern and southern sides. Pottery sherds were located in the gully line on the southern scarp.

The structure lay between two small knolls on a narrow saddle. Four gully lines on the south east had low terrace walls just below their confluence, suggesting a spring or well site. The stones in this area carried a heavy moss cover. On the northern gully line some modification to the catchment area was suggested and a large earth spoil heap was recorded.

Two late Ottoman mandras were recorded near ancient terraces above the Row Potomas. The first of these early modern mandras was possibly sited on a much older foundation as the wall material contained reuse pithoi. Several unusual pitted stone boulders were located beside the building. The adjacent terraces appeared to be very old. Another abandoned mandra was located on an adjacent ridge line opposite the kiln site. The remains of the building appeared consistent with other foundations in close proximity to kiln placements. Both mandras had westerly aspects.

A similar two-roomed stone structure was cut into a steep spur descending from Muti tou Vaterou. Only one foundation wall was standing the others were collapsed. Good stone masonry was, however, apparent. A scatter of pithoi sherds was recorded and one sherd collected.

In contrast, a four or five roomed stone dwelling or mandra was cut into a very steep spur above the river line and between two steep gullies. Very thick pine litter obscured surface visibility. Two pieces of rough tile, several large pieces of flat stone, possibly roofing material, were recorded and a few sherds. The stonework was very well laid and no re-use of material was noted. The site was approximately 50m from the river where several large natural pools occurred.

The area around Panagia Bridge was rich in artifacts. The area was consequently divided into areas falling either side of the present road. The settlement area defined as PA.A.I Settlement or Mandra lay at an altitude of 400m asl with an easterly aspect. It was divided into four loci:

Locos 1: An extensive area of abandoned agricultural land and eight roomed mandra were located on a fluvial terrace adjacent to the medieval bridge of Yephyri tis Panayias. The area covered included vineyards to the north and almond groves to the south. The mandra had solid walls of broken stone and pithoi fragment construction. One room was dug below the ground level and covered with iron sheeting. The building appeared structurally sound. An old pithoi house was recorded about 25m up hill from the main building. Although only the foundation of the pithoi house remained the holding places for six pithoi were evident. Three different broken pithoi were photographed on the slope below the out house. Other stone walls or foundations showed in the road cutting below the mandra.

Locos 2: Was an orchard area with dense grass covering. Rat middens were observed under the almond trees. One large body sherd was recovered.

Locos 3a: Was abandoned and cultivated vineyards with some diagnostic sherds. However, visibility was very poor. One red jasper flint thought to come from a threshing sled was collected.

Locos 3b: Was a roadside plantation of almonds with pottery visible under trees. Two cultivated areas provided good visibility with dense pottery scatters recorded. However, manure dumps apparently brought into the area by the cultivator, contained broken pottery and this may have created a bias in the archaeological evidence.

A second extensive settlement site (PA.A.II) was located on the fluvial terrace on the northern bank of the Phterirouni Potamos, approximately one kilometre east of the area known as Panagia Bridge. The area had been extensively bulldozed to create modern agricultural terraces. A large quantity of fineware, pithoi and tile were recovered.

A settlement or mandra (AD.A.XVI Adelphi *Poro)* was located above the ford river crossing about two kilometres south of Panagia Bridge. It appeared to be a two-roomed stone building or enclosure dug into the ground above the Row Potomos. Extensive old terraces and cultivated fields were adjacent. An old path descended from the building to the river where some modification to the river bed was observed.

A large agricultural or monastic complex (AD.A.IV Mandra tou Peraskeva) was situated on a level alluvial terrace above the gorge of the Phterirouni Potamos. Four separate structures of stone rubble construction and a large rough stone platform, possibly a threshing floor, were recorded. (see plan) A well was located within the complex and appeared to be in alignment with a small square room. The older and smaller buildings were of better construction than the larger room (locos 1) and may have been individual monk cells. Some walls were standing to a height of 30cm. The area was surrounded by a narrow curtain wall. The site was thickly overgrown with ground cover and plantation pines. Three diagnostic sherds, one pithos sherd and one undiagnostic ceramic were collected from the curtain wall area. One piece of slag and several lumps of gossan were observed.

The site was sheltered with first early morning sun exposure. A rich bird life was noted locally. Several ancient olive trees were located within the site parameter. These had stone wall surrounds. The inner diameter of one olive tree was 2.6m. The stone enclosure measured 3.6m in diameter. A dense covering of wild garlic was also noted.

The cliff face down to the river appears to have previously had a stair cut into it but it is now eroded out. A large quantity of water-worn pottery was located on the pebble beach below the cliff. Two very deep natural water-holes lie in the river bed below the site. An old ford crossing was also noted below the site and a narrow path leads from the ford up the gorge.

The Mandra sites may have represented situations of extensive agro-pastoralism from settlement bases at regional towns or large monastic centres. Each Mandra investigated appeared to be located within its own valley with access to permanent water, high altitude summer grazing and safe shelter during winter months. Most seem sited so as to obtain maximum winter sunlight and a strategic view in each direction up and down the valley. The most frequent aspect for mandra sites in the Makheras Forest was easterly away from the prevailing wind. In contrast those recorded in the Adelphi Forest generally faced west.

Mine sites.

Mine sites were positioned just below ridgelines in the south-eastern area of the forest. There was no pottery visible at either location although the area recorded as MK.A.I included seven shafts, each apparently following a gold bearing quartz seam. Well defined paths were connected, with valley areas below and nearby areas of industrial activity (Kyprovassa Chiflik and Khelidonia). No mine shafts were recorded in the Adelphi Forest.

The first archaeological site recorded in the Makheras Forest was a mine site (MK.A.I) located at an elevation of 830 m asl on the slopes of Pano Teratsotos above the spring known as Vrysi tis Skotnis. A mine shaft located about 5m from the north-west corner of MK 139.5. appeared about 3.5m deep, 2.5m long and 1 to 1.5 m wide. It seems to have been cut through a dyke of altered diabase with a heavy quartz vein. Other shafts were then located in the immediate area between this ridge and one adjacent to the west - two with associated platforms. In total, seven shafts were located down the spur, each following a quartz seam. The dominant site aspect was westerly. The condition of the site was good although some in filling had occurred on the floor. The sides of the shaft appeared sound and stable. Although it has not yet been possible to have a geologist investigate the site, it is known that gold was mined in the Makheras Forest and the quantity of quartz associated with the shafts may suggest this activity (Flood 1996 per.com.)

Figure 106: MK.A.I A Mine Site

Monastic sites.

In the road cutting opposite the current church of Ayios Onouphrios the outline of a structure with a floor, with two and possibly more walls exposed. A small quantity of medieval pottery was eroding from the exposed bank. There appeared to be a distinct floor level of large flat river cobbles making up two building areas. There was no structure visible at the surface level nor were any ceramics located at the surface. The original monastery was founded in the area in 1450 and was dissolved and fell into ruin in 1571. It is possible that the foundations exposed by the road cutting pertain to this early foundation.

The road bank and cut terraces in front of the current monastery and church of Prophetis Elias Monastery (MK.A.III) contained a vast quantity of Byzantine material including tile, pithoi, glazed and plain ware sherds. It is possible that earlier foundations of the monastery existed at a slightly lower altitude than the present structure, thus accounting for the quantity of material observed. The material is wide spread and extended down a gully line below the current spring head. On the steep slopes below the monastery, evidence of ancient vineyards, olive groves and pasture land were noted.

Monastic sites appear to have early dates and several foundations in the same immediate area. Large quantities of early Medieval pottery were recovered for analysis. The monastic sites were all located near permanent springs and most were at high altitudes. There is a possibility of early sanctuaries dedicated to Apollo may have existed in these localities prior to Christian conversion. The consistency of naming corresponding with the "sun god" and prophets or saints personified by fire seems consistent (Given per.com. 1995).

Settlement.

The Khelidonia site (MK.A.XVIII) included a large settlement and industrial complex with some associated low grade copper ore seams located in the spur above the site. It was situated on the confluence of four streams. The area extended over an area of approximately 10 hectares. To facilitate recording, the complex was divided into three "site" areas: Site A, a more recent Ottoman occupational area, Site B, an earlier occupation area with Roman and Medieval pottery, and Site C, the slag heap and industrial installation area.

Site A was situated on a spur above the main river line with creeks on either side and a steep cliff face above the southern river line. Five "fields" were instituted to describe Site A: Field one covered the extensive construction which was divided into eleven "rooms". Field 2 lay between the structures and the river line to the east. Field 3 covered the steep slope to the south where extensive erosion had occurred. Field 4 lay to the north and incorporated the creek line down to its confluence with the river. Field 5 incorporated the small area of alluvium opposite Site A.

Site B was situated on a plateau to the south of Site A. Seven fields were instituted to describe the area. Field 6 lay between a lightly marked pathway above a visible wall line opposite the slag heap. Several areas of erosion had exposed pottery fragments which were collected. Several geological samples were also collected from this area. Field 7 was an eroded peninsula which proved generally sterile of anthropogenic material with only a small number of sherds collected. One large piece of vitrified material however was located here. Field 8 covered an area of thick grassy pasture with scattered small fire places presumably built by hunters. One "T" shaped dug-out appeared to be associated with military activity. A considerable amount of finely broken slag occurred in the spoil from the dug-out. Field 9 was a cut bank above the track where a concentration of pottery fragments were observed *in situ* and some were eroding onto the track. The bank extended for about 30m. A quantity of diagnostic sherds was collected for analysis. Field 10 extended down a steep bank from the track to the river line. An old wall appears quite distinctly running parallel to the track. Field 11 included the second building complex whose walls were, to a large extent, standing and intact. This area was measured and photographed. A curtain wall or enclosure wall appeared to extend about 20m from the building to the north-east. Field 12 incorporated a platform area defined by a low retaining wall. Some pottery was recovered from under very old olive trees where the ground had been cultivated. The olives were planted in rows approximately 10m apart. Several piles of rubble were evident to the south but building foundations were not discernible, although the area was heavily grassed.

Site C was situated on the end of a long spur. An old wall and possibly old smelting installations were observed on the crown of the spur adjacent to the slag heap. The slag heap extended from the top of the spur end to the creek line below where its width was about 25m . Extensive landscape modification had occurred along the length of the spur with several deep cuttings observed. These are thought to have facilitated the movement of ore. One small outcrop of low grade ore was observed midway along the spur. The area had been used for military exercises.

A site comparable in size and importance to Khelidonia was located in the Adelphi Forest. This area (AD.A.XXV) known locally as *Lithosouri,* appeared to be an abandoned village with a kiln installation on the creek line below. It extended over a large area, probably in excess of 6 acres with dense concentrations of rubble piles and large quantities of pithoi, tile and pottery. The site extended from a ridgeline along which a forest fire break has been bulldozed, to the creek line trending north. By following the ridgeline for approximately 400m the village of Kato Moni was clearly visible.

No building foundations were observed above ground level, although visibility was very poor due to a dense pine litter cover. Areas cut for pit planting of trees and along the road cuttings exposed quantities of white slipped pottery.

Above the creek line, one kiln site was observed intact. Others may lie adjacent to it. A quantity of slag and vitrified material was concentrated near the kiln. The structure and the type of material associated with it did not suggest a pitch or charcoal kiln. It was very similar to one (or more) destroyed in the Makheras forest during post wild fire terracing.

An area within the forest enclave appeared to have a quantity of pottery, suggesting a small settlement site in the area known as Laxia tou Papa Yianni. Pottery extended over five terrace levels overlooking the ford crossing into forest Beat 3 of the Adelphi Forest.

Three other smaller settlement sites were recorded. A series of rubble piles, cut by road construction near Lakxioes tou Seraphim (AD.A.XXVI) exposed pottery and vitrified material. The quantity of material and the extent of the complex was difficult to ascertain due to the condition of the site.

A series of abandoned terraces were cut by recent road works at Kamnel tou Dhespoti Porto (AD.A.XII), Bishops Leap, to reveal walls from an ancient settlement located above the medieval bridge replaced in 1956 by a modern construction. A quantity of pottery was recovered from the cutting and the terraces adjacent. Ground visibility was poor. A path led to the top of a small hill where a rubble pile suggested the presence of another small stone structure. The area possibly covered approximately half an acre.

An abandoned mandra located on a spur overlooking the Phterirouni Potamos at Kimitirin (AD.A.XIII Mandra tou Kimitirin) consisted of two rooms and a large walled enclosure. The site extended over an area of approximately four acres with pottery from the road line to the river escarpment. A quantity of fine ware, very dense pithoi and tile fragments were recorded. The structure had standing walls with an older foundation wall visible in room 2. Some slag was located in the almond orchard on the adjacent southern slope. Very dense ground cover obscured most of the area with pottery recovered only from exposed areas under cultivated trees. The old olive trees had individual terraces or walls for protection.

Settlement areas are located near springs and creek lines or above rivers. In many situations in the Makheras, a combination of water resources are located in the vicinity of any mandra or agricultural holding. Long periods of occupation in each site were evident from the associated pottery range. The building techniques which reuse materials, such as pottery, pithos and tile in new walls, indicated long periods of settlement at many locations. Modern forestry practice, the elimination of settlers from the forest reserve and the action of wildfire has reduced the visible traces of occupation. Rubble masonry, tile and pottery therefore constituted the central evidence for occupation areas. The combination of natural resources, terracing and communication lines, however, suggested that these centres were important economic elements in a wider region. The quality of glazed pottery recovered indicated a level of affluence not readily associated with isolated farmsteads.

Round stone arrangements.

Several enigmatic circular or oval stone structures were located in both forests. The first was at an elevation of 1420m asl, just below the summit of Kionia. It appeared to be a rough oval enclosure which utilised large rocks already *in situ* and included a small terrace on the eastern side. There was not sufficient rubble or tumble to suggest a large or complete structure. However, the structure did appear to be dug into the surrounding slope. Surface covered by dense pine litter. The situation appeared sheltered and a very large *Arbutus andrachne* (approximately 15 metres in height) grew nearby. There appeared to be no immediate water resource apart from the springs at Kionia picnic area. A second square stone enclosure was located 5m to the north east and was of comparable construction but in general more similar to the square stone tower features. This structure measured approximately 2.7m^2 and a quantity of rubble tumble from its collapse extended several metres down slope. One body sherd, not diagnostic, was recovered. The location of these enclosures afforded views to the southern coastline. From the site, a bearing on Forest boundary cairn 190 was 142° true. A second circular stone arrangement or pit was located on the ridgeline within the quadrat T16 AD44 (AD.A.XVII). Another circular stone (AD.A.XV) arrangement with a door way or insert was located above a modern forest track at an altitude of 700m asl, that replaced a more ancient path recorded on Kitchener's map (1885). A long stone wall or terrace ran north from the structure. A small platform appeared to be adjoining the structure. No pottery was evident. The site was at least 400 metres from the river.

This type of round stone arrangement was suggestive of the Chalcolithic houses of the southern coastal area and these may prove to be consistent with a more mobile population of the Early Chalcolithic (Peltenberg 1981;1982; per.com. 1996). The practice of transhumance pastoralism was challenged by Halstead, and the size of the arrangements did not suggest animal enclosures. Similar structures in Israel have been determined "sanctuaries" and this possibility appears more likely given the size and location of the structures. The cult of Cybele was practised at Tamassos with rural sanctuaries associated with the cult, believed to be located in the forest (Stewart 1962).

Terracing.

Terracing for agricultural needs, and the more recent forestry terracing for erosion prevention, appeared to be long lasting. Hillside gradoli used in post-fire reafforestation may take thousands of years to remould into more natural lines. Erosion from road building was also evident, although in comparison with European forests there is less road per hectare in the Makheras forest. The geological formation, particularly the diabase, was easily fractured and weathered causing frequent landslide and roadway collapse during winter. Landslide debris provides a ready surface for the germination of seedling trees and in some instances, assists in the regeneration potential of some areas.

Terracing appears to have been a major agricultural activity in the Makheras forest. Dating terraces by the age of old olive groves, placed many terraces in the late Roman or Early Byzantine era. Most terraces were small, located in high altitude areas with paths near by. Terraces were located below saddles and most frequently have a westerly aspect. Their distance from mandras varied from one kilometre to several kilometres. Numerous pottery sherds were collected in sheltered saddles near high altitude olive groves. The presence of wild olives as an important forest sub-strata species cannot be ignored. Preference for oil production however, continued to favour the small wild spinose variety, whereas this species was grafted for edible fruit preparation.

Bridges.

Three bridges of ancient construction were recorded, all in the Adelphi Forest. The bridges were marked on Kitchener's map of 1885. One bridge was subsequently

Figure 107:A 13th Century Bridge

destroyed to allow the construction of a wider concrete structure at Bishop's Leap. The main bridge crossing at Panagia was widened and the original single span incorporated into the present structure. The bridge providing access to the Byzantine church of Stavros Dhiosmatro has been preserved.

Several crossings cut through rock to provide a descent through the gorge were observed. In some situations, the stone placed river fords were still *in situ*. It has been suggested (Demetriou 1996 per. com.) that the extant bridges can be dated to the establishment of the churches, to which their associated paths are directed.

Church sites outside the forest.

Around the perimeter of both the Makheras and Adelphi Forests lay a series of small churches, most of which are in a ruined condition. Each small church was located near a spring and bore a recorded saints' name. From south to north, the churches lying beside ancient paths were Ayios Nicoloas near Kyprovasa Chiftlik (ruin), Ayios Yeorghios located between two springs and near the junction of paths from the southern district, Lythrodondas and Prophetis Elias converge, Ayia Paraveski also located beside a spring (ruin), Ayios Demetrios (ruin) located on the path between Mandra tou Kambiou and Lythrondondas, the area of MK.A.II believed to be a former Byzantine church which lay almost equidistant between Ayios Demitrios and Mandra tou Kambiou where a small ruined church was recorded late last century. The two monastic settlements of Ayious Onouphros and Panagia tou Makhera are located in the north western forest area of the Makheras. The ruin church of Agnlotissa in the Stenokambia area is located in the valley above the deserted village of Philani. An area known as Ayios Venetos is located in the forest enclave in the Yialias River basin.

A similar pattern is observed around the Adelphi Forest with a series of churches along the southern edge of the forest. These include the historic Byzantine churches of Ayios Ioannis, Stavros tou Ayiasmati and the ruined church of Ayios Efstathios.

The importance of spring water on ancient paths or travel

routes may indicate long use of these communication lines. These sites were not investigated during the survey as they lay outside the forest perimeter but they may also act to define the extent of the forest in ancient times.

Following the trade pattern described by Harrison (1993), the spatial orginisation of sites along river valleys became apparent and this may relate to a network of economic exchange. A point of origin in the mountain areas, such as Mandra tou Kambiou or Kheledonia, using city states such as Tamassos or provincial centres like Kyprovassa Chiflik as transaction points with further intermediary centres such as Idalion or Kalavassos Ayios Dhimitrios.

At any time in the history of Cyprus, the development of a political geography has been intrinsically linked to spatial relationships and the effect of, and dependence on, an immediate environment. Geopolitical relationships can take two forms: (i) the relationship between a group, as a separate unit and the environment or (ii) as the relationship between groups as "political contacts" and the environment. The former illustrates the concentration of population and natural resource within legally demarcated boundaries of social or cultural areas. The second case deals with processes of co-operation or conflict. These were modified by characteristics of space between cultural groups in which such processes were enacted, for example, economic warfare for resource control.

In this situation, the Makheras Forest must be seen as lying between triangular points of power with the city kingdoms of Tamassos to the north, Kalavassos/Maroni to the south east, and Amathus to the south. From the higher reaches of the forest the coastline from Salamis to Kourion can be clearly seen. The topography of the afforested area in respect to the environmental barriers and corridors which facilitate exchange in material culture is again evident in surface scatters of artefacts which occur on paths running above river lines, thus presenting the valley areas as potential zones of influence. For example, to the south Kyprovassa and Mandra tou Klosmatou are within easy reach of Kalavassos while Mandra tou Kambiou lies at the head of the valley formerly dominated by Tamassos.

Figure 109: An example of wood construction translated into stone at the Tombs of the Kings, Tamassos.

Settlement location and subsequent cultural changes which have altered site values might be considered as extensions of their source of origin. The settlement sites need to be considered as extensions of original requirements of a particular power base. For example, the utilisation of a timber resource is linked not only to a particular stage of socio-economic development but also to the dynamic structure of an extended political environment with wide-ranging trading implications. The exploitation of a forest resource, particularly timber, as

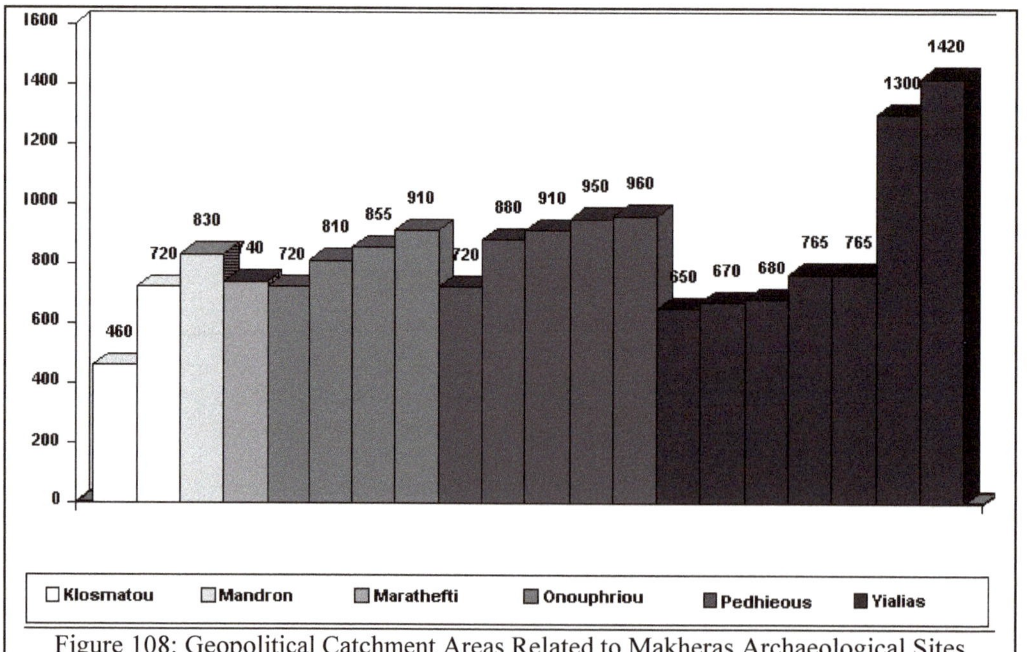

Figure 108: Geopolitical Catchment Areas Related to Makheras Archaeological Sites

either a domestic or export commodity, reaches a climax defined by population growth and available labour, wealth accumulation or attained production. The delineation of a culture, as defined by an environment, is relative to its capacity as a society to sustainably develop its habitat. It is of note that excavations at Tamassos have identified the widest range of tree species yet published from an archaeological site - almost all indigenous to the Makheras forest, while the architecture of the Royal tombs imitates in stone, wooden building techniques.

The forest has also acted to define the distribution of political and cultural energy to the limits of an active frontier, of particular note during the Lusignan Period when the mountains were sought for the beauty of the vines and the freshness of the air. The expansion of court culture and the establishment of feudal holdings at Lazania, Philani and Philcardhou, create a distinct boundary separate from the central seat of court culture in Nicosia, whilst the legendary feuding between these village centres make the location of this site of possible military significance. The second early 13th century site, with its rich layer of exposed pottery may represent a single occupation site, contemporary with the adjournment of the court to Makheras during the plague.

The forest also represents an area of contest in struggles for domination during periods of civil conflict. Isolated look-outs and military "fox holes" are located at the top of spurs. Brass bullet casings from pistols and rifles have been recovered. These artefacts testify to a continuation of forest use as a refuge in recent past history. The destruction of the Kionia Forest Station by fire has also been linked to the struggle for independence.

Societies may choose to forget their past and many build elaborate monuments around that process. However, the forest and its biotic environment is imprinted with human associations to which it is now tied. Recently a series of commemorative monuments to heroes of the Independence movement have been built in the forest. These monument coincide with the approximate area where a death occurred although some are situated in prominent public view on a ridgeline. Such monuments are an example of a created place or space as a deliberate human construction involving a new or different relationship with the natural world.

Thus, instead of creating an intellectual structure around features of natural topography, monument building is a way of establishing or enhancing the significance of a particular location. In altering the perception of the environment the emphasis is not on the sacred nature of the landscape as a whole or on particular features within it, but rather on a perceptional and altitudinal change exemplifying a definite conceptual shift in both spatial and temporal consciousness (Bradley 1993: 4-17).

Archaeological sites located within the forest must be viewed as palaeohabitat islands where segments of the past ecosystem processes are preserved. Alternatively, they may be areas of intense human activity in which elements of past ecosystems were either artificially and selectively concentrated, or where new and unique habitats were created which some elements of the local flora and fauna were able to exploit.

Chapter Ten: Transformation of Ecosystem Components over Time.

Vegetation transformation in an isolated situation such as island ecology is governed by geomorphological factors, perturbations and time, or as Fanta described it "V=f,t,u,". The ecological systems represented in the present environment of the Makheras Forest were initially isolated but as greater and more frequent contact with mainland regions to the south and east, as well as other Mediterranean islands developed, more cosmopolitan features became apparent.

The dimensions of physical space within which the evolution of the forest bioresource has occurred might, therefore, be considered as twofold. First is the geographic formation and location of the island which is sea-bound, but within sight of nearby mainland areas. Second is the geomorphology of the mountain area to which the forest has been confined since the commencement of cultural records for the island. These dimensions locate the forest system and have determined the associations developed over time between the forest bioresource and human activity.

In the tightly interlocking spatial system represented by an island, action in one ecosystem cannot be isolated from any other. Any event directly or indirectly affects each ecosystem of a cluster with the result that a modification of that ecosystem cluster occurs (Forman 1987). Each disturbance is therefore long-lasting and wide-ranging. If impact is more frequent than the environment can accommodate, a rapid transformation in the character of the environment can occur. Change can open new ecological niches readily filled by alien species or colonising invader species.

Cyclical succession at several levels is evident within the forest ecosystem dynamic. The maturation rate for various tree species governs the longest period of cyclic return. These interact with forest harvesting regimes which to date have been governed by internal and external economic need. Seasonal cycles are also of importance both for flora and faunal species, particularly migratory birds which impact on forest insect populations. The pattern of cyclic growth, energy loss and return can only be mapped over an extended time frame. The complex aggregation of species, dominated by tree types exists in close association with the physical environment. The forest has developed its own trends, features of inter- and intra-species competition, and a regeneration ecology for particular niche species.

As a result of the survey, landscape mosaics of current vegetation (plant microlandscapes) were identified. Based on these mosaics, the types of potential vegetation were defined using species associations at spatial and distributional levels. Negative impacts on species composition and forest structure were analysed. Although impacts have varied over time and space they can be grouped in a simplified way as follows:

(i) **Biotic impacts including foreign introductions.**

Biotic impacts are concentrated in the more accessible areas and generally result from the economic potential and utilisation of the forest bioresource. Environmental impacts recorded include settlements, terracing, paths, olive groves and mechanical reafforestation.

(ii) **Stand structure changes.**

An appraisal of the level of anthropogenic impact has resulted in the identification of levels of forest degradation based on the diversity of plant associations, their successional positions, faunal associations and anthropogenic evidence. No natural landscapes were observed in the Makheras Forest. Quadrats dominated by complex, uneven-aged stands, that is, with a biodiversity score >13 were limited in the Makheras to six areas. Forest associations with complex elements in even-aged stands were low as a result of mechanical reafforestation processes. However, elements in uneven-aged stands were well developed with the result that forest associations with a characteristic species mosaic in a poorly structured stand were most characteristic. Modified forest associations involving natural forest species (ie olive groves) with forest invasion were confined to higher altitude areas and they appeared to be very limited in extent. Natural associations on archaeological sites were recorded as more complex with the inclusion of various fruiting species now uncommon in the forest but recorded from ancient archaeological sites. Radical manipulation of the forest environment (e.g. mechanical terracing) was extensive and there is an urgent need for silvicultural management of sapling stands in need of immediate thinning.

Dominant mosaics in forest associations and their characteristics were as follows:

- 65 of the 93 quadrats carrying *Pinus brutia* showed some form of regeneration although in many areas *Quercus alnifolia* germination was more prevalent
- 65 of the 93 quadrats carrying *Pinus brutia* were classed as "mixed stands"
- At higher altitudes with a northern aspect, quadrats were dominated by *Quercus alnifolia*
- Less than 15% of the quadrats surveyed in the Makheras could be considered optimum stands

The production of the forest in terms of potential timber harvesting is low. Twenty eight quadrats exhibited the effects of clearfelling. The main stress on the present silvicultural management of the forest appears to be the result of production economies.

Previous management activities have led to the establishment of single species pine plantations of even age, extensive artificial terracing and the removal of native sub-strata species.

Anthropogenic threats to the forest environment.

Anthropogenic hazards were assessed mainly with regard to the potential threat to water pollution. The purity of water rising in the forests is a critical social and economic factor for the whole community. The disturbance of water patterns has occurred through the capping of most natural springs thus disturbing the natural drainage pattern. Sewage contamination of creeks downstream from picnic and recreation areas pollute streams and may contaminate ground water reserves. Solid waste from industrial or construction sites and household garbage continues to be illegally dumped from access roads. The dumping of goat carcasses also impacts on the water supply, soils, plants and animals.

Biogeocenoses of the Forest.

Research results correlating the soil types and soil fertility were the basis for an analysis of the natural conditions for forest growth. It was assumed that the geological parent material played an important role in preconditioning the spatial distribution of forest associations. The soil structure and nutrient status in conjunction with hydrogeological properties of the bed-rock would determine the soil moisture content. The synthesis of the biotic and abiotic conditions would, therefore, determine floristic characteristics.

The distribution of specific species throughout the forest indicated the dependence of certain types to edaphic factors. The range the endemic *Cyclamen* and the domination of *Quercus alnifolia* coincides with northerly aspects at an altitude over the 800 asl contour. The role of the *Acer obtrusifolia* and *Rhus coriaria* in delineating boundaries of hydrogenic areas is also obvious when the dominant plantation species, *Pinus brutia,* is eliminated from calculations.

An analysis of the changing pattern in forest associations can be attributed to two factors:

(i) changes resulting from silvicultural management induced mainly by clear felling practices

(ii) changes occurring in natural associations caused by multiple factors as a result of previous anthropogenic impact, e.g., regeneration after wild fire, grazing or harvesting).

The clear felling system of forest management and the application of artificial regeneration after clear felling or wild fire using one, two, and rarely three different species, has caused an impoverishment in species composition. Stands appear uniform in age and structure with many elements occurring in natural associations having been eliminated. The felling of species with no known economic value has further reduced the complexity of the ecosystem to the detriment of future forest health.

Arborous species influence, to a large extent, the forest environment, including site microclimates and light conditions. As well, they act to shape the species composition of associations. Monoculture in species composition and stand structure eliminates diversity, with the resulting impact on geomorphological resources creating substantial changes. These changes can influence site humidity, light availability and soil fertility. The result of ecological requirements by species and shifts in forest composition reflect changes in flora associations.

Succession strategy after perturbation is as complex as the structure of the ecosystem prior to disturbance. In some situations there may be a greater range of species growing in association after disturbance than that occurring in an old growth stand. Retrogressive succession, seen as monoculture plantations, can move, over time, into more progressive and complex forest stands. The degradation sequences observed in the forest instituted by grazing, clear felling and frequent wildfire tend to be naturally confined by the geophysical characteristics of the environment. Unstable screes inhibit passage and are colonised by *Quercus*, *Rhus* and *Acer*. Leaf fall from these species stabilises the ground surface and produces a humus level that facilitates further regeneration of sub-strata species. The process takes several centuries but given the maturation rates of the species involved, the progression of the cycle can be understood. Secondary succession species such as the *Pinus brutia* cannot survive in the dense shade generated by the more tolerant and longer living *Quercus*. The role of the *Acer*, *Rhus* and *Arbutus* as long living tolerant dominants that are fire sensitive is being assessed in the context of forest succession.

The Makheras and eastern sector of the Adelphi Forest under study covered an area dominated by geomorphological and topographical properties. The geological formations, following a north to south trend, are: Upper Pillow Lavas, Lower Pillow Lavas, Basal Group and Sheeted Dykes. Stockwork sulphides have been investigated within the forest perimeters and several large mines occur immediately outside the forest periphery. Ore carrying bodies were located near the site of Khededonia and a possible collapsed mine head in the gossan on a ridgeline behind Prophetis Elias. Seven mine shafts following quartz seams were located on a spur near the peak of Pano Teratsotos and these veins carry gold bearing deposits.

The Makheras Forest is also confined by topographical features. It is bounded along its western sector by the Pedhieous River, to the north by Onouphriou River and to the south by the mountain range dividing the Larnaca and Nicosia Districts. In the east, the forest perimeter is uneven following the upper catchment streams of the Yialias River, the Pendaskhinos River (which subsequently flows south) and the Argaki tou Ieroupha which flows north east toward the village of Lythrodhondas.

The forest is further defined by the maximum winter and summer rainfall. Overall the forest receives in excess of 400 inches of rain per annum, the higher altitudes in receipt of a little more which generally falls as snow. Aspects over the forest vary although most compartments surveyed had the more favourable north to north westerly aspects. In these positions the trans-evaporation is reduced and favours a range of floristic species.

Evolution and change through interaction.

The evolution of the forest since the early Holocene follows distinctive trends. Cyprus is a marine island of volcanic origin and has had no land bridge connection with nearby continental land masses. Its geology has influenced the biological process that have occurred. Island colonisation has been well researched (Azzaroli 1979, 1981; Diamond 1987; Gotelli and Graves 1990; Guenther 1988; Higgs 1981; Honegger 1966; Hooijer 1967; Keegan and Diamond 1987; Kuss 1973, 1975; MacArthur and Wilson 1967; Simberloff 1974; Strid 1971, Held 1993; Schule 1993). Following the principle that species on islands of marine origin will increase with time and decrease in proportion to the island's distance from a mainland resource pool (Schule 1993) the ecological composition on Cyprus has resulted in a greater similarity in species with those from the Levant (Zohary 1966) rather than species from either the Greek islands or mainland, or Asia Minor to the east. Isolation has induced the evolution of many endemic species, a high proportion of which are confined to the mountain areas. Their presence may indicate a reduction in range due to a disturbed environment on the cultivated lowlands and Messoria plain, which may have contributed to high insolation and greater trans-evaporation.

The fossil tufa beds described by Holmboe contain quantities of plant material identified as *Platanus orientalis*, *Lauris nobilis* and other mesophytic species. The presence of broad-leafed, deciduous species suggests a canopy consisting in high proportion of broadleaf types.

A fossil bed of a coniferous tree types has been located near Kambia in the foothills where the dominant geological material is upper pillow lava.

The pattern of broadleaf species in the current forest composition is consistent with the possible cyclic rotation of *Quercus* and *Acer* species. The importance of *Arbutus* as a stabilising species and *Pinus* as an opportunistic invader is evident in the forest species associations mapped during the survey.

Cyclical succession at several levels is evident within the ecosystem dynamic. The maturation rate for various tree species governs the longest period of cyclic return. These interact with forest harvesting regimes which to date have been governed by internal and external economic need. Seasonal cycles are also of importance both for flora and faunal species, particularly migratory birds which impact on forest insect populations. The pattern of cyclic growth, energy loss and return can only be mapped over an extended time frame.

Based on an analysis of the structure and complexity of mature quadrat stands with extensive canopy cover and correlating this information with analyses from European reports the following phases of forest patterning were defined.

The Fagus dominated canopy.

A *Fagus* dominated forest has a closed canopy producing deep shade over the forest floor. There is usually no shrub sub-strata and few mosses due to the dense ground litter from fallen leaves. Beech forests do not require a high ground water level and they can extend to altitudes of over 1500m asl. A grass and herb layer can include a range of species. Those listed from Greek and European forests correlate favourably with ground cover species recorded in the Makheras Forest. Mast years, that is the years when abundant fruit is produced, are variable depending on the weather conditions during the preceding summer - a warm summer usually causes a heavy flowering the subsequent year. Mast years are often followed by years when few flowers are produced. Two important associate trees in the *Fagus* forest are the *Acer* and to a lesser extent the *Sambaccus nigra* (Elderberry).

Quercus dominated forests.

A deciduous *Quercus* dominated forest allows light to infiltrate to the forest floor, facilitating the growth of a strong shrub sub-strata and an extensive herb and grass ground cover. The timber from the oak forests has had many varied uses and over exploitation may have induced a rapid decline between the Roman and Venetian periods. The impact on oak regeneration by rodent activity has further inhibited its range.

Pine invasion.

Pine is a light demanding tree that establishes on bare soil. A forest dominated by *Pinus brutia* is in its natural formation, an open forest with single trees widely spaced. Regeneration occurs in the open areas outside canopy cover. Plantation pine stands utilise the intolerant aspects of the species to produce fast growing even-aged stands. Saplings occurring under the canopy produce deformed stems. If germination occurs in a dense stand, saplings fail to reach maturity.

Evergreen oak regeneration.

As the pine produces a canopy that cannot be sustained due to its natural light requirement, a niche is created to promote the germination of the *Quercus alnifolia*. This process was probably established during the Ottoman period but the extensive demand on the species for fuel and industrial wood has confined its growth. Clear felling of the *Quercus alnifolia* to promote plantation pine growth has reduced its natural range. Its ability to with stand fire however, has ensured its role as a future forest dominant.

Magnitude of the transformation.

The complex aggregation of species, dominated by tree types, exists in close association with the physical environment. The forest has developed its own trends with features of inter- and intra-species competition, and a regeneration ecology for particular niche species replacement.

Factors that have introduced a systematic and irreversible shift in the basic characteristics of the forest's complexity have been traditionally linked to anthropogenic impact, particularly harvesting regimes. Environmental fluctuations contribute to an irregularity in seasonal conditions but no major climatic shift has been recorded since the commencement of human settlement on the island. The shift from a broadleaf dominated forest to one of a secondary conifer invasion is an assessment made at the canopy level of the forest. Sub-strata formations and ground cover species and their role in maintaining forest stability has been previously discounted. The stability of ground cover and sub-strata is most affected by wildfire and erosion although invasion following these perturbations can be rapid.

Natural Factors Determining Canopy Shifts.

The river catchment for the Pedhieos and Yialias Rivers reaches to high altitudes. The Pedhieos rises in the range west of Kionia and the Yialias immediately below Kionia's peak. The Agakis tou Mandra tou Klosmatou rises between the twin peaks of Dhyo Mouttes and flows south after its confluence with the Argaki tou Marathefti and Argaki tou Mandron near Kyprovasa Chiftlik. The Mandron and Marathefti rise in the wide basin below Prophetis Elias Monastery. Many deep natural pools have developed in river lines which, for the most part, have cut channels through solid rock. These pools retain water during prolonged summer periods.

The environmental impact of extensive spring capping has been considerable. The natural spring flow of

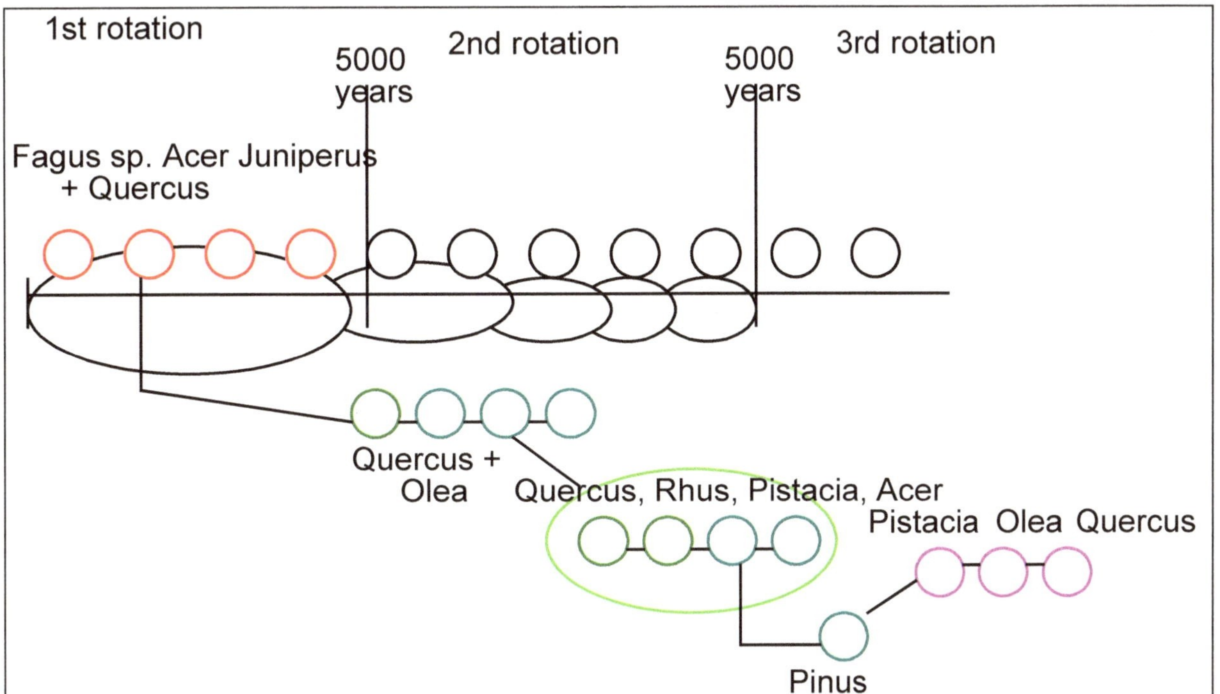

Figure 110: Natural cyclic/evolution and non-cyclic factors influencing forest composition change over an extended time frame.

subterranean water reserves during the summer months provided an even release for plant and animal use. The reduction in a summer water flow along streams has contributed to the loss of riparian cover and an increase in evaporation. Dry stream beds or their depletion to stagnant pools during the summer months has diminished the range and possible occurrence of faunal habitats.

Resource Utilisation Resulting in Disturbance.

The change in ecosystem dynamics as a response to human utilisation of forest bioresources can be tentatively mapped following predictions based on known factors such as economic need and species suitability for specific purpose use.

Factors that appear to have introduced a systematic and irreversible shift in the basic characteristics of the forest's complexity have been traditionally linked to anthropogenic impact, particularly harvesting regimes. The ecological phenomenon of deforestation leading to hill top erosion and subsequent forest degradation has been cited as the primary factor in the environmental degradation in Cyprus. Rackham (1983) challenged the conventional understanding of this process saying; "They [scholars] have too easily supposed that a succession of builders of triremes, insouciant shepherds, marauding armies, Turkish mal-administrators, and rapacious woodcutters - some of whom might otherwise be expected to undo the damage done by others - have reinforced one another and each contributed to a progressive decline". Rackham suggests this scenario to be an improbable reconstruction as it presupposes a single-minded and uninterrupted process of destruction with no compensation for natural rehabilitation processes (Snodgrass 1987:72).

Approximately half the known occupation span of Cyprus, that is, between 8,000 - 3,200 BP is the Neolithic period (T1). The forest during the Neolithic Period was probably a *Fagus* dominated closed canopy with the *Acer* as a codominant and principal sub-strata species. The natural species shift between a *Fagus* to *Quercus* dominated forest follows the normal cycle of change.

During that time impact on the forest was probably confined to browsing by animals imported as domesticates that either were released or escaped into the wild. The principal ungulate identified from archaeological deposits is the *Dama dama mesopatomia* (fallow deer). The forest would also have provided a range of forage, edible plants, fruits and a ready fuel wood supply from fallen timber. The impact of browsing would be the gradual reduction of broadleaf species at lower and mid-altitudinal range and juniper species in the alpine regions. The long term effect of browsing on the forest composition would be the gradual reduction of the deciduous broadleaf species. Lower branches of established trees would be initially eaten while seeding trees would also be reduced. The reduction of faster growing species would initially have opened the sub-strata level and allowed the creation of a dense evergreen sub-strata. The reduction in ground leaf litter and the slow depletion of a humus layer would also effect germination and regeneration. Fuel wood gathered from the forest floor reduces the forest fuel load thus reducing the intensity of wild fire. Fruit harvesting from the forest may have had some impact over the long term although most fruits occurring naturally (with the exception of the olive) form part of the riverine association. In some areas remnant riparian vegetation is dense and impenetrable.

The decline in the *Fagus* dominated forest coincided with the Neolithic occupation of the island. Woodland grazing opened the canopy so that light demanding species were able to invade and establish. The gradual removal of forest floor litter changed the soil nutrient supply and pH. A result was the decrease in nitrogen demanding species and an increase in acid indicator species. The decrease in ground litter depth also facilitated the germination of species that preferred soils without a litter layer (Falkengren-Grerup 1990).

The transition to a deciduous *Quercus* dominated forest can be correlated with a steady increase in pine invasion. This was possible because the canopy allowed light to infiltrate to the forest floor and encouraged the establishment of the *Pistacia terebinthus* as a major sub-strata species. The introduction of mosaic burning would promote the growth of fire tolerant sub-strata species such as the *Olea europea*, *Quercus alnifolia* and the *Arbutus andrachne.* Low intensity burns would increase grass regrowth and open forest areas for both better quality grazing and patch cultivation. Species with a longer maturation rate would survive and reproduce more consistently under human pressure of this type.

A more open canopy would also provide better growing conditions for the *Pistacia terebinthus*, an important economic tree of the Bronze Age. In order to produce resin in the quantities excavated from ancient ship wrecks, the *Pistacia terebinthus* would need to be a mature tree. Only two mature trees were observed on Cyprus, one located outside the Prophetis Elias Monastery in the Makheras Forest and the other outside the Hala Sultan Tekke near Larnaca. In both locations the trees stood alone and carried wide spreading crowns. The specimen at Hala Sultan Tekke had resin tapping scars in its bark. All trees of this species recorded during the survey were small and spindly, growing usually under a *Pinus brutia* canopy. These specimens were also observed to be severely damaged by feral goat browsing in areas adjacent to or where these animals were observed. Using the *Pistacia terebinthus* as an indicator for sub-strata composition it could be suggested that by the Bronze Age, large expanses of the forest had a broken or open canopy allowing sufficient light to penetrate to establish large specimens of *Pistacia terebinthus.*

Fuel wood supply for pyrotechnology has been considered the primary cause of forest depletion during the Bronze Age. However, the species required to produce the maximum heat for a furnace would be a hardwood type. These species coppice well and can be managed to produce lengths of a suitable diameter for easy cartage and rapid burn. The importance of medicinal and aromatic ground cover species also indicates a more open canopy. Cistus flourishes in full sunlight and if the canopy closes the plant declines. The continuation of an open canopy would also allow ease of passage through the forest for hunting, timber selection and herb gathering.

The Archaic and Classic periods extended the forest resource use to include large scale timber construction of ships and buildings. Building requirement specified species use. While only mature timber was harvested, the depletion of forest structure in the form of old mature or senescent trees providing deep canopy shade and seed supply would have influenced species shifts. The impact of forest utilisation on the *Juniperus spp.* and *Cedrus brevifolia* would begin to become significant, particularly in the Makheras Forest.

Intentional use of wild fire as a strategic war weapon would have completed the depletion of deciduous broadleaf species, particularly saplings or immature trees. The Roman preference for the *Quercus* species as a construction timber would also have depleted the genetic pool of this species in the forests.

The extended gap phase that is believed to have occurred during the Medieval and Ottoman periods was most probably the beginning of a natural reversal to a closed canopy forest with the evergreen *Quercus alnifolia* and *Arbutus andrachne* occupying the ecological niche formerly supplied by the *Juniperus* and *Fagus*. These species are not eaten by goats. However, they represent valuable fuel wood and industrial timbers. Evidence for heavy coppicing is still apparent and only one specimen of *Arbutus* was recorded without coppice effects. This tree was situated in an area exhibiting elements of a mature complex forest. It was in excess of ten metres in height and carried a wide spreading crown. No *Quercus alnifolia* were without extensive coppice scars. The ability of these species to have long maturation rates emphasises their importance as leading elements in the current and future forest structure providing both complex elements and long term stability.

From the survey several factors emerged which may contribute significantly to the failure of natural regeneration of the *Pinus brutia* over large areas of the forest. The colonial plantations were of *Pinus brutia* which was widespread in the mountain areas as a secondary invader species. The growth requirements of the *Pinus brutia* tend to reduce its potential for regeneration at higher altitudes and on soil types not derived from the Pillow Lavas. The role of the *Pinus brutia* as an intermediary species in the pattern of cyclic return in the forest dynamic has not hitherto been recognised.

The simplistic scenario of forest degradation attributed to goat grazing, pyrotechnology or shipbuilding is unfounded. The complexity of the forest elements must be observed as unfolding, one species attaining dominance for a period then declining as another becomes precedent. Human interaction with the forest bioresource may have contributed to a more rapid transition between forest phases. Forest utilisation appears to have had no great effect on the genetic pool present in the forest although some species appear to have declined. The loss of three major dominants, the *Fagus*, *Quercus lucitania* and the *Juniperus spp.* is being compensated for by the increase in *Quercus alnifolia* and *Arbutus*. The *Rhus* and *Acer* act as main sub-strata elements facilitating the closing of the canopy. An area believed planted with *Cupresses sempervirens* by old shepherd Hadjilukis, flourishes to dominate the canopy cover. This species also grows well in the Forest Arboretum at Kionia Science area. More importantly, there are seedling trees from these stands which may indicate a natural affinity to the area and these will in time contribute to the forest's diversity. The most critical problem facing the future of forest regeneration processes is the high rodent numbers and the effect that their eating requirement will have on acorn germination.

The new forest policies for water catchment, recreation, education and scientific investigation, emphasise the non-wood value of the forest resource. With a decline in harvesting and the income derived, it is critical that silviculture be maintained at a high level to ensure the continued rehabilitation of the forest's dynamic. The impact of picnickers on several localities has reached a critical level with effluent pollution occurring in creek lines downstream. Continued illegal grazing and the dumping of goat carcasses in creek lines impacts on the watershed areas with potential and increasing erosion and again water pollution. The extent of hunting debris and its density is likewise of concern. A correlation between the modern artefactual material recovered and its spatial relationship to archaeological sites may determine the access routes most used by those walking in the forest.

Chapter Eleven: Sustainable Development of the Forest Bioresource.

The evolution of forest management practices toward a framework of ecologically sustainable development has resulted in the adoption of controlled, responsible, environmentally safe, and socially acceptable initiatives by many Forest Departments world wide. The main goal of forest management has thus made a paradigm shift from anthropomorphic exploitation of forests solely for wood production to one of greatest sustainable long term benefit allowing restricted present use whilst maintaining a potential reserved for future generations.

Ecological sustainability is the foundation upon which modern forestry management for wood production must be based. Within its guidelines, existing species and community patterns will be maintained regardless of the complexity of those patterns. Stand development and ecosystem processes would be managed along lines consistent with those of "natural" forests, and elements of wildlife habitat and all native animals would be protected. The outcome will be the preservation of long term forest health that will be maintained despite disturbance generated by harvesting.

Principles based on forest capacity for optimum ecological robustness will require the establishment of controls based on an interpretation of a wide range of empirical data. The productivity of the Makheras forest must be ascertained not only in economic terms, but also in the assessment on non-wood values and objectives. Forest management for multiple purposes, therefore, relies on sustaining a range of values in a quality manner. The possible restriction of production harvesting to ensure the continuation of other values in a balanced and harmonious way will, however, create additional cost in providing the integration of wider community values.

Ecological sustainability has become an essential element of silvicultural management decision making in an environment where the stated Forest Policy is toward an emphasis on non-wood values for community benefit (Christodoulou 1990). Forest managers are challenged to maintain natural community patterns in a forest that is essentially a remnant plantation stand and where only a limited range of tree species are growing naturally. Emphasis on ecosystem processes and wildlife habitat potential are essential components in ensuring the continuing health and productivity of the forest and must be delineated.

An ecologically sustainable and socially acceptable pattern of resource use will be achieved only where an adequate planning process is in place. Much conflict stems from management plans that are not consistent with the objectives and priorities established at the strategic planning phase. Operations also need to comply with an established forest practices code (Florence 1990).

In determining the principles for forest management, contemporary policies developed from UNESCO guidelines for the balanced conservation and utilisation of natural resources might be adopted. These could include:

1. The integration of unique natural landforms, areas of ecological significance and cultural landscapes representing the nature of the Troodos range and human interaction over a long time span
2. The creation of diversified models of sustainable conservation and utilisation of forest bioresources
3. Conservation of native animal populations characteristic of the forest ecosystem
4. Conservation and enhancement of sustainability, naturalness and biodiversity of forest ecosystems and the retention of natural gene pools
5. Enhancing the forest aesthetic and balancing community values with economic development of the forest resource
6. Shaping preconditions for multidisciplinary research, monitoring biosphere change and ensuring ongoing appraisal of traditional impacts and modern use of the natural resource in order to protect the occurring processes (Łonkiewicz 1996).

Under a plan of management to retain the Makheras Forest as a forest entity, the following objectives are suggested: (i) the supply of forest products to the greatest extent possible constant with conservation objectives, (ii) conservation of biodiversity, and (iii) the maintenance of an aesthetically attractive forest characterised by a high level of biological and structural diversity.

In order to achieve these objectives a system of defining areas requiring conservation plans has been defined. These areas incorporate features such as high species diversity, areas of high biodiversity, areas maintaining rare endemic species and areas of cultural significance. If these areas of high aesthetic or cultural significance were to be retained within a forest reserve (with relevant preservation orders in place), a buffer zone would need to be established around them with the aim of conserving and restoring natural forest associations.

The buffer zone would act in a twofold role: (i) sustainable protection and (ii) as a linking corridor between preserved areas. The core preserved areas would carry strict conservation protection of natural resources with access for scientific research in order to monitor change subject to the approval of senior forest officers. An outer buffer zone extending around these core sensitive areas would establish and preserve forest resources and the continuity of their ecological function. The role of nature conservation determined by the outer buffer zone would be as a distinct boundary along forest compartment lines, dividing areas of conservation value from areas utilised for economic forestry. They would also act to inhibit anthropogenic impact on the ambience of sensitive ecosystems.

It is important that the central core areas of preservation are contained within natural boundaries such as those established by Chapman as compartment boundaries. This would facilitate the mapping of sensitive areas using Global Information Systems (GIS) and the monitoring of natural features occurring as the result of natural succession. Scientific research would be limited to a monitoring role with active experimentation being excluded from these specific areas (Łonkiewicz 1996). Core preservation areas would also act as refugia for protected fauna. The initial buffer zone would create an ecological safe-guard to enhance the distribution of biodiversity or high species richness across the landscape through the establishment of forest corridors. The outer buffer zone would be a transitory stage between natural systems and those under economic management. Within this zone silvicultural work should extend to stand restructuring in order to establish a complex forest composition with a high structural diversity. Study plots to establish forms of protection, species trials or arboretums would be established in the outer buffer zone. In the event of rare or endangered species establishing within the areas of buffer zones these areas should then become protected and a study undertaken to determine the ecological requirements of the species concerned. Extant areas of the forest lying outside the core preservation areas or buffer zones would require ongoing silvicultural enhancement with new plantings integrated into the current monoculture pine stands to enhance species diversity.

Potential for Sustainable Productivity.

The utilisation of the Makheras Forest as an environmental, multiple-use resource will have far reaching repercussions on its management. Environmental restraints will act to enhance or improve a wide range of values associated with the forest, making a significant contribution to ecological conservation, research and regional biodiversity.

Ecosystem management for sustainable forest production integrates a range of practices, techniques, goals and objectives that share overlapping attributes or characteristics. It needs to encompass longer time scales than traditionally allowed for production rotation, as well as consideration of the needs of future generations and cross-political coordination to achieve cultural and scientific goals. The change in emphasis from production forestry to ecosystem management does not alter the concern expressed by generations of foresters for the land or for the role as stewards of the forest resource. On the contrary, the new terminology may not reflect a change in specific management technique, but rather a shift in emphasis (More 1996).

Production Forest.

Within the context of future timber production potential for the forest, it is clear that the Makheras Forest has the capacity to produce quality softwood logs over an extended period, providing sound silvicultural practices are implemented and adhered to. While the current production potential of the forest is variable depending on site and dominant sub-strata species, no further logging on a rotational basis is planned for a further 40 years (Christodoulou pers. com). The current dominant *Pinus brutia* stands require silvicultural enrichment to provide an ongoing complexity in the forest structure. The reintroduction of endemic species such as the *Arbutus andrachne*, *Cedrus brevifolia* and *Juniperus spp.* will enhance future forest dynamics and provide natural inhibitors resistant to destructive biotic and abiotic attacks.

Very few old mature or senescent trees remained in the forest with the majority of trees representing pole or early mature stages. Three age classes were apparent in the forest: mature age dating from trees planted at the turn of the century, pole trees planted after the second World War, and saplings grown on artificial terraces post wild fire incursions during the 1980's. Limited natural *Pinus* regeneration has occurred, particularly in gaps created by logging activity, possibly due to the lack of light ground disturbance. Failed regeneration in pit planting is also apparent. Saplings germinated on artificial terraces have not been thinned in many areas causing growth stress due to overcrowding. The lack of species diversity in many locations has curtailed ongoing forest potential and the *laise-faire* policy of single tree exploitative selection and former clear-felling practises have resulted in an accumulation of restricted growth at pole size.

The Makheras Forest remains, however, visually attractive. The implementation of lighter fellings on a longer rotational basis while stocking to capacity would ensure a gradual increase in its dynamic capability. Professional responsibility to retain the forest in its optimum state of equilibrium and an aspiration toward long term production of value-added specialist timbers from endemic trees would further enhance the forest's capacity for ecologically sustained development. The structure and spatial patterning within the forest varies significantly and relates directly to the dominant forest type *in situ*. The *Quercus alnifolia*, for example, has responded well to the reduction in harvesting although there are no uncoppiced or senescent trees visible. Sapling regeneration has occurred on sheltered, northerly aspects and areas of scree at high altitudes are slowly being stabilised. Areas of *Acer obtrusifolium* and *Rhus coriaria* are also expanding following the restriction of grazing and harvesting. Also, within the hardwood range the low yield on an economic scale has meant little commercial plantation of these species. Although the *Platanus orientalis* is expanding again naturally along water courses its value as a commercial timber is under exploited. An important factor in conserving or developing stands of tolerant species is the risk of creating stagnation within dense stands. The mixed hardwood forest has a rich structural potential and aesthetic diversity, a value for which there needs to be a future emphasis on long-term productivity planning.

It is of critical importance to return all revenue generated in the short term from forest royalties in the form of research, silvicultural development and forest maintenance. The planning and establishment of small mosaic regeneration, the detailing of standing stock, removal of suppressed saplings and poles, and the deliberate marking for retention of elite trees for future stocking purposes as habitat trees for faunal conservation, all require input currently not made under the current management policy. The opportunity to conserve creek corridors for biodiversity and the enhancement of riparian cover should be regarded as a high priority, while the regeneration of clear-felled mosaics that require intensive silvicultural management could become part of an ongoing research and educational program. Remnants of ancient *Olea europea* groves should be retained for their historical value, aesthetic, habitat and structural importance. Fire and fuel bed accumulation control through head chipping for mulch post felling should be investigated in order to reduce the risk of wild-fire damage.

While the Makheras Forest can be redeveloped for timber production on a monocultural even-aged stand system, re-establishing a complex, uneven-aged forest structure for future community needs must be considered as the most viable alternative. If the Makheras Forest can be managed under the principles and philosophy of ecological sustainable development, it will regenerate and be enhanced as a vital regional resource providing even greater options for future use.

Research & educational value.

Documentary inventory lists made of the forest's resources, both natural and anthropogenic, can be directly related to its future potential and provide a basis for continuing evaluation and current utilisation. Rare or threatened endemic plants have intrinsic value and require conservation protection within a legal framework. In order to produce legislation, an authoritative list of these species must be established. The distribution and conservation status of plants at variety and subspecies level, need to be established. This list might also include genetic resources represented by wild species related to cultivars as there appear to be several important links between wild plant species and their relationship to modern domesticates.

A range of possible research projects involving (i) biodiversity conservation, (ii) environmental evaluation and climate change, (iii) biodiversity assessment and (iv) conservation management incorporating biological surveying and modelling of faunal movement, are being initiated through overseas contacts. Other projects assessing the loss of biodiversity through forest fragmentation, wild-life population management and the implementation of forest policy are also under discussion as future research projects involving overseas scholars and Cypriot forest colleagues.

Furthermore, awareness of the cultural significance of the forest environment (including the 20 archaeological sites recorded in the Makheras Forest, together with the Arboretum at Kionia), opens new avenues of research into forest bioresource utilisation during historical and prehistoric periods. Research objectives will be realised by providing ongoing data on the historical development of the forest, the projected impact of urban areas, tourism and experimental trials for tree selection. This research will enable studies to be conducted on a complex forest ecosystem.

Social objectives and community values can also be achieved by demonstrating the forest's capacity for sustained ecological development whilst producing timber, clean water catchment areas, recreational facilities, wildlife habitats and a quality environment. To this end, a portion of the forest adjacent to the Mandra tou Kambiou Forest Station could be designated and managed for school and eco-tourist groups providing signed walking tracks, tree labelling and the provision of a forest museum that traces the forest's ecological history and utilisation.

Recreation.

The Makheras Forest has potential for only limited recreational use. The pressure of a rapidly expanding, urbanised population within an hour's travelling time requires critical assessment of future human impact. Old paths linking former areas of economic importance within the forest are currently over grown and in some areas erosion of the paths has occurred. These tracks are now most frequently used by hunters. Forest based recreation such as picnicking, walking, wildlife observation, photography and bird watching are acceptable pursuits. Many picnic sites have reached their maximum capacity and impact on the surrounding forest has been recorded. To reduce future environmental degradation new areas might be opened as picnic sites provided that fire safety standards can be maintained. This may involve the setting aside of specific forest compartments as public utilities. Hunting in these areas should be prohibited.

Eco tourism can provide incentives to protect wild species from hunters, development and possible extinction. Eco tourism is based on species diversity as an integral part of forest aesthetics. It incurs a cost in marketing renewable resources in threatened ecosystems through the provision of an infra-structure capable of ensuring holist environmental management. Marketable species, such as the mouflon or chukar, fail to capture all the values of biodiversity which includes fungi or insects, and these species with uncaptured values will receive less protection unless total ecosystem protection is encouraged.

Pharmaceutical value.

It is possible that the medicinal value of the forest bioresource may be of greater value than the forest timber reserve. Pharmaceutical companies world wide are investigating and developing rights to genetic resources in regions where high biodiversity contains multiple species likely to yield valuable drugs (Montgomery and Pollack 1996).

Beekeeping.

The forest is suitably located and has excellent access for beekeeping activities. The preference for the aromatic ground cover and sub-strata plants in the production of export quality honey enables the utilisation of most species represented in the forest. Pollination of species by bee activity will enhance regeneration potential.

Scenic value: The forest as viewed by tourists provides a range of spectacular vistas. These can be greatly enhanced through the development of the forest's structure and a shift away from a monoculture of *Pinus brutia*. Areas along major roads have been planted with both exotic species and Cypriot endemics. A greater emphasis on Cypriot species will further promote the island's forested areas.

Water catchment.

The increased regional population on the coastal strip south of the forest has increased demand on the local water catchment resource. Clean water requirements and erosion are of primary concern as aspects of forest management policy. In order to achieve a secure, pure water supply from the forest catchment area, restricted development and landscape modification in water catchment areas, particularly those at higher altitudes must be legislated for. The prevention of pollution from picnic areas, road side rubbish dumping on forest land, and carcass dumping will require legal enforcement. This will require the formulation of policies: (i) to introduce and maintain riparian cover over all streams, (ii) to create special provision for drainage, sewerage, and water supply to recreational areas within the forest and (iii) to make a reassessment of spring capping in the forest.

Passive forest use: The aesthetic appreciation of the forest resource and the "non-wood value" as perceived by community needs must also be considered in relationship to long-term potential options.

Fuel wood: There is a diminished demand for fuel wood for domestic purposes with the exception of souvla pits. Management of tree heads for this purpose may be sufficient to meet current and future demands for this commodity.

Responsibilities.

The Makheras Forest can only realise its full potential for use in various sustainable ways if it is managed to ensure its rehabilitation from past disturbance and to retain, as far as possible, the greatest number of options for future use. An essential component in future forest management will be that staff appointed to maintain the forest bioresource are dedicated to the philosophy of ecological sustainable development and are familiar with the practical application of all aspects of silviculture, forest history and ecology, needed to attain this goal. The financial input into the forest to re-establish a complex, uneven-aged stand should be matched by the silvicultural needs as ascertained by senior foresters. Restricted development through poor treatment applications will inhibit the future potential of the forest and compound an already existing imbalance in the forest's structure and biodiversity.

Problems experienced in bioresource management are the result of resource use conflict that, in the forester's view, erode or impede the successful attainment of objectives. Management problems arising from the ecological rehabilitation and sustainable development of the Makheras Forest are grouped into those that relate to community activities, previous management use or abuse, and those that relate to extremes in nature, involving natural forces not directly related to human activity.

Summary.

The Makheras Forest is in a state of rapid transition. It is responding to the conservation initiatives implemented by Unwin earlier this century and it is continuing the natural cyclic rotation from a dominant coniferous stand to the evergreen oak phase. Although several dominant crown cover species have been lost at sub-strata level, there remains strong species diversity.

The future of the forest will rely on its potential to become a community based resource with multifaceted use. Its ability to provide a range of forest products, including a capacity to produce quality, value-added timber such as cedar and juniper, depends on the injection of financial and administrative resources. Management of the forest under principles of ecologically sustainable development in an increasingly urbanised region may further challenge its current capacity. Under these circumstances, the forest's importance as a research unit, a water catchment area, in wild life conservation, for recreation or a future parkland will be enhanced and will provide the impetus to rehabilitate and develop its natural resources. It is critical, therefore, that management efforts both utilise the present resource and improve its future potential through the implementation of full silvicultural practices. A program for management of the forest following objectives and policy guidelines aimed at providing the optimum combination of ecologically sustainable development and forest utilisation as a multiple use resource will be devised in consultation with European colleagues and forest research scientists within Cyprus and internationally.

Glossary

Abiotic factors. Non-living elements (factors) of the environment - that is, soil, climate, physiography.
Adaptation. Genetically determined character or feature of an organism that serves to increase reproductive potential or chance of survival.
Allelopathy. Suppression of germination, growth, or the limiting of the occurrence of plants as the result of the release of chemical inhibitors by some plants.
Autoecology. Study of the relationship between an individual organism and its environment.
Avifauna. Bird life of a given region.
Basal area. Area of the cross section of a tree stem, generally at breast height (1.3 metres or 4.5 feet) and inclusive of bark.
Bedrock. Bottom layer; lowest stratum; unbroken solid rock, overlaid in most places by soil or rock fragments.
Biomass. Quantity of biological matter of one or more species present on a unit area.
Biosphere. Part of the earth's crust, water and atmosphere where living organisms can subsist.
Biotic factors. Relation of living organisms to each other from an ecological viewpoint (as opposed to abiotic or nonliving elements).
Bole. Tree stem of merchantable thickness.
Browse. Leaves, small twigs, and shoots of shrubs, seedling and sapling trees, and vines available for forage for livestock and wildlife.
Canopy. More or less continuous cover of branches and leaves formed collectively by the crowns of adjacent trees or shrubs.
Carnivore. Organism that consumes mostly flesh.
Carrying capacity. Number of organisms of a given species and quality that can survive in, without causing the deterioration of, a given ecosystem.
Clearcutting. Silvicultural system in which the entire timber stand is cut.
Community. Unit of vegetation that is homogeneous with respect to species composition and structure and occupies a unit area of ground.
Crown fire. Fire that burns the tops of trees and brush.
DBH (diameter at breast height). Tree diameter at breast height, 1.3 meters (4½ feet) above the ground as measured from the uphill side of the tree.
Deciduous. Perennial plants that are normally leafless for some time during the year.
Dendrology. Branch of botany dealing with classification, nomenclature, and identification of trees and shrubs.
Density. Proportion of cell-wall volume to total volume of wood. The number of individuals (trees, animals) per unit area at a given time.
Desertification. Exhaustion of the soil, often because of removal of vegetative cover in semi-arid regions, leading irreversibly to an unproductive desert.
Dominant. Trees that project somewhat above the general level of the canopy, having crowns that receive direct sunlight from above and partly from the side.
Ecology. Science that deals with the relation of plants and animals to their environment and to the site factors that operate in controlling their distribution and growth.
Ecosystem. Any complex of living organisms with their environment considered as a unit for purposes of study.
Ecotone. Transition zone between two adjoining communities.
Edaphic. Pertaining to soil conditions that influence plant growth.
Elasticity. Relationship, expressed mathematically, between a percentage change in one variable and the resulting percentage change in another variable, when all other things are held constant. The price elasticity of demand (supply) is the percentage change in quantity demanded (supplied) when price changes by 1 percent, with all of the variables such as income, population, etc., held constant.
Endemic population. Natural low population level of most species native to an area.
Evapotranspiration. Combined loss of water (through evaporation and transpiration from the soil and vegetal cover on an area of land surface.
Even-aged stand. Stand in which relatively small age differences exist between individual trees, usually a maximum of 10 to 20 years.
Exotic. Not native; foreign; introduced from other climates or countries.
Fibre. Narrow cell of wood (xylem) or bast (phloem), other than vessel elements and parenchyma; includes tracheids. Or a cell material with a length to diameter (l:d) ratio greater than 20:1.
Forb. Any herbaceous plant that is not a grass or grass-like - such plants as geranium, buttercup, or sunflower.
Gall. Pronounced, localised, tumour like swelling of greatly modified structure; occurs on plants from irritation by a foreign organism.
Ground fire. Fire that not only consumes all the organic materials of the forest floor, but also burns into the underlying soil itself - for example, a peat fire.
Habitat. Immediate environment occupied by an organism. In forestry, habitat usually refers to animal habitat.
Habitat type. Unit of land capable of supporting a single community type.
Heartwood. Inner core of a woody stem wholly composed of nonliving cells and usually differentiated from the other enveloping layer (sapwood) by its darker colour.
Herbivore. Organism that consumes living plants or their parts.
Horizon, soil. Layer of soil roughly parallel to the land surface, distinguished from adjacent layers by different physical, chemical or biological characteristics.
Humus. Decomposed lower part of the soil organic layer, generally amorphous, colloidal, and dark coloured.

Increment. Increase in girth, diameter, basal area, height, volume, quality, or value of individual trees or crops.
Knot. Portion of a branch enclosed in the xylem by the natural growth of the tree.
Leaching. Removal of soluble substance (eg., from soil or timber) by percolating water.
Lignin. Non-carbohydrate (phenolic), structural constituent of wood and some other plant tissues; encrusts the cell walls and cements the cells together.
Limiting factor. Environmental factor needed by an organism but in shortest supply,
Loam. Rich friable soil containing a relatively equal mixture of sand and silt and somewhat smaller proportion of clay.
Mast. Nuts and seeds of trees, serving as food for livestock and wildlife.
Mature. Stage of tree growth when height growth slows and crown expansion and diameter increase and become marked.
Mensuration, forest. Science dealing with the measurement of volume, growth, and development of individual trees and stands and the determination of various products obtainable from them.
Microclimate. Climate of small areas, especially in so far as this differs significantly from the general climate of the region.
Monoculture. Crop of a single species, generally even-aged.
Morphology. Study of form and its development.
Nitrogen cycle. Worldwide circulation of nitrogen atoms in which certain microorganisms take up atmospheric nitrogen and convert it into other forma shat may be assimilated into the bodies of other organisms; excretion, burning, bacterial and fungal action in dead organisms return nitrogen atoms to the atmosphere.
Nitrogen fixation. Conversion of elemental nitrogen (N_2) from the atmosphere to organic combinations or to forms readily utilisable in biological processes.
Pathogen. Organism directly capable of causing disease in living material.
Phloem. Tissue of the inner bark; contains sieve elements through which carbohydrates are transported.
Photosynthesis. Synthesis of carbohydrates from carbon-dioxide and water by green plant cells in the presence of light, with oxygen as a by-product.
Pioneer community. First stage in the ecological development of a community.
Pole. Still young tree larger than 10cms DBH to about 23cms DBH.
Predator. Any animal that preys externally on other animals, that is, hunts, kills, and feeds on a succession of hosts.
Prescribed burning. Controlled use of fire to further certain planned objectives of silviculture, wildlife management, fire-hazard reduction, etc.
Primary growth. Growth originating in the apical meristem of shoots and roots.
Profile, soil. Vertical section of the soil through all its horizons and extending into the parent material.
Provenance. Natural origin of seeds or trees, usually synonymous with "geographic origin," or a plant material having a specific place or origin.
Ray. Laterally oriented, ribbon-shaped tissue extending radially in the xylem and phloem; functions in the lateral transport of water and nutrients.
Regeneration. Renewal of a tree crop, by natural or artificial means.
Release cutting. Silvicultural treatment in which larger trees of competing species are removed from competition with desired crop trees.
Remote sensing. Collection of data acquired by a device that is not in physical contact with the object, area, or phenomenon under investigation - for example, aerial photography or satellite imagery.
Reproductive potential. Ability of a species to multiply in the absence of countervailing forces.
Resin. Pitch; the secretions of certain trees, oxidation or polymerisation products of the terpenes, consisting of mixtures of aromatic acids and esters insoluble in water but soluble in organic solvents; often exuding from wounds.
Rhizome. Horizontal underground stem, usually containing stored food.
Ring-porous wood. Wood (xylem) of hardwoods in which the earlywood vessels are much larger in diameter than vessels in the latewood; the vessels generally appear as a ring in a stem cross section.
Riparian. Of vegetation growing in close proximity to a watercourse, small lake, swamp, or spring.
Rotation age. Period of years required to establish and grow timber crops to a specified condition of maturity. Applies only to even-aged management.
Sapling. Young tree at least 1 metre high, but not larger than 10cm DBH; crowns are well elevated and usually many lower branches have started to die.
Sapwood. Predominantly living, physiologically active wood; includes the more recent annual layers of xylem that are active in translocation of water and minerals.
Secondary growth. Growth derived from lateral meristem; results in increase in girth.
Secondary succession. Succession starting after the disturbance of a previously existing plant community. See
Seedling. Youngest trees from the time of germination until they reach a height of 1 metre.
Seed-tree method. Silvicultural system in which the mature timber is removed in one cut, except for a small number of seed trees left to provide a source for seed for the next crop.
Senescent. Growing old; aging stands at this stage are over-mature; losses from mortality and decay may exceed additions in volume.
Seral community. Community that follows the pioneer community in a sere but still does not represent the fully developed mature ecosystem.

Sere. Series of communities through time that develop as a result of succession.
Shade tolerance. Capacity of trees to reproduce and grow in the shade of and in competition with other trees.
Shelterwood method. Silvicultural system in which the mature timber is removed, leaving sufficient numbers of trees standing to provide shade and protection for new seedlings.
Shrub. Woody perennial plant, seldom exceeding 10 feet in height, usually having several persistent woody stems branching from the ground.
Silviculture. Manipulation of forest vegetation to accomplish a specified set of objectives; controlling forest establishment, composition, and growth.
Snag. Standing dead tree from which the leaves and most of the branches have fallen.
Stenotopic. Organisms limited to a very specific habitat niche.
Structure, soil. Combination of arrangement of primary soil particles (eg. sand, silt, clay) into secondary particles called peds.
Subsoil. Bed or stratum of earth or earthy material immediately under the surface soil.
Substrate. Underlying material; the soil beneath plants or animals; the material on which an enzyme or fermenting agent acts, on which adhesive is spread, or on which a fungus grows or is attached.
Succession. Change in community composition and structure through time.
Suppressed. Trees with crowns completely overtopped by surrounding trees so that they receive almost no direct sunlight.
Surface fire. Fire that burns only surface litter, loose debris of the forest floor, and small vegetation.
Sustained yield. Yield a forest can produce continuously, such as timber.
Symbiosis. Mutually beneficial relationship between two dissimilar living organisms. In some cases, the symbionts form a single body or organ, as in mycorrhizae or lichens.
Synecology. Study of the community and its environment.
Systems analysis. Method of analysis that deals with the movement of energy or materials to different parts or components of a complex system.
Tannins. Complex extracellular water-soluble substances, generally formed from a variety of simpler polyphenols; part of wood extractives.
Texture, soil. Relative proportion of the various mineral particles such as sand, silt, and clay, expressed as a textural class - for example, sandy loam, clay loam, etc.
Thinning. Silvicultural treatment in which stand density is reduced to accelerate diameter growth in remaining trees.
Tissue. Group of similar cells organised into a structural and functional unit in a plant or plant organ.
Tracheary element. Tracheid or vessel member.
Tracheid. Elongated, thick-walled conducting and supporting cell of xylem. Has tapering ends and pitted walls without perforations. Found in nearly all vascular plants; the main fibrous component of wood.
Tree. Woody perennial plant, typically large and with a single, well defined stem and a more or less definite crown.
Trophic, -troph, Tropho-, Pertaining to nutrition, feeding.
Trophic levels. Steps in the movement of energy through an ecosystem.
Understorey. Any plants growing under the canopy formed by others.
Uneven-aged stand. Stand in which more than two distinct age classes and a range of size classes (seedling, sapling, pole, etc.) are present.
Uptake. Amount of water and/or nutrients absorbed by vegetation.
Vascular cambium. Cylindrical sheath of meristematic cells, the provision of which produces secondary xylem and secondary phloem.
Vascular tissue. Specialised conducting tissue in plants forming a vascular system - in woody plants comprising the whole of the xylem and phloem.
Watershed. Total area above a given point on a river, stream, etc., that contributes water to the flow at that point.
Water stress. Stress or negative pressure exerted on a water column in a plant due to transpiration.
Wood. Secondary xylem.
Xerarch succession. Primary succession beginning on a substrate that is solid rock and therefor has minimal water-storing capacity.
Xylem. Tissue containing tracheary elements through which water and minerals are transported; wood is secondary xylem.

Bibliography

Åberg, Eric.
1976 Analysis of Charcoal from Trench 3. *Hala Sultan Tekke.* Göteborg: Paul Àstroms Förlag. Vol 1, SIMA XLV:1 pp 117.

Acampora, Ralph R.
1994 Using and Abusing Nietzsche for Environmental Ethics. *Environmental Ethics.* Vol 16:2 pp 187-194.

Adams, R. and Simmons, D.
1993 Botanical remain from Bronze Age Marki Alonia. *Report of the Department of Antiquities, Cyprus.* pp 66-68.

Adams, S. N. 1975 Sheep and Cattle Grazing in Forests. *Journal of Applied Ecology.* 12 pp 143-152

Adger, W. Neil et. al.
1995 Total Economic Value of Forests in Mexico. *Ambio.* Vol 24:5 pp 286-296.

Adovasio, J. M. et al 1975 Prehistoric and Historic Settlement Patterns in Western Cyprus. *World Archaeology.* 3 pp 339-564

Agnew, John A. - Duncsan, James S eds.
1989 *The Power of Place.* Boston. Unwin Hyman.

Agnew, J. A.
1987 *Place and Politics: the Geographical Meditation of State and Society.* London. Allen & Unwin.

Alcock, Susan E.
1994 Roman imperialism in the Greek landscape. *Journal of Roman Archaeology.* Vol. 7. pp.5-34.

Allison, R. J. and Thomas, D. S. G.
1993 The Sensitivity of Landscapes. *Landscape Sensitivity.* Chichester. John Wiley and Sons.

Anastasakis, G. C. and Dermiszakis, M.
1988 Pleistocene Evolution in the Central Aegean-Cyclades region: Palaeogeographic Reconstruction Based on Marine Geological Evidence as compared to Pleistocene Mammal Findings. *Early man in Island Environments.* Internation Conference. Ed. by M. Sangers. Martini, Nuoro.

ANET James B. Pritchard, *Ancient Near Eastern texts.*

Angel, J. Lawrence.
1966 Porotic hyperostosis, anemias, malarias, and marshes in the prehistoric eastern Mediterranean. *Science*, 16th March 1966, Vol 153 pp 760-763.

Angel, J. L.
1972 Ecology and Population in the Eastern mediterranean. *World Archaeology.* 4 pp 99-105

ARAB D.D. Luckenhill, *Ancient records of Assyria and Babylonia.* 2 volumes.

Arbib, Michael A and Hesse, Mary B.
1986 *The Construction of Reality.* Cambridge. Cambridge University Press.

Archambault, L., Barnes, B. V. and Witter, J. A.
1990 Landscape Ecosystems of Disturbed Oak Forests of Southeastern Michigan, USA. Canadian Journal of Forest Research. 20 pp 1570-1582

Arnold, Matthew.
1960 *Culture and Anarchy.* Ed. J. Dover. Wilson. Cambridge. Cambridge University Press.

Åström, Paul and Hjelmqvist, H.
1983 An amphora decorated with a palm-tree motif from Hala Sultan Tekke. *RDAC* pp 175-176.

Avner, Uzi.
1990 Ancient Agricultural Settlement and Religion in the Ovola valley in Southern Israel. *Biblical Archaeologist.* September, pp 125-141.

Badawry, A. 1978 *The Tomb of Nyhetep - Ptah at Giza and the tomb of 'Ankhim' ahokat Saggara.* Berkeley: University of California Press.

Baker, Richard St. barbe.
1944 *I Planted Trees.* London. Lutterworth Press.

Baker, Graeme, and Clive, Gamble.
1989 *Beyond domestication in prehistoric Europe.* Unpublished paper.

Banlon, E. K.
1993 Dynamics of Biodiversity and Mechanism of Change: A Plea for Balanced Attention to Form Creation and Extinction. *Biological Conservation.* 66 pp 5-16

Bannerman, David Armitage and Bannerman, W. Mary Jane.
1971 *Handbook of the Birds of Cyprus and Migrants of the Middle East.* Edinburgh. Oliver and Boyd.

Banuri, Tarr and Marglin, Frederique.
1993 *Who Will Save the Forests?* London. Zed Books.

Barker, Graeme.
1991 Approaches to Archaeological Survey. *Roman Landscapes: Archaeological Survey in the Mediterranean Region.* Ed. Graeme Barker and John Llyoel. London. British School at Rome.

Barlow, Jane A. et. al. ed.
1991 *Cypriot Ceramics: Reading the prehistoric record. Preceedings of an International.* Philadelphia. University of Pennsylvania University Museum of Archaeology and Anthropology.

Barnes, B. V. et al.
1982 Ecological Forest Site Classification. *Journal of Forestry.* 80 pp 493-498

Barthes, Roland. 1977 *Image, Music, Text.* Trans. by Stephen Heath. New York. Hill and Wang.

Baskin, Yvonne.
1994 Ecologists Dare Ask: How Much Does Diversity Matter? *Science.* Vol 264 8 April 1994

Bass, G.F.; Pulak,C.; Collon, D.; and Weinstein, J.
1991 The Bronze Age Shipwreck at Ulu Burun: 1986 Campaign. *American Journal of Archaeology.* Vol 93 pp 1-29.

Bass, George F.
1990 Nautical Archaeology and Biblical Archaeology. *Biblical Archaeologist.* March.

Bass, George F.
1989 Evidence of Trade from Bronze Age Shipwrecks. *Bronze Age Trade in the Mediterranean.* Ed. N. H. Gale Jonered. Paul Åströms Förlag.

Bate, D. M. A.
1903 On An Extinct Species of Genet (Genetta plesictoides, sp. n.) from the Pleistocene of Cyprus. *Proceedings of the Zoological Society.* Vol 71 pp 121-124.

Baylis, C. A.
1953 Universals, Communicable Knowledge, and Metaphysics. *Journal of Philosophy.* Vol XLVIII

Bear, L.M.
1963 *The mineral resources and mining industry of Cyprus.* Bulletin No. 1.

Bearlin, D.
1991 *Ecology as Science-Theology.* Victoria College.

Bearlin, D.
1991 *Ecology as Science.* Canada. Victoria College. MA Thesis (M.Ed) University of Manitoba.

Beaton, J. M.
n.d. Extensification and Intensification in Central Californian Prehistory. *Antiquity.* 65/249.

Bednarz, Ulrich.
1988 *Volcanological, geochemical and petrological evolution, and sub- seafloor alteration in the north-eastern Troodos ophiolite.* Unpublished PhD thesis.

Belsky, A. Joy and Canham, Charles D.
1994 Forest Gaps and Isolated Savanna Trees: an Application of Patch Dynamics in Two Ecosystems. *BioScience* vol 44 no 2. Feb 1994

Bennett, Carl Graham.
1980 *The cults of the ancient Greek Cypriots.* 2 volumes. Unpublished thesis.

Berg, Peter and Dasmann. 1980 Reinhabiting California. *Ecologist.* 7:10 pp 399-401

Bibby, J.S., and D. Mackney.
1969 *Land use capability classification.* Dorking: Bartholomew Press.

Bickford, Anne.
1981 The Palinoi of Nostalgia. *Australian Archaeology.* Vol. 13 pp.1-7.

Binford, Lewis.
1989 *Debating Acrhaeology.* San Diego. Academic Press.

Binford, Lewis.
1983 *In Persuit of the Past.* London. Thames and Hudson.

Biodiversity Assessment Workshop
1993 *Rapid Biodiversity Assessment.* Proccedings of the Biodiversity Assessment Workshop, 3-4 May, 1993. Macquarie University.

Birrel, Jean
1969 Peasant Craftsmen in the Medieval Forest. *Agricultural History Review* Vol 17 pp 91-107

Blair, Robert, and Louis E. Brunett.
1980 Seasonal browse selection by deer in a southern pine-hardwood habitat. *Journal of Wildlife Management.* Vol 44:1.

Blumer, Mark A.
1993 Successional Pattern and Landscape Sensitivity in the Mediterranean and Near East. *Landscape Sensitivity.* Ed. D. S. G. Thomas and R. J. Allison. Chichester, John Wiley and Sons.

Bobek, Bognslaw, and Ryszard Dzieciotowski.
1972 Method of browse estimation in different types of forests. *Acta Theriologia* XVII, 14, August 1972.

Bolza, Eleanor and Keating, W. G.
1972 *African timbers - The Properties, Uses and Characteristics of 700 Species.* Melbourne CSIRO. Division of Building Research.

Boserup, Ester.
1983 The impact of scarcity and plenty on development. *Hunger and History: the impact of changing food production and consumption patterns on society.* Ed. Robert I. Rotberg and Theodore K. Rabb, pp 185-209. Cambridge University Press.

Botkin, Daniel B.
1993 *Forest Dynamics: and Ecological Model.* Oxford. Oxford University Press.

Botkin, Daniel B.
1990 *Discordant Harmonies: a New Ecology for the Twenty first Century.* New York. Oxford University Press.

Boussiou, M. et. al
1991 The Origin of the Sickle Mutation in Greece; Evidence from Bs Globin Gene Cluster Polymorphisms. *Hemoglobin* 15 (6) pp 459-467

Bovill, A.K.
1920 *Report on the natural resources and the present development of the forests of Cyprus.* Nicosia: Govt. Printing Office.

Bradford, George.
1989 *How Deep is Deep Ecology?* Ojai. Times Exchange Press.

Bradley, Richard
1993 *Altering the Earth.* Edinburgh. Society of Antiques of Scotland.

Bredemeier, Michael and Ulrich, Bernhard
1991 Input / Output Analysis of Ions in Forest Ecosystem. *Responses of Forest Ecosystems to Environmental Change.* Edited by A. Leller, P. Mathy. D. J. N. R. Jeffers. London. Elsevier Applied Science.

Briggs, John C.
1987 *Biogeography and Plate Tectonics.* Amsterdam. Elesevier.

Brown, A. C. and Catling, H. W.
1986 *Ancient Cyprus.* Oxford. Ashmolean Museum.

Brown, Thomas M. and Kraus, Mary K.
1993 Time - Ttratigraphic Reconstruction and Integration of Paleopedologic Sedimentologic, and Biotic Events. *Palaios.* Vol 8 pp 68-80.

Brunsden, D.
1993 Barriers to Geomorphological Change. *Landscape Sensitivity.* Edited by D. S. G. Thomas and R. V. Allison. Chichester. John Wiley and Sons.

Buber, Martin
1970 *I and Thou.* Translated by Walter Kaufmann. Edinburgh, T and T Clark.

Buchholz, H.G.
1993 *The wood industry in ancient Cyprus.* Public lecture, Famagusta Gate, Nicosia, Cyprus. 20.11.93.

Buchler, Justus
1966 *Metaphysics of Natural Complexes.* Mew York. Columbia University Press.
Bulletin 48:14-17.

Burgess, C.
1980 *The Age of Stonehenge.* London, Dent.

Burnet, Julia Ellis.
1993 *Echoes of a Lost Greenness.* Unpublished thesis. Forestry Department. Australian National University.

Burnet, Julia Ellis.
1989 *Learning to be a Tree.* Canberra. AIDS Council of the A.C.T.

Burnham, Robyn J.
1993 Reconstructing Richness in the Plant Fossil Record. *Palaios.* Vol 8 pp 376-384.

Butzer, Karl w.
1982 *Archaeology as Human Ecology: Method and Theory for a contextual approach.* Cambridge. Cambridge University Press.

Butzer, K.W.
1961 Climatic change in arid regions since the Pliocene. *A History of Land Use in Arid Regions.* Paris: UNESCO.

Cain, Stanley A. and Oliveira Castro, G.N.de
1971 *Manual of Vegetation Analysis.* New York. Hafner Publishing Company.

Callebaut, Werner and Pinxten, Rik. eds.
1987 *Evolutionary Epistemology.* Dordrecht. D. Reidel Publishing Co.

Callicott, J. Baird.
1992 The Land Aesthetic. *Renewable Resources Journal* 10/4: 12-17.

Capra, Fritjoy.
1984 *Green Politics.* New York. Dutton.

Cassirer, Ernst
1944 *An Essay on Man: an Introduction to a Philosophy of Human Culture..* New Haven. Yale University Press.

Casson, L.
1980 Two-masted Greek Ships. *International Journal of Nautical Archaeology.* Vol 9 pp 68-69.

Casson, L.
1963 Sewn Boats. *Classical Review.* Vol 13 pp 257-259.

Casson, L.
1971 *Ships and Seamanship in the Ancient World.* Princeton: Princeton University Press.

Casson, L.
1991 *The Ancient Mariners* 2nd ed. Princeton: Princeton University Press.

Casson, L.
1980 Two Masted Greek ships. *International Journal of Nautical Archaeology* 9:68-9.

CAST/ CDD.
1994 *Strengthening the Competitivenes of Cyprus Industry.* Cyprus Forest Industries Ltd. Final report. [No imprint.]

Caswell, H.
Life-History Strategies. *Ecological Concepts.* Ed. J. M. Cherrett. Oxford. Blackwell Scientific Publications.

Catling, Hector W.
1962 Patterns of settlement in Bronze Age Cyprus. *Opuscula Atheniensia* 4:129-169.

Catling, H. W. 1982 The Ancient Topography of the Yialias Valley. *Report of the Department of Antiquities, Cyprus.* pp 227-236.

Catling, H. W.
1962 patterns of Settlement in Bronze Age Cyprus. *Opuscula Atheniensia.* 4 pp 129-169
Cesnola, Louis Palma di.
1877 *Cyprus, its Ancient Cities, Tombs and Temples.* London. J. Murray.
Chambers, F. M.
1993 *Climate Changes and Human Impact on the Landscape.* Ed. F. M. Chambers. London. Chapman and Hall. CC81.C59
Chambers, David and Pullan, Brian eds.
1992 *Venice: a Documentary History.* pp 1450-1630, Oxford. Blackwell.
Chang, Claudia.
1992 Archaeological Landscapes. The Ethnoarchaeology of Pastoral Land Units in the Geneva Province of Greece. *Space, June and Archaeological Landscapes.* Ed. Jacqueline Rossignol and Lu Ann Wandsnides. New York. Plenum Press
Chapman, G.W.
1934 *A reconnaissance report for Paphos Forest.* Unpublished.
Chapman, Esther.
1967 *Cyprus trees and shrubs.* Nicosia: Forest Department.
Chapman, G.W.
1934 *A reconnaissance report for Paphos Forest.* Unpublished.
Chapman, G.W. and Davidson, D. F.
1935 *Preliminary working plan for Paphos Forest.* Unpublished.
Chapman, G.W.
1952 *Afforestation techniques in Cyprus.* Prepared for the Third Commission of the European Commission on Forestry and Forest Products. Istanbul.
Charlesworth, Max.
1982 *Science, Non-science and Pseudo-science.* Melbourne. Deakin University Press.
Chatman, Seymour, Eco, Umberto and Klinkenberg, Jean-Marie
1979 *A Semiotic Landscape: Proceedings of the First Congress of the International Association for Semiotic Studies, Milan, June 1974.* The Hague Mouton Publishers.
Chavas, Jean-Paul.
1993 On Sustainability and the Economics of Survival. *American Journal of Agricultural Economics.* Vol 75:1 pp 72-83.
Cherry, J. F.
1981 Pattern and Process in the Earliest Civilisation of the Mediterranean Islands. *Proceedings of the Prehistoric Society.* 47 pp 41-68
Cherry, John F.
1988 Pastoralism and the Role of Animals in the pre- and protohistoric economies of the Aegean. *Pastoral Economies in Classical Antiquity.* Ed. C. R. Whittaker. Cambridge Philological Society.
Cherry, J. F., Davis, J. L. and Mantzourani, E.
1991 *Landscape Archaeology as Long-term History. Northern Keos in the Cycladic Islands.* Classical Antiquity.
Chibnall, A. C.
1965 *Sherrington: Fiefs and Fields of a Buckhamshire Village.* Cambridge. Cambridge University Press.
Chicago, 1926-7.
Christensen, N. L.
1989 Landscape History and Ecological Change. *Journal of Forest History.* Vol 33 pp 116-124
Christensen, N. L. and Peet, R. K.
1984 Convergence during Secondary Forest Succession. *Journal of Ecology.* Vol 72 pp 25-36.
Christodoulou, Demetris.
1959 *The evolution of the rural land use pattern in Cyprus.* The World Land Use Survey, Regional Monograph 2. Bude: Geographical Publications.
Christodoulou, Alecos.
1993 *Forestry in Cyprus.* Nicosia: Ministry of Agriculture & Natural Resources.
Christofides, Louis N.
1992 Economic Dimensions of the Cyprus Dispute in *Cyprus: A Regional Conflict and it's Resolution.* Edited by Norma Saley. St. Martins Press. Ottowa.
Christou, Andreas K.
1991 *The Genetic and Taxonomic Status of Cyprus Cedar, Cedrus brevifolia (Hook) Henry.* Unpublished PhD thesis.
Clarendon Press.
Clark, G. A.
1982 Qualifying Archaeological research. *Advances in Archaeological Method and Theory.* Vol 5 pp 217-273.
Clarke, D. L.
1972 A Provisional Model of an Iron Age Society and its Settlement System in *Models in Archaeology.* Ed. D. L. Clarke. London. Methuen.
Clarke, David L.
1978 *Analytical Archaeology.* 2nd ed. London. Methuen.
Cohen, R. and Service. E.
1978 *Origins of the State: the Anthropology of Political Evolution.* Philadelphia. Institute for the Study of Human Issues.
Coldstream, J.N.
1982 Early Iron Age (Cypro-Geometric): the rise of the ancient kingdoms *c.* 1100-700 B.C.

Footprints in Cyprus,. Ed. Sir David Hunt, pp 58-72. London: Trigraph.

Colledge, S.
1980 Plant species from Kissonerga *Mylouthkia* Feature 16.3. *Levant* 12: 18-20.

Colledge, S.
1981 Kissonerga *Mylouthkia* 1979: palaeobotanical report. *Levant* 13: 47.

Colledge, S.
1982 Lemba *Lakkous* 1980 - flora. *Levant* 14: 53.

Commoner, Barry
1967 Biology A Molecular Science. *The Anatomy of Knowledge*. Edited by Marjorie Grene. London. Routledge and Kegan Paul.

Comotti, Giovanni
1979 *Music in Greek and Roman Culture*. Beltimore. John Hopknis University Press.

Connell, Joseph and Slayter, Ralph O.
1977 Mechanisms of Succession in Natural Communities and their Role in Community Stability and Organisation. *American Naturalis*. Vol 111 pp 1119-1144.

Constantinou, G.
1992 Ancient Copper Mining in Cyprus. Pp 43-76 *Cyprus and the Sea.* Ed. V. Karegeorghis. Nicosia.

Constantinou, G.
1992 Geological features and ancient exploitation of the cupriferous sulphide orebodies of Cyprus. In Muhly, J. D., Maddin, R. and Karageorghis, V. eds. *Early Metallurgy in Cyprus*, 4000-500 BC. Nicosia: Pierides Foundation, Pp.13-24.

Corbet, G. B.
1966 *The Terrestrial Mammals of Western Europe*. Philadelphia. Dufour Editions.

Cosgrove, D. E. 1984 *Social Formation and Symbolic Landscape.* London. Croom Helm.

Couppis, T.A.
1949 Village fuel areas in Cyprus. Reprint. *Empire Forestry Review* 28/1.

Courtois, Jacques-Claude, and Jennifer M. Webb.
1987 *Les cylindres-sceaux d'Enkomi (fouilles françaises 1957-1970).* Nicosia: Mission Archéologique Française d'Alasia.

Cowell, C. Mark
1993 Environmental Gradients in Secondary Forests of the Geogia Piedmont, U.S.A. *Journal of Biogeography*. Vol 20 pp 199-207

Creevey, Bradley.
Unpublished MA Thesis. Classics Department. University of Newcastle.

Croft, Paul.
1991 Man and beast in Chalcolithic Cyprus. *Bulletin of the American Schools of Oriental Research* 282-283: 63-79.

Croft, Paul.
1991 Man and beast in Chalcolithic Cyprus. *Bulletin of the American Schools of Oriental Research* 282-283: 63-79.

Cummins, R. P., and Miller, G. R.
1982 Damage by Red Deer (*Cervus Elaphus*) enclosed in planted woodland. *Scottish Forestry* 36.

Cunliffe, B.
1974 *Iron-Age Communities in Britain.* London. Routledge and Kegan Paul.

Cutler, D. F.
1983 Charcoal and Silica. *Vrysi: A Subterranean Settlement in Cyprus.* Ed. by Edgar Peltenberg. Waeminster. Aris and Phillips.

Cyprus, Forest Department.
1936 *Limniti Felling Series: working plan, 1936.* Unpublished.

Cyprus, Forest Department.
1955 *The Cyprus Forests*. Nicosia.

Cyprus, Forest Department.
1963 *Cyprus tour. Oxford University Post-graduate Forest Officers' visit. Friday June 28-Sunday July 8, 1963.* Nicosia: Ministry of Agriculture & Natural Resources.

Cyprus, Forest Department.
1965 *Forest Fire Fighting*. Nicosia: Ministry of Agriculture & Natural Resources.

Cyprus, Forest Department.
1967 *General information for the 7th silvicultural study tour of the European Forestry Commission of the Food & Agricultural Organisation of the United Nations.* Nicosia: Ministry of Agriculture & Natural Resources.

Cyprus, Forest Department.
1976 *Integrated Rural Development of Pitsilia Area. Proposals on Forestry.* Nicosia. Ministry of Agriculture & Natural Resources.

Cyprus, Forest Department.
1984 *Inventory of P. nigra Forests: instructions for the enumeration of the temporary sample plot.* Nicosia: Ministry of Agriculture & Natural Resources.

Cyprus, Forest Department.
1985 *Progress Report 1980-84 prepared for the Twelfth Commonwealth Forestry Conference 1985.* Nicosia.

Cyprus, Forest Department.
1990 *National report on Forestry Tenth world Forestry Conference Paris 1991.* Nicosia.
n.d. *Heavier fellings with the application of artificial regeneration in mature stands of Pinus brutia. Compartment no.23 Ayia F. series.* No imprint.

Cyprus, Government of Cyprus.
1946 *Report of the Land Utilisation Committee 1946.* Nicosia: Government Printing Office.

Cyprus, Government of Cyprus.
1957 *A Statement of Forest Policy.* Nicosia: Government Printing Office.

Cyprus, Government of Cyprus.
1984 *A Statement of Forest Policy 1984.* Unpublished.

Cyprus, Ministry of Agriculture & Natural Resources.
1984 *Agriculture in Cyprus: an analysis and evaluation.* Nicosia. Cyprus.

Cyprus, Ministry of Agriculture & Natural Resources.
1913 *Colony of Cyprus. The Goats Law.* [no imprint]

Daniels, S. and Cosgrove, D. E.
1988 Introduction: Iconography and Landscape. *The Iconography of Landscape: Essays on the Symbolic Representation, Design and Use of Past Environments.* Ed. by D. E. Cosgrove and S. Daniels. pp 1-10. Cambridge. Cambridge University Press.

Danks, F.S.
1939 *Charcoal.* Nicosia: Forest Department.

Danthine, Hélène.
1937 *Le palmier-dattier.* 2 volumes.

Dantin, C. F. A.
1969 Organism and Environment in *The Anatomy of Knowledge.* Edited by Marjorie Grene. London. Routledge and Kegan Paul.

Danto, Arthur C.
1981 *The Transfiguration of the Commonplace.* Cambridge, Mass. Harvard University Press.

Darwin, Charles.
1959 *The Origin of Species by means of Natural Selection.* Murray. London.

Davidson, D.F.
1943 *Resin Tapping Experiments (Pinus Halepensis var. brutia).* Nicosia: Forest Department.

Davidson, D.F.
1957 *Statement prepared for the British Commonwealth Forestry Conference. 1957*Nicosia: Forest Department.

Deetz, James.
1990 Landscapes as Cultural Statements. *Earth Patterns: Essays in Landscape Archaeology.* Ed. by W. M. Kelso and R. Most. pp 1-4. Charlottesville. University Press of Virginia.

Delany, M. J.
1982 *Mammal Ecology.* Glasgow. Blackie.

Derrida, Jacques
1991 *A Derrida Kader: Between the Blinds.* Edited by Peggy Kammf. New York. Columbia University Press.

Detienne, Marcel.
1986 *The creation of mythology.* Trans. Margaret Cook. Chicago: University of Chicago Press.

Diersing, Victor.
1984 Lagomorphs. *Orders and Families of Recent Mammals of the World.* Ed. by Sydney Anderson and J. Knox Joes. New York. Wiley & Sons.

Dikaios, Porphyrios.
1961 *Sotira.* Philadelphia: The University Museum, University of Pennsylvania.

Dikaios, Porphyrios.
1953 *Khirokitia: final report on the excavations of a Neolithic settlement in Cyprus on behalf of the Department of Antiquities 1936-1946.* Oxford University Press.

Dikaios, Porphyrios.
1969-71 *Enkomi, excavations 1948-1958.* 4 volumes. Mainz: Phillipp von Zabern.

Dincauze, Dena F.
1987 Strategies for Paleoenvironmental Reconstruction in Archaeology. *Advances in Archaeological Method Theory.* Vol 11 pp 255-336.

Doak, Daniel F and Mills, L. Scott
1994 A Useful Role for Theory in Conservation. *Ecology.* Vol 75 Number 3. April 1994. pp 615-626

Druitt, D. G. et al.
1990 Altitudinal Zonation in the Mountain Forests of Mt Houhungatolia, North Island, New Zealand. *Journal of Biogeography.* Vol 17 pp 205-220

Dunbar, G.S.
1983 The Forests of Cyprus under British Stewardship. *Journal of Scottish Geography Magazine* 99.

Dunn, Christopher
1991 Methods for Analysing Temporal Changes in Landscape Pattern. *Quantitative Methods in Landscape Ecology.* Ed. Marica G. Turner and Robert H. Gardner. New York, Springer-Verlag.

Eco, Umberto
1986 *Travels in Hyper Reality.* Trans. William Weaver. San Diego. Harcourt Brace. Jovanovich.

Eco, Umberto
1990 *The Limits of Interpretation.* Bloomington, Indiana University Press.

Edlin, H. L.
1972 *Trees, Woods & Man.* London. Collins.

Egler, Frank.
1964 Pesticides - in Our Ecosystem. *American Scientist.* Vol 3:1. March. pp 112

Egler, Frank G.
1954 Vegetation Science Concepts: 1. Initial

Floristic Composition - A Factor in Old Field Vegetation Development. *Vegetatio.* Vol 4:4 pp 412-417.

Ehrlich, Paul R. and Roughgarden, J.
1987 *The Science of Ecology.* New York. Macmillan Publishing.

Eliade, Mircea
1961 *Images and Symbols.* Trans. Philip Mairet. London. Harvill Press.

Eliade, M.
1978 *The Forbidden Forest.* Trans. by mac Linscott Ricketts and Mary Park Stevenson. Notre Dame. University of Notre Dame Press.

Enger, Eldon D. et. al.
1989 *Environmental Science: The Study of Interrelationships.* 3rd Edition. Dublique, Ottowa WM.C.Brown.

Erlich, Paul R. and Roughgarden, Jonathon.
1987 *The Science of Ecology.* New York. Macmillan.

Fahn, Abraham, Werker, Ella and Baas, Pieter.
1986 *Wood Anatomy and Identification of Trees and Shrubs from Israel and Adjacent Regions.* Jerusalem. The Israel Academy of Sciences and Humanities.

Falkengren-Grerup, U.
1990 Distribution of Field Layer Species in Swedish Deciduous Forests in 1929-54 and 1979-88 as related to Soil pH. *Vegetatio.* Vol 86 pp 143-150.

Fancault, Michel.
1984 *History of Sexuality.* London. A. Lane.

Fanta, J. Editor
1986 *Forest Dynamics Research in Western and Central Europe.* Proceedings of the Workshop held 17-20 Sept 1985 in Wageningen. The Netherlands. Wageningen, Dudoe.

Fanta, J.
1986 Primary Forest Succession on Blow-out Areas in the Dutch Drift Sands. *Forest Dynamics Research in Western and Central Europe.* Ed. J. Fanta. Wageningen, Pudoc.

Fanta, J.
1986 Forest Site as a Framework for Forest Succession. *Forest Dynamics Research in Western and Central Europe.* Ed. J. Fanta. Wageningen, Pudoc.

Feinbrun - Dothan, Naomi
1960 *Wild Plants in the Land of Israel.* Tel-Aviv, Hakbrutz Hameuchad Publishing House.

Fenchel, Tom
1987 *Ecology - Potentials and Limitations.* Luhe. Ecology Institute.

Finberg, H. P. R.
1964 *Lucerna.* Leicester. Leicester University Press.

Finley, M.I.
1965 *The ancient economy.* 2nd edition. London: Hogarth.

Fisher, W. R.
1908 *Schlich's Manual of Forestry* vol. V, 2nd Edition. London. Bradbury. Agnew and Co.

Fitton, Martin.
n.d. *The Limassol Forest Study: a report to Rotary, Cyprus.* Unpublished draft.

Flaksberger, C.
1927 Wheats of the Island of Cyprus. *Cyprus Agricultural Journal.* Vol 22:3.

Florence, Ross.
1992 *Stand Dynamic Processes.* Unpublished lecture notes. Forestry Department, Australian National University.

Florence, R. G.
1994 The Ecological Basis of Forest Fire Management in New South Wales. *The Burning Continent: Forest Ecosystems and Fire Management.* pp 15-33 Current Issues. Perth. Institute of Public Affairs.

Florence, R. G.
1996 *Ecology and Silviculture of Eucalypt Forests.* Collingwood. CSIRO Publishing.

Forbes, R. J.
1964 *Studies in Ancient Technology.* Vols I-IV. Leiden. E. J. Brill.

Ford, Hugh A.
1989 *Ecology of Birds: an Australian Perspective.* Chipping Norton. Surrey Beatty and Sons.

Forman, Richard T. T.
1987 The Ethics of Isolation, the Spread of disturbance, and Landscape ecology. *Landscape Heterogenity and Disturbance.* Ed. Monica Goigel Turner. New York. Springer-Verlag

Forman, R. T. T. 1983 Corridors in a Landscape: Their Ecological Structure and Function. *Ekologia (CSSR).* 2 pp 375-387

Formann, R. T. T. and Gordon, M. 1981 Patches and Structural Components for a Landscape Ecology. *Bioscience.* 31 pp 733-740

Fox, Matthew.
1989 A Cry for a Spiritual Renaissance. *Green Letter.* Vol 5:1 Spring. pp 16-17.

Frankel, D.
1974 Inter-site relationships in the Middle Bronze Age of Cyprus. *World Archaeology* 6:23-43.

Frankel, David and Webb, Jennifer M.
1990 *Archaeological research in the Marki region, Cyprus, 1990.* Mediterranean

Archaeology Supplementary Volume 3.

Franklin, J.
1992 The Scientific Basis of New Forestry. *Watershed Management: Balancing Sustainability and Environmental Sustainability and Environmental Change.* Ed. R. M. Naiman. pp 60-90. New York. Springer-Verlag.

Franklin, Jerry F.
1994 A Conceptual Basis for FEMAT. *Journal of Forestry.* April pp 21-23.

Frazer, Alistair.
1988 *Forest Policy and Management, Cyprus.* United Nations FAO, July 1988. Unpublished.

Freeman, Peter et al.
1985 *Building Conservation in Australia.* Canberra. RAIA.

Frost, H.
1963 *Under the Mediterranean.* Englewood Cliffs, N.J.

Furumark, Arne.
1965 The Excavations at Sinda, some historical results. *Opuscula theniensia* 6: 99-116.

Galen. *De Antidotis.* Trans. C.G. Kuhn, from Ancient Greek 1821-33. Leipzig.

Gamble
1991 Biology and Behaviour in Human Evolution in *Cambridge Archaeological Journal (1991)* Vol 1 pp 215-217

Garnsey, Peter.
1988 *Famine and Food Supply in the Graeco-Roman World.* Cambridge. Cambridge University Press.

Gause, G. F. 1934 *The Struggle for Existence.* Baltimore. Williams and Wilkins.

Georgiades, C.
1992 *Flowers of Cyprus I: Plants of Medicine.* 2nd ed. Nicosia: Stavrinides.

Georgiades, C.
1987 *Flowers of Cyprus II: Plants of Medicine.* Nicosia: Cosmos.

Georgiades, C. 1992 *Flowers of Cyprus I: Plants of Medicine.* 2nd ed. Nicosia: Stavrinides.

Giesen, Bernd.
1991 Code, process and situation in cultural selection. *Cultural Dynamics.* 4/2: pp 172-185.

Gifford, John Allen.
1978 *Paleogeography of archaeological sites of the Larnaca Lowlands, Southeastern Cyprus.* UMI, 1978.

Gillon, D. and Rego, 7
1991 Forest Response to Fires and Implication in Ecosystem Functioning and Management. *Responses of Forest Ecosystems to Environmental Change.* Ed. A. Leller, P. Mathy. D. J. N. R. Jeffers. London. Elsevier Applied Science.

Given, Michael.
1991 *Symbols, power and the construction of identity in the city kingdoms of ancient Cyprus, c.750-312 BC.* Unpublished PhD thesis.

Glanville, S.R.K.
1931 Records of a Royal Dockyard of the Time of Tuthmosis III:Papyrus British museum 10056 Part I. *Zeitschrift fur Agyptische Sprache und Alterumskunde* 66:105-121.

Göbel, Barbara.
1991 The beginnings of agriculture in Mesoamerica and South America: a view from decision theory. *Cultural Dynamics* 4/2: pp 135-157.

Golley, F. B., Ryszkowski, L. and Sokur, J. T.
1975 The Role of Small Mammals in Temperate Forests, Grasslands and Cultivated Fields. *Small Mammals: Their Productivity and Population Dynamics.* Ed. F. B. Golley, K. Petrusewicz and L. Ryszkowski. Cambridge. Cambridge University Press.

Golley, F. B. ed.
1977 *Benchmark Papers in Ecology 15. Ecological Succession.* Pennsylvania, Dowden, Hutchinson and Ross.

Gombrich, E. H.
1977 *Art and Illusion.* London. Phaidon.

Goodall, D. W.
1952 Quantative Aspects of Plant Distribution. *Biological Review.* Vol 27 pp 194-245.

Gordon, W.A.
1950 Regeneration at high altitudes in Cyprus. *Empire Forestry Review. Vol 29*

Gordon Childe, V.
1944 *Archaeological ages as technological stages.* London: Royal Anthropological Institute of Great Britain and Ireland.

Gore, Al
1992 *Earth in the Balance.* London. Earthscan Publications.

Goudie, A. S.
1983 *Environmental Change.* 2nd ed. Oxford. Clarendon Press.

Grace, John
1991 Incorporating Species Attributes and Environmental Heterogenity into First Models. *Responses of Forest Ecosystems to Environmental Change.* Edited by A. Leller, P. Mathy. D. J. N. R. Jeffers. London. Elsevier Applied Science.

Graham, Angus.
1981 *Chang Izu: The Inner Chapters.* London. George Allen and Unwin.

Graumlich, Lisa J.
1993 High Resolution Pollen Analysis Provides New Perspective on Catastrophic Elm Decline. *TREE* vol 8 no 11. November 1993

Gray, Elizabeth Dodson.
1981 *Green Paradise Lost.* Welleslay Mass. Round_Table Press.

Green, Karen.
1994 Freud, Wollstonecraft, and Ecofeminism. *Environmental Ethics.* Vol 16:2 pp 117-134

Greig-Smith, P
Quantitative Plant Ecology. 3rd Edition. Oxford. Blackwell Scientific Publications. Series Studies in Ecology Vol. 9.

Greig-Smith, P.
1980 The Development of Numerical Classification and Ordination. *Vegetatio* Vol 442 pp 1-9.

Greig-Smith, Peter.
1964 *Quantitative Plant Ecology.* 2nd ed. London. Butterworth.

Greller, Andrew M.
1988 Deciduous Forest. *North America terrestrial Vegetation.* Ed. Michael G. Barbour & William D. Billings. Cambridge. Cambridge University Press.

Grene, Marjorie.
1969 *The Anatomy of Knowledge: Papers Presented to the Study Group on Foundations of Cultural Unity.* London. Routledge and Kegan Paul.

Grizzle, Raymond E.
1994 Environmentalism Should Include Human Ecological Needs. *BioScience.* Vol 44 no 4 pp 263-268

Groves, Colin P.
1989 Feral mammals of the Mediterranean islands: documents of early domestication. *The Walking Larder: patterns of domestication, pastoralism, and predation,* edited by Juliet Clutton-Brock. pp 46-58. One World Archaeology. London: Unwin Hyman.

Grub, Peter J. and Whittaker, John B.
1989 *Toward a More Exact Ecology.* Oxford. Blackwell Scientific Publications.

Gunnis, R.
1936 *Historic Cyprus.* London. Methuen.

Habermas, Jurgen.
1968 *Knowledge and Human Interests.* Trans. Jeremy J. Shapiro. London. Heinemann.

Hadjioannou, Loizos Chr.
1987 *The climate of Cyprus past and present.* Nicosia: Ministry of Agriculture & Natural Resources, Meteorological Service.

Haldane, Cheryl Ward.
1990 Shipwrecked Plant Remains. *Biblical Archaeologist.* March 1990

Haldane, Cheryl.
1993 Direct Evidence for Organic Cargoes in the Late Bronze Age. *World Archaeology.* Vol 24:3 pp 348-358.

Hallam, H. E.
1965 *Settlement and Society.* Cambridge. Cambridge University Press.

Hallen, Patsy
1994 Reawakening the Erotic: Why the Conservation Movement Needs Ecofeminism. *Habitat Australia.* Feb. 1994

Hallman, Max O.
1991 Neitzsche's Environmental Ethics. *Environmental Ethics.* Vol 13.

Halstead, Paul.
1977 A preliminary report on the faunal remains from Late Bronze Age Kouklia, Paphos. *Report of the Department of Antiquities of Cyprus.* pp 261-275.

Hamilton, David.
1925 The Forests of Cyprus. *Empire Forestry Review.* Vol 4.

Hansen, Julie.
1989 Botanical Remains. *Studies in Mediterranean Archaeology.* Vol 71:3 pp 82-93.

Hansen, Julie.
1989 Khirokitia Plant remains. *Fouilles Récentes a Khirokitia (Chypré), 1983-1986.* Ed. Alain Le Brun. Paris. Editions recherche sur les Civilisations.

Hansen, Julie M.
1978 Vasilikos Valley Project: second preliminary report: botanical remains. *Journal of Field Archaeology.* Vol 5/2: pp 181-184.

Hansen, Julie M.
1991 *The Palaeoethnobotany of Franchthi Cave.* Excavations at Franchthi Cave, Greece, Fascicle 7. Bloomington: Indiana University Press.

Harris, Larry D.
1984 *The Fragmented Forest: Island Biogeography Theory and the Preservation of Biotic Diversity.* Chicago. University of Chicago Press.

Harris, Hillman G. C. 1989 *Foraging and Farming.* London. Unwin Hyman.

Harris, Larry D.
1984 *The Fragmented Forest: Island Biogeography Theory and the Preservation of Biotic Diversity.* Chicago. University of Chicago Press.

Harrison, Timothy P.
1993 Economics with an Entrepreneurial Spirit: Early Bronze Trade with Late Predynastic

Egypt. *Biblical Archaeologist.* Vol 56:2 pp 81-93.

Hawking, Stephen W.
1988 *A Brief History of Time.* London. Bantam Press.

Hayes, W.C. 1949 Career of the Great Steward Henenv under Nebhepetre entuhotpe. *Journal of Egyptian Archaeology* 35:43-49.

Hayward, G. F. and Phillipson, J.
1979 Community Structure and Functional Role of Small Mammals in Ecosystems in *Ecology of Small Mammals.* Ed. D. Michael Stoddart. London, Chapman and Hall.

Heal, O. W.
1991 Community Patterns and Dynamics in *Responses of Forest Ecosystems to Environmental Change.* Ed. A. Leller, P. Mathy. D. J. N. R. Jeffers. London. Elsevier Applied Science.

Helbaek, Hans.
1962 Late Cypriote vegetable diet at Apliki. *Opuscula Atheniensia.* Vol 4 pp 171-186.

Held, Steve O.
1993 Insularity as a Modifier of Cultural Change: The Case of Prehistoric Cyprus. *Bulletin of American Schools of Oriental Research* no 292. November pp 25-32

Held, Steve O.
1983 *Contributions to the early prehistoric archaeology of Cyprus 1. Environmental and chronological background studies.* Unpublished.

Held, Steve O.
1992 *Pleistocene fauna and early Holocene humans: A gazetteer of paleontological and early archaeological sites on Cyprus.* SIMA 95. Jonsered: Paul Åströms Förlag.

Heltzer, Michael, and E. Lipinski ed.
1988 *Society and economy in the Eastern Mediterranean (c.1500-1000 BC).* Orientalia Lovaniensia Analecta 23. Leuven: Uitgeverij Peeters.

Hettinger, Ned.
1994 Valuing Predation in Rolston's Environmental Ethics: Bambi Lovers versus Tree Huggers. *Environmental Ethics.* Vol 16:1 pp 3-20.

Higgs, E.S.
1967 *Climate, Environment and Indistries of Stone Age Greece, Part III.* The Prehistoric Society.
1969 The Origins of Agriculture: A Reconsideration. *Antiquity.* XLIII pp 31-41

Higgs, A. J.
1981 An Island Biogeographgy Theory and Nature Reserve Design. *Journal of Biogeography.* 8 pp 117-124

Hill, *Sir* George.
1940 *A History of Cyprus.* Vol. 1. Cambridge. Cambridge University Press.

Hirschfeld, Nicolle.
1990 The Ship of Saint Paul: Historical background. *Biblical Archaeologist.* March. pp 25-30.

Hjelmqvist, H.
1970 Some carbonised seeds and fruits from the necropolis of Salamis. *Excavations in the necropolis of Salamis* II, by V. Karageorghis. pp 329-335. Nicosia: Department of Antiquities.

Hjelmqvist, H.
1973 Some economic plants from ancient Cyprus. In *Excavations in the necropolis of Salamis* III, by V. Karageorghis. pp 231-255. Nicosia: Department of Antiquities.

Hodder, Ian.
1993 *Reading the Past.* 2nd Ed. Cambridge. Cambridge University Pres.

Holmboe, Jens.
1914 *Studies in the vegetation of Cyprus, based upon researches during the spring and summer of 1905.* Bergen Museums Skrifter I/2. Bergen: John Greigs Boktrykkeri.

Hoskins, W. G.
1977 *The Making of the English Landscape.* 2nd ed. London. Hodder and Stroughton.

Howell, C.
1983 *Land, Family and Inheritance in Transition.* Cambridge. Cambridge University Press.

Huffaker, C. B.
1958 Experimental Studies on Predation: Dispersion Factors and Predator-Prey Oscillations. *Hilgardia.* 27 pp 343-383

Hulin, Linda Carless.
1989 The identification of Cypriot cult figures through cross-cultural comparison: some problems. *Early Society in Cyprus.* Ed. Edgar Peltenburg. pp 127-139. Edinburgh University Press.

Hummel, F.C.
1954 *Proposals on some aspects of Forest Management in Cyprus.* Nicosia: Forest Department.

Huntley, Brian.
1991 How Ecosystems Respond to Environmental Change: Evidence from Late-Quaternary Palaeoecological Records. *Responses of Forest Ecosystems to Environmental Change.* Edited by A. Leller, P. Mathy. D. J. N. R. Jeffers. London. Elsevier Applied Science.

Huston, M.
1979 A General Hypothesis of Species Diversity.

American Naturalist. Vol 113 pp 81-101.

Hutchins, D.E.
1909 *Report on Cyprus Forestry.* London: Waterlow & Sons. Jari Florestal e Agropecuar Ltd.

Hutchins, D.E.
1979 *Yield Projection System of Jari Florestal e Agropecuar Ltd.* Unpublished.

Hutchinson, ir Joseph.
n.d. Erosion and Land Use: The Influence of Agriculture on the Epirus Region of Greece. *Agricultural History Review.* Vol 17 pp 85-90

Hythbern, H.
1986 Methods of Forest Dynamics research. *Forest Dynamics research in Western and Central Europe.* Ed. by J. Fanta. Wageningen, Pudoc.

IJames, N. D. G.
1982 *The Forester's Companion* 3rd edition. Oxford, Basic Blackwell.

Jane, F.W.
1956 *The Structure of Wood.* 2nd ed. Completely revised by K. Wilson and D.J.B. White. London: Adam and Charles Black.

Jarman, M. R.
1972 A Territorial Model for Archaeology: A Behavioural and Geographical Approach. *Models in Archaeology.* Edited by D. C. Clarke. London. Methuen and Co.

Jarvis, D. I. and Helin, Liu.
1993 Vegetation Patterns in the Pinus yunnanensis - sclerophyllous broadleaf forests, Mianning County, Sichvan Province, China. *Journal of Biogeography.* 20 pp 505-524

Jenik, J.
1986 Forest Succession: Theoretical Concepts. *Forest Dynamics Research in Western and Central Europe.* Ed. by J. Fanta. Wageningen, Pudoc.

Jenkins, R.J.H.
1970 Cyprus Between Byzantium and Islam AD 688-965. *Studies on Byzantine History of the 9th and 10th Centuries.* London, Variorum Reprints.

Jenkins, R.J.H.
1970 The Mission of St. Demetrianus of Cyprus to Bagdad. *Studies on Byzantine History of the 9th and 10th Centuries.* London, Variorum Reprints.

Johnson, Edward A.
1992 *Fire and Vegetation Dynamics: Studies from the North American Boreal Forest.* Cambridge. Cambridge University Press.

Johnson, G. A.
in *Theory and Explanation in Archaeology.* Edited by Colin Rerfrew.

Johnson, Edward A.
1992 *Fire and Vegetation Dynamics: Studies from the North American Boreal Forest.* Cambridge. Cambridge University Press.

Johnston, R. J.
1973 *Spatial Structures.* London. Methuen.

Johnstone, P.
1972 Bronze Age Sea Trial. *Antiquity.* Vol 46 pp 269-274.

Kanawati, N. et al.
1984 *Excavations at Saqqara north-west of Teti's pyramid.* Sydney: Macquarie University Ancient History Documentary Research Centre.

Kangas, Jyriki and Kuusipalo, Jussi.
1993 Integrating Bio-diversity into Forest Management Planning and Decision Making. *Forest Ecology and Management.* Vol 61 pp 1-15

Kappelle, Maarten Kennis, Peer, A. F. and Vries, Rob, A. J. De.
1995 Changes in diversity along a successional gradient in a Costa Rican upper montane Quercus forest. *Biodiversity and Conservation.* Vol 4 pp 10-34.

Karageorghis, Jacqueline.
1977 *La grande déesse de Chypre et son culte à travers l'iconographie, de l'époque néolithique au VI`me s. a. C.* Lyon: Maison de l'Orient.

Karageorghis, Vassos.
1993 *The coroplastic art of ancient Cyprus. II. Late Cypriote II - Cypro- Geometric.* Nicosia: A.G. Leventis Foundation.

Karageorghis, Jacqueline.
1977 *La grande déesse de Chypre et son culte à travers l'iconographie, de l'époque néolithique au VI`me s. a. C.* Lyon: Maison de l'Orient.

Karageorghis, Vassos.
1976 *The civilisation of pre-historic Cyprus.* Athens: Ekdotike Athenon SA.

Karageorghis, Vassos.
1981 *Ancient Cyprus: 7000 years of art and archaeology.* Athens: Ekdotike Athenon SA.

Karageorghis, Vassos.
1985 *Excavations at Kition 5: The pre-Phoenician levels, Part 2.* Nicosia: Department of Antiquities of Cyprus.

Karageorghis, Vassos.
1991 *The coroplastic art of ancient Cyprus. I. Chalcolithic - Late Cypriote I.* Nicosia: A.G. Leventis Foundation.

Karageorghis, Vassos.
1993 *The coroplastic art of ancient Cyprus. II. Late Cypriote II - Cypro- Geometric.* Nicosia: A.G. Leventis Foundation.

Katzev, Michael L., and Katzev, Susan Womer .
1986 Kyrenia II. Research on an ancient shipwreck comes full circle in a full-scale replication. *Institute of Nautical Archaeology Newsletter* 13/3 pp 2-11.
Katzev, Michael L., and Katzev, Susan Womer .
1987 *Kyrenia II*. Piraeus: Hellenic Institute for the Preservation of Nautical Tradition.
Keating, W. G. and Bolza, Eleanor.
1982 *Characteristics, Properties and Uses of Timbers South East Asia, Northern Australia and the Pacific*. Vol. I Melbourne. Inkata Press.
Kehoe, Dennis P.
1994 Pastoralism and Agriculture. *Journal of Roman Archaeology*. Vol 7 pp 388-398.
Keller, Evelyn Fox and Grontkowski, Christine R.
1983 The Minds Eye. *Discovering Reality*. Ed. Sandra Harding and Merrill B. Hintikka. D. Reidel.
Kent, Martin and Coker, Paddy.
n.d. *Vegetation, Description and Analysis*. Boca Raton, Ann Arbor, CRC Press.
Keswani, P. S.
1994 The Social Context of Animal Husbandry in Early Agricultural Societies: Ethnographic Insights and an Archaeological Example from Cyprus. *Journal of Anthropological Archaeology*. Vol 13 pp 255-277.
Kienast, Felix, et al.
1987 Tree-ring studies on conifers along ecological gradients and the potential of single year analysis. *Canadian Journal of Forestry Research*. Vol 17.
Kitromilides, Paschalis M., and Marios L. Evriviades (compilers).
1982 *Cyprus*. World Bibliographical Series, Volume 28. Oxford: Clio Press.
Knapp, A. Bernard et. al.
1993 The Sydney Cyprus Survey Project (SCSP) - Second Preliminary Season in *RDAC*.
Knapp, A. Bernard.
1986 *Copper production and divine protection: archaeology, ideology and social complexity on Bronze Age Cyprus*. SIMA Pocket-book 42. Göteborg: Paul Åströms Förlag.
Knapp, A. Bernard.
1991 Bronze Age Trade in the Mediterranean in *Studies in Mediterranean Archaeology* vol XC. Edited by N. H. Gale. Jonsered. Paul Astroms-Forlag
Knapp, A. Bernard.
1993 Social Complexity: Incipient, Emergence and Development on Prehistoric Cyprus. *BASOR* 292 pp 85-106
Knapp, A. Bernard.
1994 Emergence, Development and Decline on Bronze Age Cyprus. *Development and Decline in the Mediterranean Bronze Age*. Ed. Clay Mathers and Simon Staddart. Sheffield. University of Sheffield Department of Archaeology and Prehistory.
Knapp, A.B. 1992 Bronze Age Mediterranean Island Cultures and the Ancient Near East. Part 1. *Biblical Archaeologist* September 1992:112-128
Knapp, A. Bernard, Held, Steve O. and Manning, Sturt W.
1994 The Prehistory of Cyprus: Problems and Prospects. *Journal of World Prehistory*. Vol 8:4 pp 377-453.
Knapp, A. Bernard.
1992 Bronze Age Mediterranean Island Cultures and the Ancient Near East, Part 2. *Biblical Archaeologist*. Sept. pp 112-128.
Knapp, A. Bernard.
1992 Bronze Age Mediterranean Island Cultures and the Ancient Near East, Part 1. *Biblical Archaeologist*. June pp 52-67.
Knapp, A. Bernard.
1979 *A re-examination of the interpretation of Cypriote material culture in the MC III-LC I period in the light of textual data*. Unpublished thesis.
Knapp, A. Bernard.
1986 *Copper production and divine protection: archaeology, ideology and social complexity on Bronze Age Cyprus*. SIMA Pocket-book 42. Göteborg: Paul Åströms Förlag.
Koestler, Arthur.
1978 *Janus: A summing up*. New York. Vintage Books.
Kolchugina, Tatyana P. and Vinson, Teel S.
1993 Carbon Sources and Sinks in Forest Biomes of the former Soviet Union. *Global Biogeochemical Cycles*. Volume 7, Number 2. June 1993
Kollman, Franz.
1951 *Technologie Des Holzes Und Der Holwerkstoffe*. Berlin. Springer-Verlag.
Koucky, Frank L., and Arthur Steinberg.
1982 Ancient mining and mineral dressing on Cyprus. In *Early pyrotechnology: the evolution of the first fire using industries*. Ed. Theodore A. Wertime and Steven F. Wertime. Washington: Smithsonian Institution Press.
Krebs, J. R. and Houston, A. I.
1989 Optimisation in Ecology. *Ecological Concepts*. Edited by J. M. Cherrett. Oxford. Blackwell Scientific Publications.

Krebs, C. J.
1972 *Ecology: the Experimental Analysis of Distributuin and Abundance.* New York. Harper and Row.

Kreutz, B.M.
1976 Ships, Shipping and the implication of change in the Early Mediterranean. *Viator* 7:79-109

Kuhn, Thomas.
1970 *The Structure of Scientific Revolutions.* 2nd ed. Chicago.

Kuhn, Thomas.
1970 *The Structure of Scientific Revolutions.* 2nd ed. Chicago. University of Chicago Press.

Kuusinen, Mikko.
1996 Cyanobacterial Marolichens on *Populus tremulla* as Indicators of Forest Continuity in Finland. *Biological Conservation.* Vol 75 pp 43-49.

Kyllo, M.
1982 The Botanical Remains. In *Vrysi: a subterranean settlement in Cyprus: excavations at prehistoric Ayios Epiktitos Vyrsi 1969-73*, by E.J. Peltenberg. Warminster: Aris and Phillips.

Kyrris, Costas P.
1985 *History of Cyprus, with an introduction to the geography of Cyprus.* Nicosia: Nicoles.

Lapin, Marc and Barnes, Burton V.
1995 Using the Landscape Ecosystem Approach to Assess Species and Ecosystem Diversity. *Conservation Biology.* Vol 9:5 Oct. pp 1148-1158

Lavender, David.
1962 *The story of Cyprus Mines Corporation.* San Marino, California: The Huntington Library.

Lavers, G. M.
1969 The Strength Properties of Timbers. *Bulletin* 50: Second ed. London: Her Majesty's Stationery Office.

Law, R. and Watkinson, A. R.
Competition. *Ecological Concepts*. Ed. J. M. Cherrett. Oxford. Blackwell Scientific Publications.

Lawrence, A. W.
1979 *Greek Aims in Fortification.* Oxford. Clarendon Press.

Lawrence, A. W. 1979 *Greek Aims in Fortification.* Oxford. Oxford University Press.

Le Brun, Alain.
1981 *Un site néolithique précéramique en Chypre: Cap Andreas-Kastros*. Mémoire no. 5. Paris: Éditions Recherche sur les civilisations.

Le Brun, Alain.
1984 *Fouilles récentes à Khirokitia (Chypre) 1977-1981.* Mémoire no. 41. Paris: Éditions Recherche sur les civilisations.

Le Brun, Alain. -- and others.
1987 Le néolithic précéramique de Chypre. *Anthropologie.* Vol 91 pp 283-316.

Leaf, Walter
n.d. The Scamander Valley - I in *ABSA* 17 (1910-11).

Leaf, Walter
n.d. The Scamander Valley - II in *ABSA* 18 1911-12.

Leippert, H.
1981 *Vegetation and its Potential Value for Landscape Protection in Akamas - case study.* Germany: Federal Ministry for Economic Cooperation.

Lenis, Martin W
1992 *Green Delusions.* Durham. Duke University Press.

Leontiades, Leontios.
1963 *Economic Aspects of Forestry in Cyprus.* Oxford: Commonwealth Forestry Institute.

Leontiades, Leontios.
1969 *Game and wildlife management in Cyprus.* Fordingbridge: Eley Game Advisory Station. *Pitsilia Regional Development Programme. Provisional Five-year Plan for Kyperounda Village 1966-1970.* Nicosia: Ministry of Agriculture & Natural Resources. Unpublished.

Leontiades, Leontios.
1977 *Report on Wetlands and Marine Parks in Cyprus, prepared for the Export Consultation on Mediterranean Marine Parks and Wetlands, convened by the UNEP in Tunisia, 1977.*

Leopold, A.
1949 *A Sand Country Almanac.* London. Oxford University Press.

Lepper, Frank and Frere, Sheppard.
1988 *Trajans Column.* Gloucester. Alan Sulton Pub.

Levick, Barbara.
1967 *Roman Colonies in Southern Asia Minor.* Oxford. Clarendon Press.

Lewis, Martin W.
1992 *Green Delusions.* Durham. Duke University Press.

Librairie C. Klincksieck.
1983 *Corpus des cylindres-sceaux de Ras Shamra-Ugarit et d'Enkomi-Alasia.* Volume I. Paris: Editions Recherche sur les Civilisations.

Likens, Gene.
1992 *The Ecological Approach: its Use and Abuse.* Oldendorf / Lune. Ecology Institute.

Lindeman, R. L.
1942 The Tropic-Dynamic Aspect of Ecology.

Ecology. 23 pp 399-418

Lloyd, John.
1991 Forms of Rural Settlement in the Early Roman Empire. *Roman Landscapes: Archaeological Survey in the Mediterranean Region.* Ed. by Graeme Barker and John Llyoel. London. British School at Rome.

Long, Douglas C.
1968 Particulars and their Qualities in *Philosophical Quarterly*. Vol. XVIII

Loret, V.
1916 Quelques notes sur l'arbre âch. *Annales du Service.* Vol 16 pp 33-51.

Lott, M.
1983 *Cluster Analysis for Social Scientists.* San Francisco. Jossey-Bass.

Lovelock, James E.
1979 *Gaia.* Oxford. Cambridge University Press.

Lowenthal, D.
1979 Environmental Perceptions: Preserving the Past. *Progress in Human Geography.* Vol 3 pp 549-559.

Lowenthal, David.
1979 Dilemmas of Appreciation. *The Interpretation of Ordinary Landscapes.* Ed. by D. W. Meinig. Oxford. Oxford University Press.

Lubsen-Admiraal, Stella M., and Joost Crouwel.
1989 *Cyprus and Aphrodite.* 's-Gravenhage: SDU Uitgeverij.

Lucas, A.
1948 *Ancient Egyptian Materials and Industries.* 3rd ed. London: Edward Arnold.

Macarthur, Robert H. and Wilson, Edward O.
1967 *The Theory of Island Biogeography.* Princeton. Princeton University Press.

Macarthur, Robert.
1955 Fluctuations of Animal Populations, and a Measure of Community Stability. *Ecology.* Vol 36 pp 533-536.

Macarthur, Robert.
1967 *The Theory of Island Biogeography.* Princeton. Princeton University Press.

Macarthur, Robert.
1972 *Geographical Ecology: Pattern in the Distribution of Species.* New York. Harper and Row.

Macdonald, David and Barrett, Priscilla.
1995 *European Mammals Evolution and Behaviour.* London. Harper Collins.

Macinnes, Lesley and Wickham-Jones, Caroline R.
1992 *All Natural Things: Archaeology and the Green Debate*. Oxford. Oxbow Books.

Madon, P.G.
1881 *Forest conservation in the Island of Cyprus.* [Reprinted Nicosia: Nicosia Govt. Printing Office, 1930]

Maier, Franz Georg.
1968 *Cyprus from earliest times to the present day*. Trans. Peter Gorge. German edition 1964. London: Elek Books.

Maisels, Fiona G.
1988 *The Feeding ecology of the Cyprus Mouflon Ovis orientalis Gmelin 1774, in the Paphos Forest Cyprus.* Unpublished thesis.

Malthus, T. R.
1788 *An Essay on the Principle of Population.* London. Johnson.

Manniche, Lise.
1989 *An Ancient Egyptian Herbal.* London. British Museum Publications.

Manniche, L.
1989 *An Ancient Egyptian Herbal.* London: British Museum Publication.

Manning, S. W. and Conwell, D. H.
1992 Masoni Valley Archaeological Survey Project: Preliminary Report on the 1990-1991 Field Season. *Report of the Department of Antiquities, Cyprus.* pp 271-283.

Maritain, Jacques.
1940 *Science and Wisdom.* Trans. by Bernard Wall. London. Geoffrey Bleo.

Mather, Alexandar S.
1990 *Global Forest Resources.* Portland. Oregon. Timber Press.

Matthäus, Hartmut.
1985 *Metallgefäße und Gefäßuntersätze der Bronzezeit, der geometrische und archaischen Periode auf Cypern.* Prähistorische Bronzefunde II 8. München: C.H. Beck.

Mattingly, David J.
1993 Understanding Roman Landscapes. *Journal of Roman Archaeology.* Vol 6 pp 359-366.

May, Robert M.
1972 Will a Large Complex System be Stable? *Ecological Succession.* Ed. by Frank B. Golley. Stroudsbug. Dowden, Hutchinson & Ross.

May, R. M.
1986 The Search for Patterns in the Balance of Nature: Advances and Retreats. Ecology. 67 pp 1115-1126

May, B. M.
1974 Biological Populations with Non-overlapping Generations: Stable Points, Stable Cycles and Chaos. *Science.* 186 pp 645-647

McBryde, Isobel ed.
1985 *Who Owns the Past?* Melbourne. Oxford University Press.

McLaren, K.
1983 *The Colour Science of Dyes and Pigments.*

Bristol. Adam Hilger Ltd.
McNeal, Ann P.
1994 Engendering a New Science. *Bioscience.* Vol 44:3. March. pp 183-184.
McNeely, Jeffery A.
1994 Lessons from the past: Forests and Biodiversity. *Biodiversity and Conservation.* Vol 3 pp 3-20.
Mee, Christopher *et al.*
1991 Rural Settlement Change in the Mmethana Peninsula, Greece. *Roman Landscapes: Archaeological survey in the Mediterranean Region.* London. British School at Rome.
Megan, J. V. S. and Simpson, D. D. A.
1979 *Introduction to British Prehistory.* Leicester. Leicester University Press.
Meiggs, Russell.
1982 *Trees and Timber in the Ancient Mediterranean World.* Oxford: Clarendon Press.
Meikle, R.D.
1977 *Flora of Cyprus.* Vol.1. Kew, Royal Botanic Gardens, The Bentham- Moxon Trust.
Meikle, R.D.
1985 *Flora of Cyprus*, vol. 2. Kew, Royal Botanic Gardens.The Bentham- Moxon Trust
Meing, D. W. ed.
1979 *The interpretation of Ordinary Landscapes.* Oxford. Oxford University Press.
Merchant, Carolyn.
1992 *Radical Ecology.* New York. Routledge.
Merleau-Ponty, Maurice.
1964 *Signs.* Trans. Richard C. McCleary. Northwestern University Press
Merrillees, R.S.
1968 *The Cypriote Bronze Age Pottery found in Egypt.* Lund: Studies in Mediterranean Archaeology.
Merrillees, R.S.
1987 *Alashia revisited.* Cahiers de la Revue Biblique. Paris: J. Gabalda.
Merrillees, R.S.
1975 Problems in Cypriote History. *The Archaeology of Cyprus: recent developments*, pp15-38. Ed. Noel Robertson. Park Ridge,N.J., Noyes Press.
Michaelides, E.D.
1962 *The production of improved strains of tree.* Oxford: Commonwealth Forestry Institute.
Michaelides, E.D.
1967 *Essential oil extraction from certain forest plants of Cyprus.* Nicosia: Ministry of Agriculture & Natural Resources.
Michaelides, E.D.
1983 *Provisional stand management rules for Pinus brutia forests for Cyprus.* Unpublished.
Mighall, T. M. and Chamber, F. M.
1993 Early Mining and Metal Working: Its Impact on the Environment. *The Journal of the Historical Metallurgy Society.* Vol 27 no2 pp. 71-83
Mighall, T. M. and Chambers, F. M.
1995 Hologene Vegetation History and Human Impact at Bryn Y Castell, Snowdonia, North Wales. *New Phytologist.* Vol 130 pp 299-321
Miller, G. Tyler, jr.
1975 *Living in the environment, concepts, problems and alternatives.* Belmont: Wadsworth.
Miller, Naomi F.
1990 Clearing land for farmland and fuel: archaeobotanical studies of the ancient Near East. *Economy and settlement in the Near East: analyses of ancient sites*,Ed. Naomi F. Miller, p. 70-78. Philadelphia: University Museum, University of Pennsylvania.
Mitford, T. B.
1980 *Roman Cyprus.* Berllin, Walter de Gruyter.
MKE Meghali Kypriaki Engiklopedhia (Great Cypriot Encyclopedia)
Milsicek, C. H.
1988 *in Karagheorghis V. + Demas M Excavations at Moa-Polaeokostro.* 1979-1986 pp 467-470
Moore, P. D.
1976 Site History in *Methods in Plant Ecology.* Edited by P. D. Moore and S. B. Chapman. 2nd Edition. Oxford. Blackwell Scientific Publications. pp 525-556
Moran, William L. (editor and translator)
1992 *The Amarna letters.* French edition 1987. Baltimore: The Johns Hopkins University Press.
Morris, Jan.
1980 *The Venetian Empire: A Sea Voyage.* London. Faber and Faber.
Muhly, James D., Robert Maddin, and Vassos Karageorghis (editors).
1982 *Acta of the International Archaeological Symposium, 'Early Metallurgy in Cyprus 4000-500 B.C.,' Larnaca, 1-6 June 1981.* Nicosia.
Myres, John L. and Ohnefalsch-Richter, Max.
1899 *A Catalogue of the Cyprus Museum.* Oxford. Clarendon Press.
Myres, John L.
1914 *Handbook of the Cesnola Collection of Antiquities from Cyprus.* New York: Metropolitan Museum of Art.

Naess, Arne.
1990 Sustainable Development and Deep Ecology. *Ethics of Environment and Development.* Ed. Ronald Engel and Joan Gibb Engel. London. Belhaven Press.

Nash, Roderick Frazier
1989 *The Rights of Nature: A History of Environmental Ethics*. Madison. University of Wisconsin Press.

Naveh, Zev
1989 Mediterranean Europe and East Mediterranean Shrublands. *The Biology and Utilisation of Shrubs*. Ed. Cyrus M. McKell. San Diego. Academic Press.

Negbi, Moshe
1989 Theophrastus on Geophyles. *Botanical Journal of the Linnean Society*. 100 pp 15-43

Nibbi, A.
1994 Some Remarks on the Cedar of Lebanon. *Discussions in Egyptology* 28:35-52

Nicolaou, Ino.
1979 Endurance for the Cult of Cybele in Cyprus. *Studies Presented in memory of Porphyrios Dikaios.* Ed. V. Karageorghis et. al. Nicosia, Lious Club of Nicosia.

Nicolaou, Kyriakos. 1968 *Ancient Monuments of Cyprus*. Nicosia: Department of Antiquities.

Niering, William A. and Lowe, Charles H.
1984 Vegetation of the Santa Catalina Mountains: Community Types and Dynamics. *Vegetatio.* Vol 58 pp 3-28.

Noble, I. R. and Slayter, R. O.
1977 Post Fire Succession of Plants in Mediterranean Ecosystems. *Proceedings of the Symposium on the Environmental Consequences of Fire and Fuel Management in Mediterranean Ecosystems.* Ed. H. A. Mooney and C. E. Courad. USDA. For. Ser. Gen. Tech. Rep. WO-3.

Noble, I. R. and Slayter, R. O.
1979 The Effect of Disturbances on Plant Succession. *Proceedings of the Ecological Society of Australia.* Vol 10 pp 135-145.

Noble, I. R. and Slayter, R. O.
1980 The Use of Vital Attributes to Predict Successional Changes in Plant Communities Subject to Recurrent Disturbance. *Vegetatio.* Vol 43.

O'Hara, Kvin L. et. al.
1994 Silviculture and Our Changing Profession: Leadership for Changing Paradigms. *Journal of Forestry*. 92(1) pp 8-13.

Odum, E. P.
1969 The Strategy of Ecosystem Development. *Science.* Vol 164 pp 262-270.

Odum, H. T.
1971 *Environment, Power and Society.* New York. John Wiley.

Ogden, John.
1995 The Long-term Conservation of Forest Diversity in New Zealand. *Pacific Conservation Biology.* Vol 2:1 pp 77-90.

Ohnefalsch-Richter, Max.
1893 *Kypros, the Bible and Homer: Oriental civilization, art and religion in ancient times. Elucidated by the author's own researches and excavations during twelve years' work in Cyprus.* London: Asher.

Oldeman, R. A. A. and Schmidt, P.
1986 Silvicultural Design, and Exercise for Students. *Forest Dynamics Research in Western and Central Europe.* Ed. J. Fanta. Wageningen, Pudoc.

Olszewski, Deborah I.
1993 Subsistence Ecology in the Mediterranean Forest: Implications in the Epipaleolithic Southern Levant. *American Anthropologist*, 93. pp 420-435

Outer, R. W. Den Veenendaal, W. L. H. Van and Versteech, Chr.
1988 *Determination Keys for Important West European Woods and Tropical Commercial Timbers.* Wageningen. Agricultural University.

Pantelas, Vassos S.
n.d. *Summary of Results of the Forest Inventory and Management Recommendations for the Pinus brutia Productive Main State Forests for the Period of 1982-1991.* Nicosia: Department of Forests. No imprint.

Pantelas, Vassos S., Takis Ch. Tsintides, and Antonis Antoriou.
1986 *The Akamas Peninsula Provisional Action Plan (summary).* Nicosia: Ministry of Agriculture & Natural Resources.

Pantin, C. F. A.
1969 Organism and Environment. *The Anatomy of Knowledge*. Edited by Marjorie Grene. London. Routledge and Kegan Paul.

Pantin, C. F. A.
1954 The Recognition of Species. *Science Progress.* 43:168 pp 578-598

Papageorgiou, Nicolaas K.
1978 Food preferences, feed intake and protein requirements of red deer in Central Greece. *Journal of Wildlife Management* 42/4.

Papageorgiou, Nicolaas K.
1978 Food preferences, feed intake and protein requirements of red deer in Central Greece. *Journal of Wildlife Management* 42/4.

Parker, A. J.
1992 *Ancient Shipwrecks of the Mediterranean and the Roman Provinces.* Oxford. (BAR

International Series: 580)

Parks, Peter J.
1991 Model of Forested and Agricultural Landscapes: Integrated Economics. *Quantitative Methods in Landscape Ecology.* Ed. Monica G. Turner and Robert H. Gardner. New York. Springer-Verlag.

Paschalis, Piotr.
1996 *Protection of Forest Ecosystems Biodiversity of Biakowieza Primeval Forest.* Warsaw. GEF 05/21685P01 Grant Office.

Patrick, Richard A.
1976 *Political Geography and the Cyprus Conflict.* 1963-1971. University of Waterloo, Department of Geography, Faculty of Environmental Studies. Ontario.

Pearsall, Deborah.
1989 *Paleoethnobotany: a handbook of procedures.* San Diego: Academic Press.

Peerenboom, R. P.
1991 Beyond naturalism: S Reconstruction of Dauist Environmental Ethics. *Environmental Ethics.* Vol 13 Summer.

Peet, R. K.
1974 The Measurement of Species Diversity. *American Review of Ecological Systems.* Vol 5 pp 285-307.

Peltenberg, E.J.
1985 Rammeside Egypt and Cyprus. Pp150-179 in *Acts of the International Archaeological Symposium: Cyprus between the Occident and the Orient.* ed. V. Karegeoghis. Nicosia.

Peltenberg, E. D.
1991 Kissonerga - Mosphilia: A major Chalcholithic Site in Cyprus. *Bulletin of the American Schools of Oriental Research.* Vol 282/283 pp 17-35.

Peltenberg, E. J.
1983 *VRYSI: A Subterranean Settlement in Cyprus.* Waeminster. Aris and Phillips.

Peltenburg, Edgar J.
1982 *Recent developments in the later prehistory of Cyprus.* SIMA pocket-book no. 16. Göteborg: Paul Åströms Förlag.

Peltenburg, Edgar J.
1985 *Lemba Archaeological Project, Volume 1: Excavations at Lemba Lakkous, 1976-1983.* SIMA 70:1. Göteborg: Paul Åströms Förlag.

Peltenburg, Edgar J.
1989 *The beginnings of religion in Cyprus.* In *Early Society in Cyprus*, edited by E. Peltenburg. pp 108-126.

Peltenburg, Edgar J. -- and others.
1988 Kissonerga-Mosphilia 1987: ritual deposit, Unit 1015. *RDAC* 1988 I: pp 43-52.

Perlin, John
1989 *A Forest Journey: the Role of Wood in the Development of Civilisation.* New York. WN Norton.

Peters, R.
1986 Natural Regeneration of Beech Forest in Japan. *Forest Dynamics Research in Western and Central Europe.* Ed. by J. Fanta. Wageningen, Pudoc.

Petit, Paul.
1976 *Pox Romana.* Trans. by James Willis. London. B. T. Batsford.

Petrie, W.M.F.
1898 *Syria and Egypt from the Tell El Amarna Letters.* London: Metheun.

Piaget, Jean.
1971 *Insights and Illusions of Philosophy.* Trans. Wolfe Mays. New York. World Publishing Co.

Pickett, S. T. A. and White, P. S. eds. 1985 *The Ecology of Natural Disturbance and Patch Dynamics.* New York. Academic Press.

Pliny the Elder 1938 *Natural History* in Ten volumes. Trans H. Rackham & W.H.S Jones, from Latin. Loeb Classical Library. Cambridge Mass: Harvard University Press.

Plumwood, Val.
1991 Ethics and Instrumentalism: A Response to Janna Thompson. *Environmental Ethics* vol. 13 Summer

Pollard, John
1977 *Birds in Greek Life and Myth.* London. Thames and Hudson.

Polunin, Oreg.
1987 *Flowers of Greece and the Balkans.* Oxford, Oxford University Press.

Polunin, Oleg, and Anthony Huxley.
1981 *Flowers of the Mediterranean.* London: Chatto & Windus.

Polycarpou, A.
1954 *The production of 'elite' forest tree seed in Cyprus.* Nicosia: Forest Department.

Polycarpou, A.
1956 *Inventory investigations in the Pinus brutia forests of Cyprus.* Oxford: Imperial Forest Institute. Unpublished thesis.

Polycarpou, A.
1954 *Classification of forests into crop types and the preparation and interpretation of stock maps in Cyprus.* Nicosia: Forest Department.

Polycarpou, A.
1957 *Report on a reconnaissance survey of the minor State Forests and recommendations on their future use.* Nicosia: Forest Department.

Polycarpou, A.
1965 *Provisional Development Plan for the Akamas Forest*. Nicosia: Ministry of Agriculture & Natural Resources.

Polycarpou, A.
1956 *Inventory Investigations in the Pinus brutia Forests of Cyprus 1953-55.* Oxford: Imperial Forestry Institute.

Polycarpou, A.
1969 *Forestry in Cyprus: an overview.* FAO/SIDA/Cyprus seminar on Forestry Public Relations. Unpublished.

Porada, Edith.
1986 Late Cypriote cylinder seals between East and West. In *Acts Occident & Orient* 289-299.

Porada, Edith.
1948 The Cylinder Seals of the Late Cypriote Bronze Age. *American Journal of Archaeology.* Vol 52 pp 178-199.

Porada, Edith.
1972 On the complexity of style and iconography in some groups of cylinder seals from Cyprus. In *The Mycenaeans in the Eastern Mediterranean.* pp260-273.

Porada, Edith.
1986 Late Cypriote cylinder seals between East and West. In *Acts Occident & Orient.* pp 289-299.

Post, G.E.
1899 Fir. *Dictionary of the Bible.* Ed. James Hastings. Edinburgh: T & T Clark.

Post, G.E. 1899 Fir. *Dictionary of the Bible.* Ed. James Hastings. Edinburgh: T & T Clark.

Powlesland, D.
1983 Pots, Pits and Portables. *Practical Computing.* June 1993 pp 144-146.

Pred, Allan
1990 *Marine Histories and Constructing Human Geographies*. Boulder. Westview Press.

Pred, A.
1983 Structuration and Place: on the Becoming of Sense of Place and Structure of Feeling. *Journal for the Theory of Social Behaviour.* Vol 13 pp 45-68.

Prentice, I. C.
1986 Some Concepts and Objectives of Forest Dynamics Research. *Forest Dynamics Research in Western and Central Europe.* Ed. by J. Fanta. Wageningen, Pudoc.

Prentice, I. C.
1986 The Design of a Forest Successional Model. *Forest Dynamics Research in Western and Central Europe.* Ed. by J. Fanta. Wageningen, Pudoc.

Pret, Robert K.
1980 Ordination as a Tool for Analysing Complex Data Sets. *Vegetatio.* Vol 42 pp 171-174

Price, H. H.
1964 *Perception.* London. Methuen.

Prince, B. J.
1978 Secondary State Formation: An Explanatory Model. *The Origins of the State: The Anthropology of Political Evolution.* Ed. R. Cohen and Z. Service. Philadelphia. Institute for the Study of Human Issues.

Pritchard, J.B.
1950 *Ancient Near Eastern Texts.* Princeton: Princeton University Press.

Proctor, James D.
1996 Will the Real Land Ethic Please Stand Up? *Journal of Forestry.* Vol 94:2 Feb. pp 39-43.

Prodon, R.
1991 Animal Communities and Vegetation Dynamics: Measuring and Modelling Animal Community Dynamics along Forest Succession. *Responses of Forest Ecosystems to Environmental Change.* Edited by A. Leller, P. Mathy. D. J. N. R. Jeffers. London. Elsevier Applied Science.

Puigdefabregas, Juan.
1995 Desertification: Stress Beyond resilience, Exploring a Unifying Process Structure. *Ambio.* Vol 24:5 Aug. pp 311-313.

Quattrochi, Dale A. and Pelletier, Rammonae.
1991 Remote Sensing for Analysis of Landscapes: An Introduction. *Quantitative Methods in Landscape Ecology.* Ed. Monica G. Turner and Robert H. Gardner. New York. Springer-Verlag.

Rackham, Oliver.
1983 Observations on the Historical Ecology of Boeotia. *Annula of the British School at Athens.* Vol 78 pp 345-346.

Rackham, Oliver.
1980 *Ancient Woodland.* London. Arnold.

Rackham, Oliver.
1976 *Trees and Woodland in the British Landscape.* London, Dent.

Rackham, Oliver.
1978 The Flora and Vegetation of Thera and Crete Before and After the Great Eruption. *Flora and the Aegean World.* Vol 1. Ed. by C. Doumas. pp 755-764. London. Thera and the Aegean World.

Rauthman, Marcus
1990 Field Survey in Northern Keos. *Journal of Roman Archaeology*. Vol 3 pp 421-425.

Ravensdale, J. R.
1974 *Liable to Floods.* Cambridge. Cambridge University Press.

Record, S.J. & Hess, R.W.
1943 *Timbers of the New World.* London: Oxford University Press.

Renfren, Jane.
1970 Carbonised Seeds and Fruits from the Necropolis of Salamis. *Excavations in the Necropolis of Salamis II.* Vassos Karageorghis. Nicosia. Department of Antiquities.

Renfren, A. C. ed.
1974 *British Prehistory.* London. Duckworth.

Reyes, A. T.
1994 *Archaic Cyprus.* Oxford. Clarendon Press.

Richardson, David M. and Cowling, Richard M.
1993 Biodiversity and Ecosystem Processes: Opportunities in Mediterranean-type Ecosystems. *TREE* vol 8 no3 March 1993.

Riddle, J.M.
1985 *Dioscorides on Pharmacy and Medicine.* Austin: University of Texas Press.

Riddle, John M.
1985 *Discorides on Pharmacy and Medicine.* Austin. University of Texas Press.

Rife, Joseph L. and Giesen, Myra J.
1994 Regional Population Diversity and the Archaic to Early Byzantine Cranial Remains from Isthmia, Greece. *Beyond the Site.* Ed. P. Nick Kardulias. University Press of America.

Rimmer, Joan
1969 *Ancient Musical Instruments of Western Asia.* Lonson. Trustees of the British Museum.

Rindos, David.
1984 *The origins of agriculture: an evolutionary perspective.* London: Academic Press.

Risser, Paul G.
1995 Biodiversity and Ecosystem Function. *Conservation Biology.* Vol 9:4 Aug. pp 742-746.

Roberts, David W.
1987 A Dynamical Systems Perspective of Vegetation Theory. *Vegetatio.* Vol 69 pp 27-33.

Rolsten, Holmen IVI.
1988 *Philosophy Gone Wild.* Buffalo. Prometheus Books.

Rorty, Richard.
1980 *Philosophy and the Mirror of Nature.* Oxford. Basil Blackwell.

Rorty, Richard.
1980 *Philosophy and the Mirror if nature.* Oxford. Basil Blackwell.

Rosen, Steven and Avni, Gideon.
1993 The Edge of the Empire: The Archaeology of Pastoral Nomads in the Southern Negev Highlands in Late Antiquity. *Biblical Archaeologist.* Vol. 56:4.

Rossignol, J. 1992 Concepts, Methods, and Theory Building: a Landscape Approach. *Space, Time and Archaeological Landscapes.* Ed. by J. Rossignol and L. Wandsnider. pp 3-16. New York. Plencem Press.

Rouner, Leroy S.
1984 *On Nature.* Indiana. University of Notre Dame Press.

Rowe, J. S.
1992 The Ecosystem Approach to Forest Land Management. *Forestry Chronicle.* Vol 68:N01 pp 222-224.

Runciman, Steven.
1951-4 *A history of the Crusades.* 3 volumes. Reissue: Harmondsworth: Peregrine Books, 1965.

Rupp, D. N. et al. 1986 The Canadian Palaipaphos (Cyprus) Survey Project: Third Preliminary Report, 1983-1985. *Acta Archaeologica (Copenhagen).* 57 pp 27-45

Salem, Norma
1992 *Cyprus: A Regional Conflict and its Resolution.* Ottowa. Canadian Institute for International Peace and Security - St. Martins PRTLS?

Sallares, Robert
1991 *The Ecology of the Ancient Order World.* London. Duckworth.

Salleh, Ariel
1992 The Ecofeminism / Deep Ecology Debate: A Reply to Patmarchoal Reason. *Environmental Ethics* vol 14.

Samuels, Marwyn S.
1979 Cause and Culpability. *Interpretation of Ordinary Landscapes.* Ed. by D. W. Meinig. Oxford. Oxford University Press.

Sauer, Carl Ortwin
1965 *Land and Life.* Ed. John Leighly. Berkely University of California Press.

Schaeffer, Claude F.-A.
1983 *Corpus des cylindres-sceaux de Ras Shamra-Ugarit et d'Enkomi-Alasia.* Volume I. Paris: Editions Recherche sur les Civilisations.

Schaeffer, Claude F.-A.
1952 *Enkomi-Alasia I: Nouvelles missions en Chypre 1946-1950.* Paris:

Schoch, Werner H.
1988 Holzkohlenanalytische Untersuchengen des Grabungsplatzes Tamassos auf Zypern. *Ber.net.med. Verein Innsbruck.* Suppl. 2 pp 9-12.

Schoener, T. W.
1989 The Ecological Niche in *Ecological Concepts*. Edited by J. M. Cherrett. Oxford. Blackwell Scientific Publications.

Schugart, H. H. and Urban, D. L.
n.d. *Factors affecting the Relative Abundances of Forest Tree Species*.

Schule, Wilhelm.
1993 Mammals, Vegetation and the Initial Human Settlement on the Mediterranean Islands: A Palaeoecological Approach. *Journal of Bibliography*. 20 pp 399-412.

Schule, W. 1990 Landscape and Climate in Prehistory: Interaction of Wildlife, Man and Fire. *Fire in the Tropical Biota.* Ed. by J. G. Goldammer. pp 273-318. Heidelberg, Springer.

Schwartz, Jeffery H.
1973 The palaeozoology of Cyprus: a preliminary report on recently analysed sites. *World Archaeology* 5/2: 215-220.

Schwartz, Stephan A.
1978 *The secret vaults of time: psychic archaeology and the quest for man's beginnings*. New York: Grosset & Dunlap.

Scott, M. H.
1953 *Utilisation Notes on South African Timbers*. Union of South Africa. Department of Forestry.

Scott, Andrew C. and Jones, Timothy P.
1994 The Nature and Influence of Fire in Carboniferous Ecosystems. *Palaeo.* Vol 106:1-4.

Semple, E.C.
1931 *The Geography of the Mediterranean Region: its relation to ancient history*. New York: Holt.

Serephim, G.M.
1961 *Report on the Effect of Goat Grazing on the Forests of Cyprus.* Nicosia: Forest Department.

Serephim, G.M.
1962 *Forests, Forest Products and Forest Industries in Cyprus*. Nicosia: Forest Department.

Serephim, G.M.
1957 *The regeneration of the main conifers of the Southern Range forests of Cyprus.* Oxford: Imperial Forestry Institute.

Sessions, George.
1985 Ecological Consciousness and Paradigm Change. *Deep Ecology.* Ed. by Michael Fabias. San Diego. Avant Books.

Shanks, M. and Tilley, C.
1987 *Social Theory and Archaeology.* Cambridge. Polity Press.

Sharpe, D. M. et. al.
1987 Vegetation Dynamics in a Southern Wisconsin Agricultural Landscape. *Landscape Heterogenity and Disturbance.* Ed. Monica Goigel Turner. New York. Springer-Verlag.

Shawcross, W.
1972 Energy and Ecology: Thermodynamic Models in Archaeology. *Models in Archaeology.* Ed. D. L. Clarke. London. Methuen.

Shepard, Paul.
1969 Ecology and Man: a Viewpoint. *The Subversive Science: Essays Toward an Ecology of Man.* Ed. by Paul SHepard and Daniel McKinlecy.

Sherratt, Susan and Andrew.
1993 The Growth of the Mediterranean Economy in the Early First Millenium B.C. World Archaeology. Vol 24:3 361-378

Shmida, A. and Ellner, S.
1984 Coexistence of Plant Species with Similar Niches. *Vegetatio.* Vol 58 pp 29-55.

Shugart, Herman H.
A Theory of Forest Dynamics. New York. Springer-Verlag.

Sidebotham, Steven E.
1986 *Roman Economic Policy in the Erythra Thalassa 30 BC - AD 217.* E. J. Brill.

Silberman, Neil Asher.
1989 *Between past and present: archaeology, ideology and nationalism in the modern Middle East.* New York: Henry Holt.

Simmons, Alan H.
1991 Preliminary report of the interdisciplinary excavations of Akrotiri- Aetokremnos (Site E): 1987, 1988, 1990. *RDAC* 1991: pp 7-14.

Simon, H. A.
1962 The Architecture of Complexity. *Proceedings of the American Philosophical Society.* Vol 106 pp 467-482.

Sinos, S.
1990 *The Temple of Apollo Hylates at Kourion and the Restoration of it's south-west corner.* Athens: Leventis Foundation.

Sklar, Fred H. and Costanza, Robert
1991 The Development of Dynamic Special Models for Landscape Ecology: A Review and Prognosis. *Quantitative Methods in Landscape Ecology* . Ed. Monica G. Turner and Robert H. Gardner. New York. Springer-Verlag.

Smart, J. J. C.
1989 *Our Place in the Universe a metaphysical Discussion.* Oxford. Basil Blackwell.

Smith, W.S.
1965 *Interconnection in the Ancient Near East.* Newhart & London.

Smith, Peter.
1981 *Realism and the Progress of Science.* Cambridge. Cambridge University Press.

Snodgrass, Anthony M.
1987 *An Archaeology of Greece.* Berkeley.

University of California Press.
Sopher, Charles D.
1978 *Soils and Soil Management.*reston. Reston Publishing Co.
Sophocleous, Sophocles.
1985 *Atlas des représentations chypro-archaiques des divinités.* SIMA Pocket-book 33. Göteborg: Paul Åströms Förlag.
Soproni, Sándor.
1980 *The Archaeology of Roman Panneuia.* Ed. by A. Lengyel and G. T. B. Radan. Budapest. Akademiai Kiado.
Soren, David.
1987 *The Sanctuary of Apollo Hylates at Kourion, Cyprus.* Tucson: University of Arizona Press.
Sorensen, L. W. and Rupp, D. W. eds. 1993 The Land of the Paphian Aphrodite 2. *Studies in mediterranean Archaeology.* 104:2. Goteberg. P. Astroms Forlag.
Soule, M. E. Ed.
1987 *Viable Populations for Conservation.* Cambridge. Cambridge University Press.
Soule, H. E. ed.
1986 Patterns of Diversity and Rarity: Their Implications for Conservation. *Conservation Biology. The Science of Scarcity and Diversity.* pp 117-121. Sutherland, Mass. Sinaues Associates.
Spellberg, Ian F and Hardes, Steve.
1992 *Biological Conversation.* Cambridge. Cambridge University Press.
Sprugel, Douglas G.
1991 Disturbance, Equilibrium and Environmental Variability: What is 'Natural' Vegetation in a Changing Environment? *Biological Conservation* 58 pp 1-18
Spyridakis, C.
1964 *A Brief History of Cyprus.* Nicosia. Greek Communal Chamber.
Stager, Lawrence E., and Anita M. Walker, editors.
1989 *American Expedition to Idalion, Cyprus 1973-1980.* Oriental Institute Communications 24. Oriental Institute of the University of Chicago.
Stanley Price, N.P.
1979 *Early prehistoric settlement in Cyprus: a review and gazetteer of sites, c. 6500-3000 B.C.* Oxford: British Archaeological Reports, International Series 65.
Stanley Price, N.P.
1979 *Early prehistoric settlement in Cyprus: a review and gazetteer of sites, c. 6500-3000 B.C.* Oxford: British Archaeological Reports, International Series 65.
Stearn, W. T.
1976 From Theophrastus and Discorides to Sibthorp and Smith: the Background and Origin of the Flora Graeca. *Biological Journal of the Linnean Society.* Vol 8 pp 285-298.
Steffy, J. Richard.
1985 The Kyrenia Ship: an interim report on its hull construction. *American Journal of Archaeology.* Vol 89/1: pp 71-101.
Steverson, Brian K.
n.d. Ecocentrism and Ecological Modelling in *Environmental Ethics.* Vol. 16.
Stewart, James R.
1962 *The Early Cypriote Bronze Age.* Lund: [Gleerup].
Stewart, Robert T.
1974 Paleobotanic investigation: 1972 season. In *American Expedition to Idalion, Cyprus: first preliminary report: seasons of 1971 and 1972*, edited by Lawrence E. Stager, Anita M. Walker, and G.E. Wright, pp123-129. Supplement to the Bulletin of the American Schools of Oriental Research No. 18. Cambridge, Massachusetts.
Stockton, E. D.
1968 Pre-Neolithic remains at Kyrenia Cyprus. *Report of the Department of Antiquities, Cyprus.* pp 16-19
Storrs, Sir Ronald, and Bryan Justin O'Brien (editors). 1930 *The Handbook of Cyprus.* 9th issue. London: Christophers.
Strabo.
1929 *The Geography* in Eight Volumes. Trans H. L. Jones, from Latin. Loeb Classical Library. Great Britain.
Streets, R. J.
1962 *Exotic Forest Trees in the British Commonwealth.* Oxford. Clarendon.
Stylianou, Andreas.
1969 *The Old Castography of Cyprus.* ΠΡΑΚΤΙΚΑ ΤΟΥ ΠΡΟΤΟΨ ΔΙΕΟΝΟΨΣ ΚΨΠΡΟΛΟΓΙΚΟΨ ΣΨΝΕΔΡΙΟΨ. ΛΕΨΚ ΟΣΙΑ. ΚΨΠΡΙΚΩΝ ΕΠΟΨΔΟΝ ΕΤΑΙΡΕΙ Α.
Stylianou, A. & J.
1980 *The History of the Castography of Cyprus.* Nicosia. Publication of the Cyprus Research Centre. III.
Swiny, Stuart.
1989b Prehistoric Cyprus: a current perspective. *Biblical Archaeologist* 52: 178-189.
Swiny, Stuart.
1988 The Pleistocene fauna of Cyprus and recent discoveries on the Akrotiri Peninsula. *RDAC* 1988I: pp 1-14.
Swiny, Stuart.
1989a From round house to duplex: a re-assessment of prehistoric Cypriot Bronze Age society. *Early Society in Cyprus. Ed.*

Edgar Peltenburg, 14-31. Edinburgh University Press.

Swiny, Stuart.
1989b Prehistoric Cyprus: a current perspective. *Biblical Archaeologist* Vol 52: pp

Tapper, Richard L.
1988 Animality, humanity, morality, society. *What is an Animal?*, edited by T. Ingold. London: Unwin Hyman.

Theophrastus.
1916 *Historia Plantis* in Two Volumes. Trans Sir Arthur Hart, from Ancient Greek. Cambridge Mass: Howard University Press.

Theophrastus.
1916-1926 *Inquiry into Plants.* Trans. Sir Arthur Hart Vol 1 1916; Vol 2 1926. Cambridge.Mass. Harvard University Press.

Thirgood, J.V.
1981 *Man and the Mediterranean forest: a history of resource depletion.* London: Academic Press.

Thirgood, J.V.
1956 *Technical Pamphlet no. 19.* Nicosia: Cyprus Forest Department.

Thomas, J. W. et. al.
1993 *Viability Assessments and Management Considerations for Species Associated with Late-successional and Old-growth Forests of the Pacific Northwest.* Washington. USDA Forestry Service.

Tilley, Christopher
1993 *Interpretative Archaeology.* Berg. Proudence.

Tilly, Louise A.
1983 Food entitlement, famine and conflict. *Hunger and history: the impact of changing food production and consumption patterns on society.* Ed. Robert I. Rotberg and Theodore K. Rabb, p.135-151. Cambridge University Press.

Tivy, Joy.
1982 *Biogeography: a Study of Plants in the Ecosphere.* London. Longman.

Todd, I.A.
1987 *Excavations at Kalavasos-Tenta.* Vasilikos Valley Project 6. SIMA 71:6. Nicosia: Paul Aströms Förlag.

Torjesen, K.
1985 Hermaneutical Procedure and Theological Method. *Origin's Exegesis.* Berlin. Walter de Gruyter and Co.

Trabaud, L. and Lepart, J.
1981 Changes in the Floristic Composition of a *Quercus Coccifera* L. garrique in relation to Different Time Regimes. *Vegetatio.* Vol 46 pp 105-116.

Trigger, B. G.
1989 *A History of Archaeological Thought.* Cambridge. Cambridge University Press.

Trigger, B. G.
1978 *Time and Traditions: Essays in Archaeological Interpretation.* New York. Columbia University Press.

Tsoumis, George
1968 Wood as Raw Material. Oxford. Pergamon Press.

Tsoumis, G.
1991 *Science and Technology of Wood.* New York: Chapman & Hall.

Tuan, Yi-Fu.
1979 The Eye and the Mind's Eye. *The Interpretation of Ordinary Landscapes.* Ed. D. W. Meinig. Oxford. Oxford University Press.

Tubbs, C.R., and J.W. Blackwood.
1971 Evaluation of Land for Planning Purposes. *Biological Conservation* 3. UNESCO/FAO.

Tubbs, C.R., and J.W. Blackwood.
1970 *Ecological Study of the Mediterranean zone.* Paris.

Turner, Monica G. and Dale, Virginia H.
1991 Modelling Landscape Disturbance. *Quantitative Methods in Landscape Ecology.* Ed. Monica G. Turner and Robert H. Gardner. New York. Springer-Verlag.

Turner, Monica Goigel ed.
1987 *Landscape Heterogeneity and Disturbance.* New York. Springer-Verlag.

U. S. Department of the NAVY.
1945 *Wood: A Manual for its Use as a Ship Building Material.* Washington.

Van Zeist, W.
1981 *Plant remains from Cape Andreas Kastros.* [off-print]

Verwijst, T. and Cramer, W.
1986 Age Structure of Woody Species in Primary Succession on a Risong Bothnian Sea-shore. *Forest Dynamics Research in Western and Central Europe.* Ed. by J. Fanta. Pudoc. Wageningen.

Viktorov, S. V., Vostokovia, Ye. A. and Vyshivkin, D. D.
1964 *Short Guide to Geo-botanical Surveying.* Trans. J. M. Maclennan. Oxford. Pergamon Press.

Vinson, Steve.
1990 Ships in the Ancient Mediterranean. *Biblical Archaeologist.* March pp 13-18

Vita-Finzi, C.
1978 *Archaeological Sites in Their Setting.* London. Thames and Hudson.

Wagstaff, J. M.
Landscape and Culture: Geographical and Archaeological Perspectives. Oxford. Basil

Blackwell.

Wagstaff, J.M.
1988 The Tenta catchment resource exploitation in Neolithic Cyprus. *Conceptual issues in environmental archaeology*. Ed. Binford et al.

Waines, J.G., and N.P. Stanley Price.
1975-7 Plant remains from Khirokitia in Cyprus. *Paléorient* 3: 281-284.

Walker, D.
1989 Diversity and Stability in *Ecological Concepts*. Edited by J. M. Cherrbett. Oxford. Blackwell Scientific Publications.

Walker, Ernest P.
1964 *Mammals of the World.* Ballimore. The John Hopkins Press.

Wallace, Paul W. and Orphanides, Andreas.
1990 *Sources for the History of Cyprus.* Vol 1. University of Albany, State University of New York. Institute of Cypriot Studies.

Wallace, Paul and Orphanides, Andrease G.
1990 *Sources for the History of Cyprus.* Vol 1. Albany. State University of New York.

Wallace, Alfred.
1881 *Island Life.* New York. Harper & brothers.

Ward, W.A. 1971 *Egypt and the East Mediterranean World 2200-1900 BC.* Beirut: American University of Beirut.

Ward Haldane, C.
1990 Shipwrecked Plant Remains. *Biblical Archaeologist* March 1990 pp 56-60.

Ward Haldane, C.
1990 Shipwrecked Plant Remains. *Biblical Archaeologist* vol. 53 (1), March 1990: 55-60.

Waring, R. H.
1989 Ecosystems: Flukes of Matter and Energy. *Ecological Concepts*. Ed. J. M. Cherrett. Oxford. Blackwell Scientific Publications.

Warren, Karen.
1991 Ecological Feminism. *Hypatia*. Vol 6:1.

Waterer, R.R.
1949 *The problem of soil erosion and its control in Cyprus.* Misc. Pub. no. 140.

Waterer, R.R.
1943 *Report on land utilisation and conservation in Cyprus*. Misc. Pub. 120.

Waterer, R.R.
1947 *The place of forestry in land utilisation with special reference to soil and water conservation.* Nicosia: Government Printing Office.

Waterer, R.R.
1949 *The problem of soil erosion and its control in Cyprus.* Misc. Pub. no. 140.

Watson, J. S.
1951 *The Rat Problem in Cyprus.* London. His Majesty's Stationary Office.

Watt, Alex S.
1947 Pattern and Process in the Plant Community. *Journal of Ecology.* Vol 35:1&2 pp 1-22.

Welch, Bruce L.
1989 Nutriture Value of Shrubs. *The Biology and Utilisation of Shrubs*. Ed. Cyrus M. McKell. San Diego. Academic Press.

Wenger, Karl F
1984 *Forestry Handbook* 2nd Edition. New York. John Wiley and Sons.

Wertime, T.
1978 *Prime culprit: the pyrotechnological furnace or the goat?* Conference on deforestation, erosion and ecology, April 1978. Smithsonian.

Westbroek, Peter et. al.
1993 World Archaeology and Global Change: Did our Ancestors Ignite the Ice Age? *World Archaeology*. Vol 25

Whitakker, Robert J.
1992 Stochasticism and Determinism in Island Ecology. *Journal of Biogeography*. 19 pp. 587-591

Whitakker, C. R.
1988 *Pastoral Economics in Classical Antiquity.* Cambridge. The Cambridge Philological Society. pp 6-34.

Whitakker, R. H.
1972 *Evolution and the Measurement of Species Diversity.* Takon. Vol 21 pp 213-251.

White, Lyn Jr.
1970 The Historical roots of our Ecological Crisis. *The Environmental Handbook.* Ed. by G. DeBell. Ballontine Books, Inc.

Whyle, R.D.
1948 The fodder resources of Cyprus. Nicosia: Government Printing Office.

Wiegers, J.
1986 Succession in Fen-Woodlands in the Dutch Hap District. *Forest Dynamics Research in Western and Central Europe.* Ed. by J. Fanta. Waseningen, Pudoc.

Wild, A.E.
1879 *Report on the forests of the South and West of the Island of Cyprus.* London: Harrison & Sons.

Wild, J.P.
1970 Borrowed names for borrowed things? *Antiquity* 44: pp 125-130.

Wiley, G. R. 1991 Horizontal Integration and Regional Diversity: and Alternating Process in the Rise of Civilisation. *American Antiquity.* Vol 56 pp 197-215

Wilson, S. D. and Keddy, P. A.
1988 *Species Richness, Supervisorship and Biomass Accumulation along Environmental Gradient.* Oikos. Vol 53 pp

375-380.
Wittgenstein, Ludwig
1974 *Philosophical Investigations*. Trans. G.E.M. Anscombe. Oxford. Basil Blackwell.
Wittgenstein, Ludwig
1964 *The Blue and Brown Books*. Oxford. Basil Blackwell.
Wittig, Rudiger
1991 Patterns and Dynamics: The Example of the European Beech (Fagus sylvatica) Forests. *Responses of Forest Ecosystems to Environmental Change*. Ed. A. Leller, P. Mathy. D. J. N. R. Jeffers. London. Elsevier Applied Science.
Wolf, Edward C.
1987 (June) *On the Brink of Extinction: Conserving the Diversity of Life*. (World Watch Paper 78) Washington D.C. World Watch Institute.
Wolseley, Garnet.
1878 *Cyprus 1878: the journals of Sir Garnet Wolseley*, edited by Anne Cavendish. Nicosia: Cultural Centre of the Cyprus Popular Bank.
Wright, H.E., jr.
1955 Geological aspects of the archaeology of Iraq. *Sumer* 11/2: 83-91.
Yadin, Yigael.
1963 *The art of warfare in biblical lands*. 2 volumes. New York: McGraw-Hill.
Young, James A. et al.
1989 Use of Shrubs for Fuel in *The Biology and Utilisation of Shrubs*. Edited by Cyrus M. McKell. San Diego. Academic Press.
Zeist, Willem Van and Roller, Gerrit Jan De.
1994 The Plant Industry of Aceramic Cayouu, SE Turkey. *Prehistoria.* Vol 33/34 pp 65-96.
Zohary, M.
1978 *The Plant World.* Tel Avivd

Appendix 1: Makheras Forest Database

Transect	Quadrat	Date	Time	Visibility	Compartment Number	Access to the site	Geology	Topography	Slope	Elevation
1	1	29/09/1994	10:00	1	102	Kaphedes - prophet Elias monastery road	diabase	2	4	580
1	2	29/09/1994	12:10	1	103	Kaphedes - prophet Elias road	diabase	2	4	570
1	3	06/10/1994	08:15	4	107	Lithrodhondas - prophet Elias monastery road	diabase	4	4	600
1	4	06/10/1994	11:00	1	108	Lithrodhondas - prophet Elias monastery road	diabase	1	4	620
1	5	29/09/1994	16:55	0	109	Lythrodhondas - prophet Elias monastery road	diabase	2	5	840
1	6	06/10/1994	13:30	1	135	prophet Elias - Lythrodhondas rd. via Petromouti	diabase	4	4	580
1	7	09/10/1994	13:33	2	143	prophet Elias - Lyhtrodhondas rd. via Petromouti	diabase	2	4	460
1	8	28/10/1994	07:55	2	142	along ridge line N-S from Kyprovasa - prophet Elias road	diabase	2	5	510
1	9	28/10/1994	10:10	2	138	along transect line from Kyprovasa - prophet Elias road	diabase	2	4	705
1	10	28/10/1994	14:55	2	139		diabase	2	5	860
1	11	28/10/1994	13:55	3	140		diabase	2	4	850
2	12	16/10/1994	11:25	1	101	from Kaphedhes to Ayious Dhemitrious	diabase	2	3	580
2	13	16/10/1994	12:45	3	100	from Kaphedhes to Ayious Dhemitrious	diabase	2	4	560
2	14	29/10/1994	07:30	2	99	road from Kaphedes -> mandra tou Kambiou	diabase	1	5	590
2	15	29/10/1994	09:45	2	98		diabase	1	3	640
2	16	26/10/1994	12:50	1	104	forestry road going Lythrodhondas -> prophet Elias	diabase	2	3	770
2	17	29/10/1994	14:55	3	105	on westerly road bearing from forest road	diabase	4	3	710
2	18	30/10/1994	08:30	1	106	ridge line from prophet Elias monastery	diabase	2	5	890
2	19	30/10/1994	10:45	1	111	above Prophetis Elias mon. on transect line	diabse	4	3	830
2	20	30/10/1994	14:45	0	132		diabase	4	4	760
2	21	30/10/1994	15:40	3	133	bearing cross country on transect line	diabase	2	4	725
2	22	04/11/1994	12:40	0	145	Vavatsinia to Lythrodhondas road	diabase	2	5	790
2	23	04/11/1994	10:45	1	144	Vavatsinia to Lythrondas road	diabase	2	4	760
2	24	04/11/1994	08:40	2	141	forest road from Kyprouassa-Chiflik	diabase	2	4	830
3	25	02/11/1994	07:10	0	95	Kaphedes -> Makheras road.	diabase	2	4	740
3	26	02/11/1994	09:15	1	94	abandoned road just above junction	diabase	2	4	660
3	27	02/11/1994	10:35	2	93	on bearing from forestry road	diabase	4	3	660
3	28	02/11/1994	15:10	3	76	mule track from Prophetis Elias mon.	diabase	2	4	765
3	29	02/11/1994	16:15	3	114	forest road from Prophetis Elias	diabase	4	3	770
3	30	06/11/1994	10:25	1	115	forest road from 5 ways junction	diabase	2	4	800
3	31	06/11/1994	12:15	3	116	gully line from forestry road	diabase	1	5	760
3	32	06/11/1994	14:15	3	119	on bearing	diabase	2	4	765
3	33	07/11/1994	07:55	2	129	forest road from Kyprousa & Vausatsinh jnt.	diabase	2	5	765
3	34	07/11/1994	13:30	2	130	ridge line from forest road	diabase	2	5	890
3	35	07/11/1994	11:45	1	146	along ridge line to N->S bearing between BC 168 & 169	diabase	2	5	915
4	36	09/11/1994	08:05	3	88	forest road from Kambiou picnic area	diabase	2	4	665
4	37	09/11/1994	09:35	1	87	forest road from Kambiou picnic area	diabase	2	2	800
4	38	09/11/1994	10:55	1	86	bearing below picnic area at Kambiou	diabase	3	3	580
4	39	09/11/1994	16:15	3	90	forest road from Kambiou forest station	diabase	2	4	700
4	40	10/11/1994	07:40	2	91	Kambiou - Prophetis Elias road	diabase	2	4	640
4	41	10/11/1994	08:55	1	81	Kambiou - Prophetis Elias road	diabase	2	4	650
4	42	10/11/1994	11:10	0	74	on bearing up ridge line from Kambiou prof.elias	diabase	2	4	810
4	43	10/11/1994	15:15	3	75	mule path from Prophetis Elias monastery	diabase	2	4	960
4	44	10/11/1994	13:55	2	118	mule track from Prophetis Elias	diabase	2	4	960
4	45	13/11/1994	15:15	3	119	ridge line from yellow stone corner	gabbro	2	4	950
4	46	13/11/1994	12:35	1	120	forest road from yellowstone corner#	gabbro	2	4	960
4	47	13/11/1994	08:10	2	121	forest road from Prophetis Elias	diabase	2	5	840
4	48	13/11/1994	10:20	2	127	forest road from Prophetis Elias	diabase	1	4	820
5	49	16/11/1994	09:00	2	13.5	on bearing from ay.Onouphrious - mandra tou Kambiou road	diabase	2	5	780
5	50	16/11/1994	10:20	0	84	bearing from MK 13a	diabase	2	4	720
5	51	18/11/1994	13:00	0	83	forest road 3 from mandra to Kambiou	diabase	2	3	880
5	52	20/11/1994	08:30	2	82	on bearing from Skordokephallas	diabase	2	4	1000
5	53	18/11/1994	09:40	2	79	on bearing	diabase	2	4	1140
5	54	18/11/1994	11:45	0	77	forest road from mandra tou	diabase	2	4	960
5	55	20/11/1994	08:30	3	72	forest road from mandra tou Kambiou	diabase	2	4	860
5	56	28/11/1994	08:40	1	71	on bearing from forest road	diabase	2	5	800
5	57	28/11/1994	14:20	3	68	goat track above river from forest road	gabbro	2	5	820
5	58	28/11/1994	12:05	3	66	along old goat path parallel with river	gabbro	4	4	840
5	59	04/12/1994	15:00	3	59	track - Prophetis Elias to Kionia	diabase	2	4	1120
5	60	04/12/1994	14:00	2	124	track - Prophetis Elias to Kionia	diabase	2	4	1050
5	61	04/12/1994	10:50	2	125	along ridge line from Prophetis Elias - Makheras path	gabbro ?	2	5	1420
6	62	31/12/1994	08:30	2	6	ridgeline	diabase	2	5	740
6	63	31/12/1994	09:15	2	9	on bearing from forest road	diabase	2	4	740
6	64	31/12/1994	11:00	3	10	bearing from forest road	diabase	2	4	780
6	65	31/12/1994	12:20	3	11	on bearing from forest road	diabase	2	4	750
6	66	31/12/1994	13:45	3	16	on bearing from forest road	diabase	4	4	950
6	67	31/12/1994	16:00	0	15	Kambia - Makheras monastery road	diabase	2	4	1080
6	68	06/01/1995	07:30	2	14	Kapathes -> Makheras mon. road	diabase	2	3	880
6	69	06/01/1995	08:30	2	25	Kapathes -> Makheras mon. road	diabase	2	4	900
6	70	06/01/1995	10:00	2	24	Kapathes -> Makheras mon. road	diabase	2	3	1020
6	71	06/01/1995	11:15	2	23	Kapathes -> Makheras mon. road	diabase	2	5	940
6	72	06/01/1995	12:45	2	22	Kapathes -> Makheras mon. road (walked)	diabase	2	4	960
6	73	06/01/1995	14:50	2	29	on bearing from forest road	diabase	2	4	1060
6	74	08/01/1995	19:50	2	30	:= 300m from upper ridge road	diabase	2	3	1220
6	75	08/01/1995	08:40	2	63	up slope of quad 74 and over ridge top	diabase	2	3	1240
6	76	08/01/1995	11:05	2	64	road leading to forest tower (2nd peak east) := 500m e off road	diabase	6	5	1320
6	77	08/01/1995	12:05	2	61	by footpath from Kaphedes - Kionia road	diabase	2	3	1300
6	78	15/01/1995	08:25	3	57	path to nearby spring	diabase	2	4	1360
6	79	15/01/1995	10:05	2	56	path to spring	diabase	2	5	1410
6	80	15/01/1995	11:50	4	55	along ridge line from path	diabase	2	4	1490
7	81	05/02/1995	08:40	2	1	forest road & ridgeline	diabase	2	4	1120
7	82	05/02/1995	11:15	3	2	old forest road	diabase	2	5	940
7	83	05/02/1995	14:45	4	3	forest road	diabase	2	4	880
7	84	05/02/1995	15:45	3	4	forest road	diabase	2	4	870
7	85	15/01/1995	15:05	2	20	on bearing from Makheras mon. -> Philini road	diabase	2	5	860
7	86	22/01/1995	08:30	4	21	forest road to Kionia	diabase	2	4	910
7	87	22/01/1995	10:30	3	38	forest road	diabase	2	3	880
7	88	22/01/1995	12:30	4	36	forest road	diabase	2	5	900
7	89	22/01/1995	13:40	4	34	forest road to Kionia	diabase	2	5	1080
7	90	29/01/1995	08:00	2	62	public road to Kionia	diabase	2	3	1440
7	91	29/01/1995	09:15	1	52	on bearing from MK 62	diabase	2	3	1430
7	92	29/01/1995	12:20	3	51	on bearing from 52	diabase	2	3	1360
8	93	05/02/1995	17:05	3	5	Lazania -> politiko road	diabase	2	4	1140
8	94	12/02/1995	08:00	2	41	Makheras mon -> Lazania	diabase	2	5	740
8	95	12/02/1995	08:55	1	40	Makheras mon -> Lazania	diabase	2	4	730
8	96	12/02/1995	12:30	2	49	by eroded track from Makheras mon -> Lazania road	diabase	2	5	970
8	97	12/02/1995	14:55	3	46	on bearing from MK 49	diabase	2	4	930
8	98	12/02/1995	17:05	3	45	along ridgeline	diabase	4	4	1260

Appendix 1: Makheras Forest Database

Aspect	Wind Direction	Catastophic Weather	Climatic Features	Soil Type	Soil Fertility	Soil Depth	Soil Moisture	Soil Temperature
NW	NE	F	2	brown	3	3	4	0
W		F	2	brown	2	2	4	0
NE	N	F	2	red	2	3	4	0
N-S	NE	F	2	red	1	1	2	0
W	S	F	2	brown	2	3	4	0
W	S	F	2	brown, orangey	2	2	3	0
N	N	F	1	brown earth	1	2	3	0
E		F	2	brown loam	1	2	2	0
W		F	2	brown	1	1	2	0
N		F	1	brown loam	2	1	1	0
S		F	1	red loam	2	2	2	0
E		F	1	brown	3	3	4	0
SE	N	F	1	brown	3	3	4	0
NNE		F	2	brown loam	2	1	2	0
SSW		F	2	brown loam	2	1	2	0
W	W	F	2	brown loam	2	0	2	0
N		F	2	brown (sandy?)	2	2	3	0
NE		F	2	red	0	3	2	0
SSE		F	2	yellow clay	2	2	3	0
E		F	1	red	3	3	3	0
SSW		F	2	brown loam	2	2	2	0
NE		F	1		2	3	2	0
E	NE	F	2	red gravel/sand	3	3	0	0
NNW	NNW	F	2	brown	2	3	2	0
NW		F	2	brown	2	2	2	0
SE		F	2	brown	2	2	2	0
N-S		F	2	brown clay	2	2	2	0
NNE		F	2		3	0	0	0
		F	2	brown clay	2	2	3	0
SSW	SW	F	1	orange sand, gravel	2	3	3	0
S	SW	F	3	red-brown loam, sandy with some gravel	1	2	2	0
E	NE	F	3	brown, sandy	2	3	2	0
NE	WSW	F	2	brown	2	3	3	0
N	NNW	F	2	brown	2	2	2	0
N	N	F	2		2	3	2	0
NW	E	F	2	brown	2	2	2	0
E	E	F	2		2	2	3	0
E	E	F	2	brown	2	3	2	0
W	---	F	2	brown	2	3	2	0
S	---	F	2	brown	2	3	3	0
N	NE	F	2	brown	2	3	3	0
SW	N	F	2	brown	3	3	4	0
NNE	---	F	2	bronn	2	3	3	0
E	SSE	F	1	grey	1	1	2	0
NE	SE	F	1	dark brown	2	1	2	0
SSE	E	F	2	red - sandy orange	2	1	2	0
S	SW	F	2	brown, gravelly	2	3	3	0
N-S	NE	F	2	brown	1	0	1	0
W	NW	F	2	brown loam	1	1	2	0
SW	---	F	1	brown loam	1	1	2	0
N-S	SE	F	3	brown	2	2	2	0
NE	NE	F	2	brown	2	2	2	0
SE	E	F	1	brown	2	2	3	0
N-S	N	F	1	brown	3	3	3	0
SE	S	F	1	brown	2	1	2	0
SE	NE	F	2	brown	2	2	2	0
E	N	F	2	brown ?	3	3	2	0
N-S	N	F	1	brown	2	2	2	0
		T	2	brown loam	1	2	2	0
SE		T	2	brown	2	2	2	0
NE	N	F	2	brown loam	2	3	3	0
SE	NE	F	2	none	3	4	4	0
N	NE	F	2	red loam	2	3	2	0
SW	E	F	1	brown loam	1	2	2	0
N-S		F	1	brown loam	2	2	3	0
W	N	F	1	brown	2	2	3	0
SSE	NW	F	1	red loam	2	1	3	0
S	NW	F	1	brown	2	1	2	0
NE	N	F	1	brown loam	1	1	2	0
SE	N	F	1	brown loam	1	1	2	0
N	S	F	1	brown	3	3	2	0
SE	SW	F	1	red	2	3	3	0
NW		F	2	brown loam	1	1	2	0
NW	N	F	2	brown	3	3	2	0
SE	N	F	2	brown	1	1	2	0
E		F	2	brown	2	3	2	0
SSE		F	1	brown loam	2	1	2	0
N		F	2	grey scree	2	3	2	0
NNE		F	2	brown loam	2	3	2	0
S		F	1	brown	2	2	3	0
E		F	1	brown	2	2	2	0
NE		F	1	brown clay	3	3	2	0
S		F	3	brown loam	2	2	3	0
S		F	3	brown	2	2	3	0
NNW		F	2	grey sand	2	3	3	0
SSE		F	1	red loam	1	1	3	0
E		F	1	red	1	1	2	0
E		F	3	brown	2	1	2	0
E		F	3	brown	3	3	2	0
NNE	SE	F	2	dark brown. (clay?)	1	1	1	0
E	WNW	F	2	dark brown	2	2	2	0
S		F	1	brown	2	2	3	0
SE		F	2	red loam	2	2	3	0
N		F	2	brown	2	3	2	0
E	E	F	2	brown	2	2	2	0
W	N	F	2	brown	3	3	2	0
NE	NW	F	2	brown	1	2	2	0
E-W	NE	F	2	brown	3	3	3	0

Appendix 1: Makheras Forest Database

Soil Conductivity	Soil Structure	Drainage	Snow	Riparian Buffer Vegetation Present	Riparian Vegetation Density	Potential Forest Type	Site Class	Seed Bed Condition
0	1	3	0	T	1	2	3	2
0.5	1	3	0	T	1	2	3	1
5	1	0	0	F	0	5	4	3
1	1	2	0	T	2	5	2	1
3	1	3	0	T	0	2	4	3
5	1	0	0	F	0	2	3	2
	0	3	0	T	1	1	1	3
7	1	2	0	F	0	1	2	1
6.5	1	3	0	F	0	1	2	1
8	1	4	0	F	0	4	2	2
10	1	0	0	F	0	5	3	2
1	1	3	0	F	0	1	3	2
1.25	1	3	0	F	0	2	3	3
8	2	3	0	F	0	2	3	3
6	2	3	0	F	0	2	2	2
6	1	0	0	F	0	2	3	2
7	1	3	0	F	0	2	3	2
7	1	3	0	T	2	2	4	3
8	1	5	0	F	0	2	2	3
8	1	5	0	F	0	2	3	2
0	1	3	0	F	0	2	2	2
0	1	3	0	T	2	2	2	2
9	1	3	0	F	0	2	3	3
10	1	4	0	F	0	2	3	2
6	1	3	0	F	0	2	3	2
9	1	3	0	F	1	2	2	2
8	1	5	0	F	0	2	3	2
0	1	3	0	F	0	2	3	3
0	0	5	0	F	0	2	3	1
7	1	5	0	F	0	2	3	3
7	2	2	0	T	2	2	1	2
6	1	5	0	F	0	2	2	2
7	1	3	0	T	2	2	2	2
2	1	5	0	F	0	2	2	2
2	1	5	0	F	0	1	3	2
8	1	3	0	F	0	2	3	2
4	1	5	0	F	0	2	2	1
6	1	3	0	F	0	2	2	1
4	1	2	0	F	0	2	3	2
8	1	3	0	F	0	0	3	2
2	1	5	0	F	0	0	3	3
9	1	5	0	F	0	1	4	3
6	1	3	0	F	0	1	3	2
4	1	3	0	F	0	0	2	1
3	4	4	0	F	0	2	3	2
5	2	5	0	F	0	2	3	2
4	1	4	0	F	0	2	3	3
1	1	2	0	T	0	1	1	1
3	1	0	0	F	0	1	2	2
9.5	1	5	0	F	0	1	2	1
0	1	5	0	F	0	2	2	2
4	1	5	0	F	0	1	3	3
4	1	5	0	F	0	2	3	2
4	1	5	0	F	0	2	3	3
8	1	4	0	F	0	2	2	2
3	1	5	0	F	0	1	3	3
3	1	5	0	F	0	1	4	3
7	1	5	0	F	0	2	2	2
0	1	5	4	F	0	0	2	2
0	1	5	0	F	0	1	3	1
0	1	4	4	F	0	1	4	3
0	3	5		F	0	5	4	3
2	1	5	0	F	0	2	3	2
4	1	5	0	F	0	2	3	2
8	1	3	0	F	0	2	3	3
8	1	5	0	F	0	2	3	3
8	1	5	0	F	0	2	4	3
0	1	5	0	F	0	2	3	2
0	1	3	0	F	0	2	2	2
0	1	5	0	F	0	2	2	2
0	1	5	0	F	0	1	4	3
0	1	5	0	F	0	1	4	2
0	1	5	0	F	0	2	2	2
6	1	5	0	F	0	2	4	3
5	1	5	0	F	0	2	2	1
0	0	5	0	F	0	1	4	3
5	1	5	0	F	0	2	3	2
0	1	3	0	F	2	2	3	2
0	1	3	0	T	1	2	3	2
0	1	5	0	F	0	2	4	2
0	1	5	0	F	0	2	3	3
0	4	5	0	F	0	2	4	3
0	1	5	0	F	0	2	4	2
0	1	2	0	T	2	2	3	3
0	2	5	0	F	0	2	3	3
0	1	5	0	F	0	2	3	1
0	1	5	0	T	0	2	2	1
0	1	5	0	F	0	2	2	2
0	1	5	0	F	0	2	4	3
6	4	5	0	F	0	2	2	1
6	1	5	0	F	0	2	3	2
0	1	5	0	F	0	2	3	2
0	1	5	0	F	0	2	4	3
4	1	0	0	F	0	1	4	3
8.5	1	0	0	F	0	2	3	2
6.5	1	5	0	F	0	0	4	3
6.5	1	0	0	F	0	2	2	1
6	1	0	0	F	0	2	4	3

Appendix 1: Makheras Forest Database

Understorey Vegetation	Ground Litter Depth	Stand Composition	Ratio of Tolerant Species	Dominant Age	Stand Structure	Canopy Cover	Crown Condition	Flora Associations
2	3	1	1	6	2	2	2	4
2	2	1	1	6	2	2	3	4
2	2	5	2	1	1	1	0	4
1	2	5	2	1	2	1	2	1
3	2	1	1	4	1	1	3	4
1	1	5	1	4	1	1	3	4
2	1	5	2	2	2	1	2	5
2	1	5	2	1	2	1	2	5
2	1	5	2	4	2	1	2	5
1	1	5	3	1	2	1	2	5
2	3	1	1	1	1	1	1	4
3	3	1	1	6	1	1	2	4
3	3	1	1	4	1	1	2	6
3	0	5	1	6	1	1	2	4
2	2	5	2	4	1	1	2	1
3	1	5	2	4	1	1	3	1
2	0	5	2	4	1	2	0	5
0	2	5	3	3	1	1	3	5
2	2	5	1	3	1	1	2	5
3	0	5	1	0	1	1	3	5
3	2	5	1	3	1	1	2	5
1	2	5	3	3	2	2	2	5
2	0	5	1	1	1	1	2	4
1	1	5	2	1	2	1	2	4
3	3	1	1	4	1	2	2	4
1	0	5	2	4	2	2	2	5
3	2	5	1	4	1	1	2	0
2	1	5	3	4	2	1	2	1
2	3	5	1	4	1	1	2	4
3	2	5	2	3	1	1	2	4
2	1	5	2	4	2	1	2	4
3	2	5	1	6	2	2	2	4
1	1	5	2	6	2	2	2	5
2	1	5	2	4	2	1	2	5
1	1	4	3	6	2	3	3	5
2	0	5	1	6	2	3	2	1
2	1	5	1	1	1	3	2	4
1	2	5	1	2	1	3	2	4
2	2	5	1	4	1	1	2	4
2	3	5	1	4	1	1	2	4
3	3	5	3	3	1	1	2	4
3	3	5	3	4	1	1	3	4
2	3	5	3	3	2	1	2	1
2	1	5	2	6	2	2	2	5
2	1	4	3	3	2	1	2	5
2	2	5	3	3	2	1	2	4
2	3	5	3	4	2	1	2	5
1	1	5	3	3	2	3	2	5
1	1	5	2	6	2	2	2	1
1	1	5	3	3	2	1	2	1
2	2	5	2	2	2	1	2	4
3	3	5	2	5	2	1	3	1
2	1	5	1	3	2	1	2	4
2	2	5	1	6	2	2	3	4
2	1	5	1	4	2	2	2	4
2	2	5	2	6	2	1	2	5
3	3	5	2	4	2	1	2	4
1	1	5	3	4	2	1	2	5
1	1	5	2	6	2	2	0	5
1	1	5	3	6	2	3	3	5
2	3	5	3	6	2	1	3	5
4	4	0	1	0	1	1	0	0
2	2	5	2	4	2	1	2	4
2	1	5	1	4	2	2	1	4
2	2	5	3	3	1	1	2	4
2	0	5	2	4	2	1	2	4
2	2	5	2	4	1	1	2	4
2	2	5	2	6	2	1	2	4
2	1	5	1	4	2	1	2	4
2	2	5	1	4	2	2	2	4
3	3	5	1	5	1	1	3	4
2	3	4	3	1	1	1	2	4
1	1	5	1	6	2	3	2	4
1	3	5	3	4	2	2	3	5
1	2	5	3	6	2	2	2	5
1	3	5	3	4	2	2	3	5
1	2	5	2	6	2	3	2	5
1	3	5	3	4	2	3	3	5
1	2	5	3	4	2	2	3	5
1	2	5	3	4	2	2	3	5
2	2	5	3	4	2	1	2	5
3	3	5	3	4	2	1	2	5
2	2	4	1	1	2	1	2	4
2	2	5	3	1	2	1	2	4
2	3	5	2	4	2	1	3	4
3	3	5	1	1	1	1	2	4
1	2	5	2	1	2	1	2	4
2	3	5	2	4	2	2	2	5
3	2	5	2	4	2	1	2	0
1	1	5	3	4	2	3	3	5
2	1	5	2	4	2	2	3	5
2	1	5	3	4	2	2	2	5
0	0	5	1	1	2	1	2	4
3	3	5	3	4	2	1	2	5
2	0	5	3	4	2	1	2	5
3	3	4	3	4	2	1	2	5
1	2	5	3	4	2	3	2	5
3	3	5	0	0	0	0	0	0

Appendix 1: Makheras Forest Database

Notable injuries	Management Techniques	Stage of Ecological Succession	Site Disturbance
grazing, erosion	1	4	
	1	4	habitation, abandoned, terracing
wild fire	1	2	mechanical reafforestation
wild fire	1	4	mechanical reafforestation
wild fire	1	2	
	1	4	
clear felling	1	4	
clear felling	1	4	
wild fire	1	4	
wild fire	1	4	
wild fire	1	4	mechanical reafforestation
grazing, erosion, clear felling	1	4	grazing
grazing, erosion	1	4	
clear felling	1	4	
	1	4	abandoned road
wild fire?. grazing?	1	4	terracing
grazing (past)	1	4	
erosion	4	4	
	2	4	terracing
clear felling?	1	4	
clear felling	1	4	
clear felling	1	4	
wild fire, clear felling	1	4	
wild fire, grazing	1	3	
grazing	1	4	
storm damage	1	4	
grazing, clear felling	1	4	
clear felling, storm damage	1	4	
clear felling	1	4	road
clear felling	1	4	road
clear felling	1	4	
clear felling	1	4	
clear felling	1	4	
wild fire? clear felling	1	4	
wild fire	4	3	
	1	4	
	1	4	mechanical reafforestation
	1	4	mechanical reafforestation
	1	4	mechanical reafforestation
	1	4	
grazing?	1	3	
	1	3	
	1	3	
	1	4	
clear felling	1	3	
wild fire? clear felling	1	3	terracing
clear felling	1	4	
storm damage?	1	4	
grazing	1	4	
clear felling	1	4	
clear felling	1	4	mechanical reafforestation?
storm damage	1	4	
clear felling, storm damage	1	4	
storm damage	1	4	
storm damage	1	4	
storm damage	1	4	
storm damage	1	4	
	1	4	old path
storm damage	1	4	
storm damage	1	4	
storm damage	1	4	
mining	5	0	mining
clear felling	1	4	
storm damage	1	4	terracing
wild fire?	1	3	
	1	3	abandoned, habitation
	1	3	mechanical reafforestation
clear felling, storm damage	1	4	mechanical reafforestation
clear felling, storm damage	1	3	terracing
clear felling	1	3	terracing, road
storm damage	1	3	
wild fire	1	2	
storm damage	1	4	
wild fire?, erosion, storm damage	1	4	
	1	4	terracing (abandoned?)
storm damage	1	4	
	0	3	
storm damage, erosion	1	4	
storm damage	1	4	terracing
storm damage	1	4	
wild fire	1	3	
wild fire, stom damage	1	4	road
wild fire	1	4	machanical reafforestation, habitation, abandoned, terracing
wild fire, erosion	1	4	mechanical reafforestation, habitation, abandoned, terracing
storm damage	1	4	path
wild fire	1	4	mechanical reafforestation
wild fire	1	4	mechanical reafforestation
wild fire	1	4	mechanical reafforestation
erosion, storm damage	1	4	
	1	4	
	1	4	abandoned walking path
disease?	1	4	terracing
wild fire	1	4	mechanical reafforestation
storm damage	1	4	high tension transmission lines
clear felling	1	4	abandoned, terracing
wild fire, clear felling	1	4	
resin tapping	1	4	
	0	0	

Appendix 1: Makheras Forest Database

Anthropogenic Evidence	Biotic Associations.	Fauna Species Identified													Sum of Flora Species
	birds, goats/sheep, lizard, snake	9	1	6	62	18	19	20	42	46	49				10
hunting debris	rabbit burrow - abandoned	21	9	1	18	22	73	0	0						8
garbage, hunting debris	birds, snake-skin collected, small rodents	9	4	6	23	22	20	52	73						8
hunting debris	birds, small rodents	4	73												2
	rodents	9	1	21	73										4
	birds, rodents	1	6	4	73										4
	birds, rodents, butterflies	1	6	4	74	73	65								6
garbage: silver foil partially buried	birds, rodents, fox	2	1	26	6	74	31	27	11						8
	birds, hares, fox, rodents	1	9	2	7	74	31	42							7
hunting debris	mouflon? birds, hares?	23	7	2	5										4
	birds, hares?, fox, goat-montlon spoar	1	31	42											3
hunting evidence	birds, goats	9	1	7	10	42									5
hunting debris	birds, goats, lizards	1	9	18	42	48	53								6
garbage-tinfoil, hunting debris	birds	9	1	2	7										4
	birds, butterflies	1	9	2	7	62	73	65							7
	birds, other (unspecified!)	1	2	11	73	8	31								6
	birds, small viper	5	6	1	2	24	45	73							7
	birds, hares, rodents	1	7	2	73										4
garbage, landscape modification, hunting debris	birds	1	2	7	6	31	42	52	59						8
	birds	1	9	2	7	6	11	24	76	0					9
	birds	5	53	2	7										4
hunting debris	birds	2	7	9	30	73	74								6
	birds, fox, rodents	2	1	6	7	4	11	74	73	31					9
goat dumps	birds	2	1	7	74	42	31								6
garbage, industrial waste	birds, goats	1	7	9	10	11	31	42	55	74					9
	birds, fox, (other:fox adjacent)	1	2	7	9	6	74	10	31						8
	birds, fox?	2	1	9	7	12	10	4	42	0					9
	birds	2	1	6	23	4	73								6
landscape modification?	birds	2	1	6	7										4
hunting debris	birds, snake skin	2	1	7	6	52									5
hunting debris	birds	2	1	6	7	5	73	74	54						8
		1	5	32	74	73									5
	birds	2	1	5	7	24	71	74	73	30					9
	birds	2	1	5	6	53	31	74							7
	birds, hares, fox	2	1	5	6	31	32								6
	birds, goats	1	2	7	5	73	42								6
landscape modification	birds, goat	1	2	7	4	10	42								6
garbage, landscape modification	birds, fox?, rodents	2	1	4	5	48	73	53	51						8
	birds	2	1	7	10	31	54								6
	birds, hares	2	7	5	1	4	32	63							7
	birds, hares, goats?	2	1	6	5	42	32								6
hunting debris	birds	2	7	4	1	6	31	53							7
pottery (adjacent), hunting debris	birds, (other unspecified!)	2	1	6	4	30	31								6
	birds, hares, rodents	1	2	6	4	32	30	73	74						8
hunting debris	rodents	7	1	2	6	4	74	30	73						8
	birds, rodents	2	1	5	7	73	30	63	67	48					9
	birds, fox	2	6	5	7	4	1	73	74						8
garbage	birds	2	1	6	5										4
	birds, rodents, hares	6	2	4	73	53									5
	birds, rodents	2	6	73	74	67	53								6
garbage	birds	2	54	74											3
hunting debris	birds	2	6	4	73	53									5
	birds, hares, fox, rodents	2	6	4	31	32	73								6
hunting debris	birds, hares, rodents	2	4	6	74	57									5
	birds	2	4	51	30	63									5
	birds, rodents, small brown snake	2	13	48	67	74									5
		2	4												2
pottery, garbage, hunting debris	birds, rodents, small black spider	2	6	4	74	73	57								6
hunting debris	birds. (other unspecified)	6	31	62	74	30									5
hunting debris	birds, rodents	7	2	5	6	4	74	30	73	31					9
	birds, hares	7	4	10	32										4
landscape modification															
garbage nearby	birds	6	1	7	2	5	3	54	73	30	70				10
garbage	birds	1	2	7	6	5	42	4	54	31	53				10
	birds	6	2	1	8	74									5
pottery, hunting debris, masonary?	birds	25	5	1	6	2									5
landscape modification	birds	16	6	5	7	73	30								6
garbage, hunting debris	birds, rodents	7	2	1	5	6	73	4	54	53	55				10
garbage	birds, hares, rodent	1	9	2	5	6	32	30	73	54	57	53			11
	birds, rodent	1	5	4	30	70	45	73	48	56					9
		1	2	5	7	54	30	74							7
hunting debris	birds, hares, fox, rodent	2	9	5	1	4	30	72	53	66					9
	birds, rodents	1	7	2	6	4	74	30							7
	birds	1	9	7	5	6									5
	birds, fox, rodents	2	7	1	31	30	74	73	56	53					9
	birds	1	5	7	4	73	31	54							7
	birds, fox, rodents	1	9	2	5	4	6	7	73	30	53				10
hunting debris	birds, rodent	1	2	6	7	74	33								6
landscape modification	birds	6	1	12	2	73									5
	rodents	1	5	30	74	73									5
hunting debris	birds, hares, rodents	2	1	3	74	54	73	32	31	4					9
	birds, hares, rodents	2	5	6	32	31	74	58							7
pottery, hunting debris	birds, hares, fox, (other unspecified!)	2	1												2
pottery, hunting debris	birds	5	9	2	7	66									5
	birds, rodents	2	5	7	11	74	30	73							7
pottery	birds, fox, rodents	2	1	7	29	6	73	30							7
pottery, slag?, hunting debris	birds, rodents	2	1	10	3	73	30	53	60						8
	birds, rodents	1	2	32	73	30	74								6
hunting debris	birds, rodents	6	1	7	73	75									5
garbage, hunting debris	birds, rodents, large mammal (hare or fox)	6	10	1	2	7	9	73	30						8
	birds, rodent	2	6	10	5	1	30	32	48						8
hunting debris	birds, hares, rodents	2	5	7	6	1	30	59	73	74	32	48	61	58	13
	birds, hares	32	7												2
	birds, hares?	6	2	1	7	57	73								6
	birds, rodents, hares	2	6	1	11	10	30	48	53	61	64	66			11
	birds	5	1	2	6	7	55	58	66						8
	birds, rodents	1	6	73	74	65									5
		7	4	6	73	30									5

Appendix 1: Makheras Forest Database

DBH																						Basal Area	Count of Measured Trees
24.83	20.05	30.88	23.24	42.34	31.83	27.37	18.14	20.37														239.05	9
50.93	30.88	28.65	23.87	22.92	20.69	33.42		22.92	24.51													258.79	9
23.87																						23.87	1
																						0	0
47.75	41.38	41.38	36.61	23.87	31.83																	222.82	6
44.56	38.2	30.88	31.19	18.3																		163.13	5
22.28	39.79	49.97	41.38	29.28	27.06																	209.76	6
28.97	23.87	48.7	36.61																			138.15	4
39.79	40.43	44.88	41.06	41.06	27.37																	234.59	6
36.61	59.21																					95.82	2
																						0	0
41.38		48.38	16.71	19.1	35.01	22.28	30.88	25.15														238.89	8
33.74	26.74	33.42	35.65	35.01	28.01	34.06	27.37	26.42	42.02													322.44	10
	17.83	17.83	26.1	29.28	28.01	10.98	21.01	19.1														170.14	8
51.25	23.55	33.1	20.37	18.14	20.05	20.69	17.83	22.28	22.92	11.3												261.48	11
39.15	35.65	47.11	27.69	38.2		28.97	37.56															254.33	7
		28.01		42.02	44.56	33.74																148.33	4
27.06	38.52																					65.58	2
40.11	44.88	37.24	38.2	38.2	32.79	32.47	33.1	29.28	26.1													352.37	10
28.33	36.29	32.79	30.24	43.29	36.61	41.06	35.65	37.56	27.37	32.15	31.19											412.53	12
36.61	17.83	41.7	38.2	40.11	29.92	35.65	27.69															267.71	8
42.65	45.2	35.97	51.25																			175.07	4
	31.51	32.47	27.37	36.29	23.55																	151.19	5
48.06	45.52	43.93	40.74		35.65	39.79																253.69	6
	29.92	47.75	51.25	46.79	36.29	30.24	48.38	39.15	24.51	35.33		35.97	31.19	38.83		39.79						535.39	14
29.28	48.7	22.28	34.7	35.65		36.29	17.83	28.65	19.1	30.56												303.04	10
			34.06			37.24	34.38															105.68	3
																						0	0
57.61	43.29	17.83	39.79	26.74	25.46	17.83	23.24															251.79	8
	28.97	30.88	17.83	21.65	32.79	37.24	27.37	41.38	35.65	33.42	21.01	25.46										353.65	12
	43.29																					43.29	1
30.56	48.06	49.38	45.84	49.02	39.47	28.97	46.79															338.09	8
38.52	30.56	31.83	35.01	28.65	19.74	29.28	32.79	21.01														267.39	9
	28.65	35.33	28.01		39.15	37.56	36.61	33.42														238.73	7
27.37																						27.37	1
	42.02	40.11		31.19	46.15	30.88	25.15	42.02	30.56	47.11	43.29	38.2										416.68	11
43.61																						43.61	1
24.51																						24.51	1
29.92	37.24		47.11																			114.27	3
39.79	37.24	33.74	41.7																			152.47	4
47.11	21.96																					69.07	2
32.47	54.43																					86.9	2
73.21																						73.21	1
60.48	57.93	38.2	60.48		55.07	43.61	52.84	23.87														392.48	8
63.03																						63.03	1
	21.01	17.83	36.61																			75.45	3
50.61																						50.61	1
39.79	60.8																					100.59	2
29.28	19.74		44.25	59.21	40.11																	192.59	5
	36.61	24.19																				60.8	2
25.15	52.84	31.83	36.61	44.88	31.19																	222.5	6
		53.48																				53.48	1
84.7	21.65	31.51	23.87	28.01	70.73																	260.47	6
28.33	42.97	37.56	40.74	28.65	30.88	25.46		40.74	33.74	39.15	39.79		44.56									432.57	12
67.48	40.11	30.88	36.92	52.52	24.19	26.42	39.47		39.79	36.61	28.65											423.04	11
45.52	49.34	30.88	39.79	31.19																		196.72	5
35.97	60.8	31.51	33.42																			161.7	4
	40.43	49.34	24.19	52.2	48.06																	214.22	5
31.19	31.51	33.74	27.37	34.7	27.37	22.28	36.29	35.65	37.56	26.74	28.33											372.73	12
		44.25	26.42	37.88																		108.55	3
33.1	28.97																					62.07	2
																						0	
38.83	37.24	40.11																				116.18	3
	22.92	16.55	21.96	38.2		31.19	48.7		52.84		36.29	51.25	28.97									348.87	10
78.3	27.37																					105.67	2
67.48																						67.48	1
46.15	30.56	46.15	44.56	48.06																		215.48	5
76.39	29.28																					105.67	2
52.52	23.87	36.92	36.29	23.24	22.6	31.19	30.88	21.8	31.51	26.58	26.74	19.58	34.38	32.47	45.2	21.33	27.06	28.97	28.33	28.33	34.38	664.17	22
42.34	34.38	35.65		29.6	47.43	86.26	56.02		36.92													368.6	8
84.03	18.46	63.66	76.71	46.47																		289.33	5
																						0	0
79.58	25.78	20.05		23.55	45.2	35.97	38.2	31.19	17.83	23.55	27.69	33.42	30.56	32.15	25.46	18.14	43.61	43.93	36.29			632.15	19
47.11	33.74	22.6																				103.45	3
36.61	34.06	39.15	16.23	17.19	49.97	43.29																236.5	7
34.38																						34.38	1
42.97	55.39	43.93	32.47	29.6																		204.36	5
39.47																						39.47	1
49.02	68.44																					117.46	2
34.06	33.74	47.75	35.01	41.38																		191.94	5
38.2	76.71	38.2	36.61	38.52																		228.24	5
29.6	44.56																					74.16	2
																						0	0
																						0	0
39.47	64.94	38.2	43.29	46.47	57.3	23.87	53.16															366.7	8
																						0	0
																						0	0
																						0	0
																						0	0
38.83	46.15	35.65	40.74	39.15	30.88	39.15	35.33	25.78	35.01	31.83	40.11											438.61	12
27.06	27.06	29.92	29.6	35.78	31.19	29.6	35.01	39.15	29.28	34.7	42.65	30.24	39.47	36.92	21.65	21.65						540.93	17
	45.52	26.42	37.24	42.65	34.7	33.42																219.95	6
																						0	0
																						0	0
44.56	45.2	51.56																				141.32	3
																						0	0
54.11	47.43	33.1	35.01	50.29	83.4	22.92	24.19	39.79														390.24	9
																						0	0

Appendix 1: Makheras Forest Database

Average dbhs	Number of Saplings	Flora Species Identified																				Count of Flora Species Identified	Soil Group	Diversity Index
26.56	8	65	58	17	67	15	48															6	Y2	14
28.75	17	65	15	48	35	67																5	D3	15
2.65	16	65	67	58	74	15	48	5	62	91	111											10	R3	9
0.00	29	65	58	17	74	67	55	15	48													8	B2	9
24.76	2	65	67	58	15	48																5	B3	4
18.13	2	65	58	17	67	15	48															6	Y3	5
23.31	6	102	65	74	3	1	15	48	82	61	49	87	26	55								13	Y2	10
15.35	0	65	67	58	74	3	15	48	61	92	5	6										11	R2	6
26.07	4	65	74	3	67	15	32	48	34	92	82	87	93	54								13	Y2	7
10.65	0	65	74	3	15	48	61	35	66													8	B2	4
0.00	124	65	66	67	34	15	12	48	35													8	D3	9
26.54	0	65	66	48	15	92																5	Y2	8
35.83	2	65	66	48	15																	4	Y2	5
18.90	7	65	67	48	15	92																5	Y2	10
29.05	9	65	58	67	15	92	48	81	55	32	6											10	B2	7
28.26	3	65	17	67	58	15	48	70	61													8	G2	6
16.48	5	65	74	67	34	15	48	5														7	D3	8
7.29	0	65	74	67	48	15	42	58	17													8	B3	4
39.15	4	65	74	58	67	48	15															6	Y2	6
45.84	7	65	74	15	48																	4	B3	6
29.75	7	65	74	15	48	67																5	B2	5
19.45	0	65	74	17	67	3	15	11	90	96	86	44	32	84	82							14	B2	10
16.80	8	65	66	67	17	15	48	31	108													8	B3	7
28.19	11	17	65	74	67	102	66	3	15	82	48	34	58	92								13	D3	8
59.49	8	65	15	82																		3	D3	8
33.67	14	65	67	74	9	55	15	48	92	58	32	6										11	B3	11
11.74	6	65	74	67	15	48																5	G2	6
0.00	0	74	65	66	58	32		42														6	MISSING	5
27.98	7	65	17	74	67	15	48															6	Y2	6
39.29	7	65	67	74	15	48	5															6	B3	6
4.81	0	65	74	68	1	55	66	15	48	90	28	51	102	56	10	69						15	B3	8
37.57	9	65	67	74	61	92	15	5														7	D3	13
29.71	7	5	65	74	3	102	15	75	82	92	32											10	R2	14
26.53	5	65	74	3	15																	4	B2	5
3.04	6	74	3	65	15	48																5	B2	15
46.30	8	65	58	67	17	15	32	92	93	35												9	Y3	16
4.85	24	65	66	9	34	15	48	5														7	D3	14
2.72	16	65	17	67	15	48	5															6	D3	13
12.70	4	65	66	67	15	61	5	32	9	70	17											10	Y2	6
16.94	0	65	67	58	15	48	71	5	92	35	9	32										11	Y3	6
7.67	2	67	58	65	15	48	5	61	92	32		93	108									11	B3	7
9.66	0	65	17	67	15	5	48	100	35	70	32											10	D3	5
8.13	1	74	67	58	17	15	82	5	41		69	70	100									11	G3	7
43.61	0	65	74	67	15	82	5															6	G3	12
7.00	2	74	67	15	92	70	65															6	G2	6
8.38	0	65	74	67	66	15	31	5	92	75	32	41	35									12	B2	6
5.62	0	65	74	3	1	75	9	15	48		92	32	55	31	67	69	5	41				16	R3	7
11.18	0	65	68	47	1	67	3	74	55	15	41	75			16	20	81	58	32	82	6	18	G3	10
21.40	1	65	74	17	58	9	26	15	31	70	5	35	93	54	92	48	6					16	R1	14
6.76	0	65	58	17	67	9	87	41	15	32	66											10	B2	5
24.72	2	65	17	19	67	15	5															6	B2	5
5.94	2	65	58	74	67	15	48	92	5	17	9	60	69	70	93	54	28					16	B2	9
28.94	5	65	67	17	15	48																5	Y1	5
48.06	11	65	74	67	5	35	92															6	Y1	14
47.00	4	65	74	67	15	5	32	48	92													8	G3	7
21.86	2	65	67	74	17	15	32	5	92	41	1											10	D3	10
17.97	1	65	74	67	58	1	15	20	5	28	92	62										11	R2	8
23.80	0	41	65	67	66	17	74	58	15	48	5	32	92									12	G2	6
41.41	3	65	74	3	64	15	82															6	R3	12
12.06	0	65	74	3	66	67	15	82	5	92	41	64										11	Y2	15
6.90	0	65	74	1	17	66	15	48	28	49	3	41	82									12	R3	12
	0																						G2	1
12.91	1	65	74	17	9	15	48	32	16	92												9	R1	6
38.76	1	65	67	74	58	9	15	32	5	92												9	G2	8
11.74	0	65	17	67	58	9	32	15	5	69	70											10	B3	4
7.50	0	65	58	74	67	15	5															6	R2	4
23.94	0	65	17	67	15	92	70	5	35													8	MISSING	5
11.74	5	65	17	15	48	5	32	70	7	90	93	26	35									12	B3	14
73.80	1	65	58	17	15	70	5	32	90	7	35											10	D3	9
40.96	2	65	67	17	74	15	76	6	62	35	32											10	D3	8
32.15	1	65	74	58	15	5	62	90														7	G3	8
0.00	1	58	15	5	62	26	35															6	B3	6
70.24	12	65	74	1	15	82	62	5	20	28	90											10	G3	19
11.49	0	65	74	67	58	1	15	103	5	62	28	90										11	R2	9
26.28	0	65	74	17	67	15	7	5	62	32	28	90	82									12	B2	15
3.82	0	65	74	58	17	15	16	62	28	90	2	4	81	32	82	5						15	G3	10
22.71	0	65	17	74	67	15	5	82														7	G2	14
4.39	1	65	74	1	3	15	41	32	81	28												9	G3	12
13.05	0	65	74	1	91	41	81	28														7	G3	8
21.33	1	65	74	67	15	41	82															6	Y1	7
25.36	3	65	74	67	17	15	79	5	31	92	53	35	32	12	6	70	48					16	B2	8
8.24	0	65	74	67	15	17	5	7	48	32	92											10	D3	5
0.00	15	17	67	9	74	15	5															6	D3	7
0.00	11	65	68	1	8	67	58	55	81	7	5	15										11	D3	8
40.74	1	65	74	58	17	67	15	5	67	4	90	35	62	32								13	D3	9
0.00	0	65	67	17	74	15	92	5	31	35												9	Y3	5
0.00	0	65	74	1	47	67	91	15	5	87	81	13	43	19	20							14	Y3	6
0.00	0	65	67	74	1	15	82	4	20	5	32	33	17	6	35							14	G3	8
0.00	0	65	74	1	67	17	28	90	62	15	16	32	41	5	110	93	20	35				17	G3	8
48.73	3	65	74	1	41	15	82	5	93	49	28	90	67									12	Y1	13
60.10	5	65	74	15	5	28	90	49	35	82												9	B2	10
24.44	12	65	74	19	15	82	5															6	B2	11
0.00	13	65	58	67	15	5	7	35	92	70												9	B3	7
0.00	0	65	74	67	58	1	16	62	6	63	15	82	28	90	32	20	92					16	Y1	9
15.70	0	65	74	17	58	1	67	16	87	32	15	7	113	5	27	110	112					16	Y1	8
0.00	0	74	67	58	1	15	33	16	5													8	G3	5
43.36	2	65	74	1	15	16	20	28	90	32	82											10	G3	11
0.00	0	65	74	67	15	5	82	20	41													8	G3	3

Appendix 2: Adelphi Forest Database

Transect	Quadrat	Date	Time	Visibility	Compartment Num	Access to the site	Geology	Topography	Slope	Elevation
13	104	08/04/1995	7:12	3	85	Forest road from Mitsero	1	2	3	740
13	105	08/04/1995	9:51	3	86	On bearing from Forest road	1	1	2	700
13	106	08/04/1995	11:30		90	Forest road	1	2	3	540
13	107	08/04/1995	12:45	2	89	Forest road	1	2	4	580
14	110	08/04/1995	3:35	2	82	Forest road	2	2		540
14	111	13/04/1995	11:50	3	83	On bearing from Forest road	1	2	4	580
14	112	14/04/1995	10:18	3	78	On bearing from Forest road	1	4	3	640
14	113	14/04/1995	9:07	2	76	On bearing from Forest road	1	2	3	640
14	114	13/04/1995	14:11	3	75	On bearing from Forest road	1	2	3	420
14	115	20/04/1995	9:50	2	74	Forest road	1	1	3	620
14	116	26/04/1995	13:40	2	94	On Bearing.				
14	117	14/04/1995	15:53	2	93	On bearing from Forest road	2	4	3	560
14	118	17/04/1995	8:19	2	97	On bearing from Forest road	1	4	3	580
14	119	17/04/1995	12:40	2	100	On bearing from Forest road	1	2	4	600
14	120	17/04/1995	11:22	1	96	On bearing from Forest road	1	4	3	570
14	121	19/04/1995	13:26	2	101	Forest road	1	2	4	560
14	122	19/04/1995	11:33	2	102	On bearing from Forest road	1	2	5	680
14	123	19/04/1995	10:00	2	103	On bearing from BC 102	1	2	5	630
14	124	22/04/1995	16:48	3	104	Path from bc 102	1	2	5	530
14	125	22/04/1995	17:54	3	105	On bearing from forest path	1	2	4	480
14	126	23/04/1995	10:37	2	106	On bearing from Ridgeline	2	2	5	680
14	127	23/04/1995	8:38	2	107	On bearing from Ridgeline	2	2	5	780
15	128	23/04/1995	14:24	1	81	Ridgeline from Village road	1	4	4	360
15	129	23/04/1995	15:40	3	79	Bearing from Forest road	1	2	3	540
15	130	02/05/1995	8:54	2	77	Forest road	1	2	4	520
15	131	26/04/1995	8:50	2	73	Forest road	1	2	4	500
15	132	26/04/1995	10:40	2	95	On bearing from Forest road	1	2	3	530
15	133	26/04/1995	12:45	1	94	On bearing from Forest road	1	2	5	500
15	135	05/05/1995	11:30	3	70	Forest road - ridgeline	2	2	5	560
15	136	05/05/1995	10:20	3	69	Gully line from abandoned forest road	2	2	5	580
15	137	04/05/1995	12:23	2	68	Ridge line from forest road	2	2	5	760
15	138	04/05/1995	13:30		67	Forest road -> path	2	2	5	720
15	139	04/05/1995	9:53		66	Path from forest road	1	2	5	800
15	140	08/05/1995	8:58	1	65	Abandoend forest road	2	2	4	560
15	141	08/05/1995	12:00	2	58	Ridgeline from old path	2	2	5	400
16	154	11/05/1995	10:30	2	55	On bearing from Forest road				
16	155	11/05/1995			56	no data				
16	153	11/05/1995	12:25	2	59	Spur line				
16	152	11/05/1995	14:10	2	62	On bearing from Forest road				
16	142	06/07/1995	7:00	1	41	Ridgeline from forest road	2	4	4	510
16	143	06/07/1995	8:47	1	40	Ridgeline from forest road	2	2	5	480
16	144	06/07/1995	10:28		38	Ridgeline from forest road	2	2	4	440
16	145	06/07/1995	11:46	1	37	Forest road	2	4	4	450
16	146	09/07/1995	9:10	2	42	Ridgeline from forest road	1	2	5	580
16	147	09/07/1995	10:23	2	44	Ridgeline from forest road	1	4	4	560
16	148	09/05/1995	13:16	2	45	Gully line from forest road	1	2	5	580
16	149	10/05/1995	8:40	2	46	Forest road	1	2	5	580
16	150	10/05/1995	10:26	1	64	Forest road	1	2	3	620
16	151	10/05/1995	12:30	2	61	Gully line from forest road	1	2	4	610
16	154	02/05/1995	8:33	2	96	On bearing from Forest road	1	2	3	500
17	156	18/12/1995	7:50	3	21	Spur line between forest roads	1	4	4	620
17	157	18/12/1995	10:03	3	38	Old forest path from road	2	2	5	680
17	158	15/12/1995	11:04	3	36	Spur line from forest road	1	2	3	640
17	159	18/12/1995	13:10	2	33	Spur line	2	2	5	720
17	160	19/12/1995	7:56	2	32	Spur line from forest road	2	4	4	680
17	161	19/12/1995	9:14	2	31	Forest road	2	4	4	640
17	162	19/12/1995	13:13	2	30	Gully line from forest road	2	2	5	700
17	163	19/12/1995	10:52		29	Forest road	1	2	5	660
17	164	02/07/1995	12:59	1	48	Ridgeline from forest road	2	4	4	740
17	165	02/07/1995	17:57	2	54	Ridgeline from forest road	2	2	5	720
18	166	20/12/1995	10:08	3	17	Path from Ayia Marina	1	2	2	470
18	167	20/12/1995	11:51	2	18	On bearing from Forest road	1	2	3	500
18	168	20/12/1995	13:12	1	15	Forest road	1	4	2	620
18	169	27/12/1995	7:26	3	14	Forest road	1	2	3	740
18	170	27/12/1995	9:06	2	23	Spur line from forest road	1	2	5	870
18	171	27/12/1995	10:38	2	25	Spur line from forest road	1	2	4	860
18	172	27/12/1995	13:06	3	26	Spur line from forest road	1	2	5	830
18	173	15/01/1996	13:29	3	27	Forest road	1	2	3	800
18	174	15/01/1996	12:45	3	6	Spur line from old forest path	1	2	4	700
18	175	15/01/1996	11:54	3	5	Spur line from old	1	2	4	830
18	176	17/01/1996	10:44	4	4	Spur line from old forest path	1	2	4	870
18	177	17/01/1996	9:44	3	3	Spur line from old forest path	1	2	4	930
19	178	10/01/1996	9:57	3	10	Fire break	1	2	4	900
19	179	10/01/1996	14:24	4	8	Old forest path	1	2	4	780
19	180	10/01/1996	13:12	3	7	Old forest path	1	2	5	840
19	181	10/01/1996	12:04	2	2	Old forest path	1	2	4	1000

Appendix 2: Adelphi Forest Database

Aspect	Wind Direction	Catastophic Weather	Climatic Features	Soil Type	Soil Fertility	Soil Depth	Soil Moisture
SSW		2	1	Brown gravel	2	2	3
E		2	1	Brown gravel	2	2	4
NNW		2	1	Red	2	2	3
SW		2	1	Red gravel	3	3	4
NE	N	2	1	Brown	2	3	3
E			1	Red	2	3	4
S		2	1	Red-Orange	2	3	4
SE		2	1	Red-Orange	2	3	4
SSW		2	3	Red	3	3	4
N-S		2	2	Brown-Orange	2	3	2
SSW		2	2	Red-Orange	2	2	3
SW		2	1	Orange	2	2	4
N		2	2	Brown	2	2	3
SW		2	2	Orange	2	3	3
W	W	2	2	Brown-Orange	2	3	3
N	W	2	2	Browm	2	3	2
S	W	2	2		2	3	4
N	E	2	2	Browm	2	3	3
W	W	2	2	Red	2	2	3
W	NW	2	2	Brown Gravel	3	3	4
N	NW	2	2	Browm	2	3	3
SW	W	2	2	Brown-Orange	2	3	4
SW	W	2	2	Browm	2	2	3
N		2	2	Brwon loam	2	1	2
E	S		2	Brown-Orange	2	2	2
W	-	2	2	Brown	2	2	2
SE	S	2	2	Brown-Orange gravel	3	3	4
N	-	2	1	Brown	2	2	3
N-S	-	2	1	Brown gravel	3	3	2
SE	-	2	1	Brown gravel	2	3	3
SE			3	Brown gravel	2	3	4
	-	2	2	Brown	2	3	4
SSE	-		2	Orange gravel	2	3	4
NW-SE	N	2	2	Brown loam	2	2	2
				no data			
NW	-	2	2	Brown sand	3	3	4
N	N	2	2	Brown	2	2	3
S	-	2	2	Orange gravel	2	3	4
S	N	1	2	Orange gravel	2	2	3
NE		2	2		3	3	2
S		2	2	Orange sandy	2	2	2
SW		2	2	Orange gravel	2	3	4
S	-	2	2	Orange gravel	2	3	4
W	N	2	2	Brown	1	2	3
S		2	2	Brown	2	3	2
NNW	-	2	2	Orange gravel	2	3	3
N	W	2	2	Brown	2	2	3
N	W	2	2	Brown	1	2	1
S	W	2	1	Podsolic	2	3	4
W		2	2	Orange gravel	3	3	4
N		2	2	Brown gravel	2	3	2
E		2	2	Brown gravel	2	3	3
N		2	2	Brown loam	3	3	2
N			2	Brown gravel	2	3	3
N	N	2	2	Orange gravel	2	3	4
NE	N	2	2	Orange gravel	3	3	4
SE	N		1	Orange gravel	2	3	3
W	N	2	2	Brown	2	3	2
E	N	2	2	Brown	2	3	3
E		2	2	Brown	2	2	3
E	NE	2	2	no data	2	3	2
SE	NE	2	2	Orange gravel	3	4	4
N	-	2	2	Orange gravel	2	3	2
S	N	2	1	Brown gravel	2	3	3
N	N	2	2	no data	2	3	2
S	N	2	1	Brown	2	3	3
E	-	2	4	Brown	2	2	2
SE	-		3	Brown	2	3	2
E	E	2	1	Brown	2	3	1
E	N	1	2	Brown	2	3	3
N	N	2	1	Brown	3	3	1
SSE	N	2	1	Brown	1	2	2

Appendix 2: Adelphi Forest Database

Soil Temperature	Soil Conductivity	Soil Structure	Drainage	Riparian Buffer Vegetation Present
	1.5	3	3	2
	0	3	2	2
	3	3	5	2
	4	3	3	2
	0.5	3	5	2
	0	1	3	2
	2.5	1	5	2
	0.1	1	5	2
	0.5	1	5	2
	4	1	3	2
	1.5	1	5	2
	0.5		5	2
	0.1	1	5	2
	0.1	1	4	2
	0.1	1	3	2
	0.2	1	5	2
	4.5	1	5	2
	3	1	5	2
	2	1	5	2
	2	1	5	2
	3	1	5	2
	7	1	3	2
	3	1	5	2
	2	1	5	2
		1	3	2
		1	5	2
		2	5	2
	0.5	1	5	2
	1	1	3	1
	0.1	1	4	2
		1	5	2
		1	5	2
		1	3	2
	4	1	5	2
	0.1	2	5	2
	0.2	1	3	2
	0	1	5	2
	0.1	1	3	2
		1	3	2
	5	1	5	2
	0.5	1	3	2
	0.5	1	3	2
	0.5	1	5	2
	1.5	1	3	2
		1	5	2
		2	5	2
		1	3	2
		1	3	2
		1		2
		1	5	2
		1	5	2
		1	5	2
		1	5	2
	0.2	1	5	2
	0		5	2
		1	2	1
		1	3	2
		1	5	2
		1	4	2
		1	5	2
		1	5	2
		1	5	2
		1	5	2
		1	3	2
		1	5	2
		1	5	2
		1	5	2
		1	5	2
		1		2
		1	5	2
		1	5	2

Appendix 2: Adelphi Forest Database

Riparian Vegetation Density	Potential Forest Type	Site Class	Seed Bed Condition	Understorey Vegetation	Ground Litter Depth
	2	3	2	3	3
	2	2	2	2	2
	2	2	2	2	1
	2	3	3	3	
	2	3	3	3	2
	2	3	2	3	
	2	3	2	3	3
	2	3	3	3	3
	2	3	3	3	
	2	3	3	3	3
	2	3	2	3	2
	2	3	2	3	
	2	2	1	3	1
	2	3	2	3	2
	2	3	2	3	2
	2	3	3	2	1
	2	3	2	2	1
	2	2	3	2	3
	2	3	3	3	
	2	4	3	2	3
	2	4	3	3	3
	2	3	3	3	3
	2	3	2	3	
	2		2	3	1
	2	3	3	3	2
	2	2	2	3	2
	2	4	3	3	3
	2	3	2	3	
1		3	3	3	3
	2	3	3	3	2
	4	3	2	3	
	2	3	3	3	
	2	2	2	3	3
	2	2	2	3	2
	2	3	3	3	2
	2	3	2	3	2
	2	3	3	3	3
	2	3	3	3	3
	2	3	3	3	2
	2	2	2	3	2
	2	3	2	3	3
	2	3	3	3	3
	2	2	2	3	
	2	3	2	3	2
	2	2	2	3	1
	2	2	2	3	2
	2	2	2	3	2
	2	3	3	3	3
	1	4	3	3	3
	2	2	2	3	2
	2	3	2	3	2
	2	3	3	3	3
	2	3	2	3	3
	2	2	2	3	2
	2	3	3	3	2
1	2	2	2	3	3
		3	3	3	3
	2	3	2	2	2
		3	2	3	2
	2	4	3	3	3
	2	4	3	3	3
	2	2	2	3	2
	2	3	2	3	3
	2	3	2	3	3
	1	3	3	3	3
	2	3	2	3	1
	2	3	2	3	1
	2	3	2	3	3
	2	2	3	3	1
	1	4	3	3	1
	1	2	2	2	2

Appendix 2: Adelphi Forest Database

Stand Composition	Ratio of Tolerant Species	Dominant Age	Stand Structure	Canopy Cover	Crown Condition	Flora Associations
1	1	6	1	2	1	4
5	1	4	1	2	3	4
1	1	6	1	2	2	4
1	1	4	1	1	3	4
1	1	6	1	1	2	4
5	1	2	1	2	2	4
1	1	6	2	1	2	4
5	1	2	1	2	2	6
5	1	3	1	1	2	4
1	1	6	2	1	2	4
				1		
1	1	6	2	2	2	4
1	1	1	1	1	2	4
1	1	1	1	3	2	4
1	1	6	2	1	2	4
1	1	4	1	2	2	4
1	1	2	1	1	2	4
1	1	6	2	2	2	4
5	1	2	2	1	2	4
5	1	4	1	1	3	4
5	1	5	2	1	3	1
5	1	4	1	1	3	4
1	1	2	1	1	2	4
1	1	4	1	2	3	4
1	1	4	1	2	3	4
1	1	4	1	2	3	4
1	1	6	2	2	2	4
1	1	4	1	1	3	4
1	1	6	2	2	2	4
5	1	1	1	1	2	4
1	1	3	1	1	2	4
1	1	3	1	1	2	4
1	1	4	1	2	2	4
1	1	6	1	1	2	4
1	1	4	1	2	2	4
				1		
				2		
				1		
				1		
1	1	2	1	1	2	4
5	1	2	1	1	1	4
5	1	4	1	1	3	4
1	1	3	1	2	2	4
5	1	2	1	1	2	4
5	1	6	2	2	2	4
1	1	4	1	2	2	4
1	1	4	1	1	2	4
1	1	2	2	2	2	4
1	1	4	1	2	2	4
1	1	4	2	2	2	4
1	1	3	1	2	2	4
1	1	2		2	2	4
1	1	4	1	1	2	4
1	1	2	1	1	2	4
5	1	1	1	1	2	4
1	1	3	1	1	2	4
1	1	4	1	1	2	4
1	1	3	1	1	2	4
5	1	2	1	1	2	4
5	1	4	1	1	2	4
1	1	1	1	2	2	4
1	1	2	1	2	2	4
1	1	1	1	1	2	4
1	1	1	1	1	2	1
5	1	4	1	1	3	4
5	1	6	1	2	2	
5	2	2	1	1	2	2
1	1	2	1	1	2	4
5	1	4	1	1	3	4
1	1	6	1	2	2	4
1	1	2	1	2	3	
1	1	4	1	2	3	4
5	1	3	1	1	2	4
1	1	4	1	1	3	4
5	2	3	1	1	2	2
5	1	6	2	2	2	4

Appendix 2: Adelphi Forest Database

Notable injuries	Management Texhniques	Stage of Ecological Succession	Site Disturbance
Storm damage	1	3	Mechanical Reafforestation
Insects	1	3	road line
	1	3	
	1	3	
Clear felling	1	3	
Clear felling	1	3	Mechanical Reafforestation
Clear felling, Storm damage, Wild fire?	1	4	Mechanical Reafforestation
Storm damage	1	3	Mechanical Reafforestation
Clear felling	1	3	Mechanical Reafforestation
Storm damage	1	4	
Clear felling	1	4	
Clear felling, Storm damage	1	4	
	1	4	Mechanical Reafforestation
Clear felling	1	4	Logging tracks
Clear felling, Storm damage	1	3	
Storm damage	1	3	Path
Storm damage	1	4	
Storm damage, Erosion	1	3	
Clear felling, Storm damage	1	4	
Storm damage	1	3	
Clear felling, Storm damage	1	3	
Erosion	1	3	
	1	3	
	1	4	
	1	3	Path (disused)
	1	4	
Erosion	1	3	
Wild fire, Clear felling	1	3	Old goat paths
Wild fire, Clear felling	1	3	
Storm damage, Clear felling	1	3	
Wild fire, Clear felling	1	4	
Clear felling	1	3	Path (disused)
Clear felling	1	4	
Storm damage, Clear felling	1	4	
no data			no data
Clear felling	1	3	Road or fire break - abandoned
Clear felling	1	4	Fox hole
Clear felling	1	3	
Clear felling	1	3	
Clear felling, Storm damage	1	4	
	1	4	Circular stone arrangement
Clear felling	1	4	
Erosion	1	4	
Clear felling	1	4	
Clear felling	1	4	
Clear felling	1	3	Mechanical Reafforestation
Storm damage	1	4	
Storm damage	1	4	
Clear felling, wild fire	1	4	
Clear felling	1	4	
Clear felling	1	4	Mechanical reafforestation
Wild fire, Clear felling	1	4	
Clear felling	1	4	
Clear felling	1	4	
Clear felling	1	4	
Storm damage	1	4	
Clear felling	1	4	Mechanical reafforestation
	1	4	Path
Clear felling	1	4	
Clear felling	1	4	Mechanical reafforestation
Clear felling	1	4	
Clear felling, Storm damage	1	4	
Clear felling, Storm damage	1	4	
Clear felling	1	4	
Clear felling	1	4	
Clear felling	1	4	
Wild fire	1	4	
Wild fire	1	4	
Clear felling	1	4	
Clear felling	1	4	
Clear felling	1	4	
WIld fire, Storm damage	1	4	

Appendix 2: Adelphi Forest Database

Anthropogenic Evidence	Biotic Associations.	Fauna Species Identified									Sum of Flora Species
		32	44	1	7	22	53				6
hunting debris		10	9	1							3
		30	86	10	17	74	72				6
Hunting debris		7	57	32	64						4
Hunting debris		76	74	72	66	31	10	73			7
Garbage		15	30	32	74	54					5
		1									1
		32	30	1	16	72	61	47	74		8
Garbage		1									1
Garbage		30	9	1	10						4
		32	42	31							3
Garbage		53	49	1	32						4
Hunting debris - One cartridge		32	27	1	8						4
Garbage		30	74	1	4	71					5
		65	64	32	1	15					5
		32	31	1	42						4
Hunting debris		32	31	1	6						4
Hunting debris		32	31	9							3
		1	9	7	8						4
		8	70								2
Hunting debris		30	22	7	8	17	9	15	23		8
Hunting debris		69	7	11	8	10	64	51	1		8
Hunting debris		67	65	49	74	31	1	11			7
Hunting debris		54	65	31	9	10	1	7			7
		50	16	12							3
Garbage		32	12	8	10	31					5
		32	22								2
		1	32	31							3
Hunting debris		4	31	32							3
		31	30	71	8	17	15				6
		23	65	8	17	22					5
		32	4								2
Pottery (outside quadrant)		32	65	71	16	15	12				6
Hunting debris		32	64	17	8	21	15	76			7
Hunting debris, Garbage		32	31	4	8						4
		32	11	8							3
no data		12	15	31	4						4
		17	74								2
		32	50	74							3
Garbage		30	59	17	22						4
		31	59	1	22	76					5
Hunting debris		4	59								2
Garbage		4	59								2
		1	46	32	31						4
Hunting debris		17	31	49	76						4
		17	32								2
		22	32	31							3
		17	32	8	15	26					5
		17	32								2
Garbage											0
		31	30	61	4	1					5
		1	30	8	74	73					5
Hunting debris		1	32	31	30	49	55	54	59	2	9
		1	32	31	30	2					5
Hunting debris		1	32	31	42	30	61	2	4		8
	Tin foil	1	32	31	57	30	2	4	74	75	9
		1	32	30							3
Hunting debris		1	30	32	73						4
		16	31	30	67	1	22	23	73		8
		29	32								2
Hunting debris, Garbage		1	32	31	30	61	54	53	2	73	9
Industrial waste		1	53	30							3
		1	32	30	2						4
		2	32	1	9						4
Pottery		2	32	31	54	56	16	76			7
		4	32	31	49	1					5
		1	32	30	74						4
Pottery, Hunting debris		5	32	49	53	1	74				6
		32	73								2
		1	32	31	30	9	74				6
											0
		1	30	32	71	55	4				6
Garbage		5	32	31	30	53	1	74	73		8
		32	30	73							3
		30	74	73	1						4
Garbage, Hunting debris											

Appendix 2: Adelphi Forest Database

DBH																		
46.15	31.19	21.96	28.65	29.92														
	35.65	36.61	26.74	22.28	29.60	35.97												
29.60	34.06	41.06	29.92															
27.69	27.37	31.83	31.51	20.37	29.60													
28.33																		
60.48	52.20																	
36.29		19.42	34.06	35.65	28.97													
34.06	34.70	35.65																
35.01	46.47	30.56	33.74	36.29	21.33	26.42	24.83											
21.65	32.79	21.33	35.97	33.10	30.88	40.74	22.28	29.60	25.46	22.28	24.51	27.69						
32.15	30.88	27.37	34.70	35.97	23.24	36.92												
38.20	24.83	27.37	47.43	23.87	26.74	46.47	54.43	50.29										
52.20	44.56	36.61	33.42	30.56	38.52	39.15	47.75	31.51	24.83	29.60	23.87							
52.52	53.79	45.84	36.61	41.06	30.88	40.74												
63.66	42.02	42.02	32.47	35.01	45.20	41.70	37.88											
63.34	42.65	39.47	74.48	53.48	38.83	55.07												
53.48	75.76	48.38	43.61															
44.88	40.11	49.02																
26.74	53.48	47.75																
29.60	44.88	25.78	26.74	29.28	25.46	29.92												
57.93	44.56	47.43	50.93	35.33	31.83													
71.30				35.65	33.10	33.10												
49.66	49.97	47.11	36.29	29.92	33.10	40.74	24.51	38.83										
53.16	39.79	38.83		31.83	35.33	37.88	43.61	36.29	41.70	31.83	31.19	45.52	35.97					
45.52	39.15	54.11	37.88	38.20	47.43	41.70												
30.88	31.51	29.28	37.24	44.56	33.42	38.20	49.66	30.24										
35.97	37.56	36.29																
44.56	35.97	31.51	37.56	41.38														
36.92	38.83	30.56	33.10	30.56	32.47	29.92												
28.65	32.47	27.37	39.79	51.57														
30.88	27.37	29.60	40.43	37.24	32.15	25.78												
35.65	55.70	51.25	53.48	32.47	36.61	22.60	8.59	26.42	24.83	40.11			28.65	38.20	39.15	18.46		
34.38	47.75																	
36.29	30.24	38.52	34.06	35.65	35.97													
31.83	28.97	39.79	26.10	32.15	28.97													
35.01	28.97	38.20																
33.42	41.38	37.62	33.74	52.20	30.24	30.56												
35.01	26.42	37.56	26.74	30.56	30.24	35.65	36.92	37.88	27.37	28.65								
38.83	25.78	25.46	34.38	39.15	29.60	45.20	42.02											
44.88	39.47	34.06	32.47	29.92			29.28	22.92	32.15									
50.61	37.88	36.92	32.15	22.60	28.97	29.60	40.43	41.38	35.65	40.43								
42.02	29.60	41.38	58.25	28.65	41.06	35.97												
35.01	21.33	27.37	24.83	28.01	25.46	26.42	22.28	25.46	22.28	16.55	21.65	21.96	23.24	25.78	24.51	17.83	25.78	24.83
21.65	15.92	16.87	28.97	16.55	26.74	25.15	23.55	19.10	31.19									
54.11	23.55	24.83	30.24	29.92	39.79	34.06	19.10	19.42										
44.56																		
30.88	21.65	48.70	43.61	42.34														
38.20	35.97	21.65	32.47	31.83	17.83	17.19	20.37	25.15	22.92	20.05	26.74	21.65	18.46					
19.74	27.37	28.33	39.47	27.06	32.79	21.96	27.69	20.05	35.97									
34.06	39.15	24.19	24.83	28.33	22.92	21.65	24.19	39.47										
28.65	37.88	40.11	35.65															
25.46	45.20	40.43	39.15	37.88	43.29													
45.84	27.69	23.24	22.60	37.24	40.74													
40.43	36.92	37.24	28.97	19.74	23.87	30.88	35.97	39.47	22.60	25.46								
39.15	33.74	26.42	39.79	38.20	29.60													
1.27	26.10	38.52	33.74	32.47														
43.29																		
35.97	31.51	29.92	34.70	27.06	18.46													
44.56	0.32	28.97	29.28	22.92														
44.56	36.61	30.88	27.69	28.33	29.92													
38.20	38.20	47.43																
42.65	18.46	20.37	21.33	40.74	51.88	22.28	30.56	32.15	25.46									
20.37	34.70	36.92	30.88	46.47	47.75													
49.66	27.69	42.02	40.43	35.33	29.92	38.52	37.56		48.38	33.10	42.02	31.51	19.74					
49.97	24.19	37.56	26.42	35.33	31.51													
36.29	26.42	40.74	37.88	30.56														
36.92	39.79	21.01	28.01	35.65														
35.33	40.11	31.51	28.97	49.02														

Appendix 2: Adelphi Forest Database

Basal Area	Count DBH	Average DBHs	Number of Saplings	Flora Species Identified.															Total Flora	Diversity Index
157.88	5.00	31.58	1	65	15	17	116	21											5	12
186.85	6.00	31.14	1	65	17	8	55	15	16	67	109	11							9	7
134.65	4.00	33.66	19	65	15														2	15
168.39	6.00	28.06	4	65	116	11	111	72	67	17									7	5
28.33	1.00	28.33	21	65	15	101													3	14
112.68	2.00	56.34	10	65	17	151	11	67	109										6	8
154.38	5.00	30.88	23	65	15														2	13
0.00	0.00		40	65	17	15	5	6											5	11
104.41	3.00	34.80	4	65	15	5	116	17											5	4
254.65	8.00	31.83	15	65	15	5													3	11
			8	65	15	92	5												4	n/a
368.28	13.00	28.33	17	65	15	5	106	30											5	14
221.23	7.00	31.60	5	65	15	5	72												4	4
0.00	0.00		0	65	15	23	76	106											5	8
339.64	9.00	37.74	11	65	15	106	72												4	11
432.58	12.00	36.05	12	65	15	5	72												4	8
301.44	7.00	43.06	14	65	15	106	23	30											5	6
339.95	8.00	42.49	7	65	15	5	11												4	12
367.33	7.00	52.48	5	65	67	75	1	15	101	30	23								8	5
221.23	4.00	55.31	5	65	75	58	15	5	101										6	5
134.01	3.00	44.67	7	65	75	58	67	8	15	5	70								8	7
127.96	3.00	42.65	6	65	75	67	32	15	61	101	72	5	8						10	7
211.68	7.00	30.24	6	65	8	15	115												4	6
268.02	6.00	44.67	6	65	8	15	31	5	115	6	11								8	9
173.16	4.00	43.29	5	65	15	61	92	5											5	6
350.14	9.00	38.90	5	65	15	55	5												4	6
502.93	13.00	38.69	11	65	15	5	72	106											5	13
303.99	7.00	43.43	9	65	15	115	31	5											5	5
324.99	9.00	36.11	2	65	15	5	94	72											5	11
109.82	3.00	36.61	5	65	58	67	15	55	6	115	112								8	6
190.99	5.00	38.20	3	65	15	5													3	4
232.37	7.00	33.20	10	65	15	11	5												4	5
179.85	5.00	35.97	10	65	15	72	5												4	8
223.45	7.00	31.92	8	65	15	5													3	11
512.16	15.00	34.14	1	65	15	72	115	94	61										6	7
			3	65	74	35	39												4	
		no data	5	65	74	15													3	
			3	65	15	112													3	
			0	49	28	65	6	15											5	
82.12	2.00	41.06	12	65	15														2	6
210.72	6.00	35.12	17	65	67	15													3	7
187.80	6.00	31.30	6	65	67	15													3	5
0.00	0.00		6	65	15	111	55	58											5	7
102.18	3.00	34.06	5	65	67	15	23	11											5	4
597.79	7.00	85.40	14	65	17	15	5	72	61										6	14
353.01	11.00	32.09	7	65	15	5	11												4	7
280.43	8.00	35.05	6	65	15	115													3	5
265.15	8.00	33.14	16	65	15	61	5												4	9
396.61	11.00	36.06	9	65	15	94	72	5											5	7
276.93	7.00	39.56	12	65	15	115	5	101	94	23									7	9
460.59	19.00	24.24	5	65	15	101	72	23											5	6
225.68	10.00	22.57	8	65	15	23	72												4	7
275.02	9.00	30.56	3	65	15	31	5	8	67										6	6
44.56	1.00	44.56	12	65	11	5	67	111											5	6
187.17	5.00	37.43	7	65	75	15	5	101	6	23	53								8	7
350.46	14.00	25.03	15	65	15	115	5	112	61	53	72	23							9	8
280.43	10.00	28.04	3	65	15	23	11	101											5	4
258.79	9.00	28.75	5	65	15	5	112	72	17	1	61	114	31						10	5
142.28	4.00	35.57	30	65	74	75	67	78	15	61	114	6							9	10
231.41	6.00	38.57	2	65	67	74	15	116											5	4
197.35	6.00	32.89	6	65	15	55	8	76	34	5	11	58	87	6	72				12	10
341.55	11.00	31.05	9	65	15	67	55	8	87										6	8
206.90	6.00	34.48	14	65	15	5	72												4	6
132.10	5.00	26.42	11	65	15	11	72	111											5	6
43.29	1.00	43.29	8	65	75	17	15	6	31	23	112								8	7
177.62	6.00	29.60	12	65	67	75	26	111	32	11	5	8	31	23	63	100	72	69	15	15
126.05	5.00	25.21	10	65	75	55	1	15	87	32	63	6	23						10	6
197.99	6.00	33.00	8	65	15	5	69	72	115										6	7
123.82	3.00	41.27	0	65	67	75	15	5	61	116	32	23	63						10	5
305.90	10.00	30.59	9	65	15	5	23	114	31	116	69	72							9	14
217.09	6.00	36.18	4	65	15	116													3	6
475.87	13.00	36.61	8	65	15	72													3	8
204.99	6.00	34.17	8	65	17	67	15	5	63	112	11	72	23	111	4				12	8
171.89	5.00	34.38	2	65	15	61													3	4
161.38	5.00	32.28	7	65	75	15	5	114	23										6	6
184.94	5.00	36.99	8	65	17	15	5	11	111	72	82	114							9	13

Appendix 3: Archaeological Database

Forest Name	Site Number	Site Description	Site Altitude	Aspect	Site Topography	Structure
PA	1	SETTLEMENT - MANDRA PANAGIA	400	E	VALLEY	YES
PA	2	MANDRA / SETTLEMENT PANAGIA	510	W	VALLEY	NIL
MK	1	MANDRA		S	RIDGE	YES
MK	1	MINE SHAFT	830	W	RIDGE	NIL
MK	2	FOREST HUT/BYZANTINE CHURCH?	680	NW	RIDGE	NIL
MK	3	MONASTERY	740	E	HILL SIDE	NIL
MK	4	MANDRA	460	SE	RIDGE	YES
MK	5	MANDRA / FOREST HUT	670	NE	HILL TOP	YES
MK	6	STONE STRUCTURE	950	W	HILL SIDE	YES
MK	7	MANDRA	720	SE	RIDGE	NIL
MK	8	MANDRA?/VILLA	910	E	HILL SIDE	YES
MK	9	CHURCH/MONASTERY	270	S	RIDGE	YES
MK	10	STONE STRUCTURE ii	765	SW	HILL SIDE	YES
MK	12	MANDRA (LOOM WEIGHT)	910	S	RIDGE	NIL
MK	13	STONE STRUCTURE (ENCLOSURE?)	880	S	HILL SIDE	YES
MK	14	STONE STRUCTURE (FORT)	960	E	HILL TOP	YES
MK	15	FOREST HUT	1300	SE	HILL SIDE	YES
MK	16	STONE ARRANGEMENT / WATCHTOWER	1420	SE	HILL SIDE	YES
MK	17	MANDRA TOU ANAYIOU	720	S	SPUR	YES
MK	18	SETTLEMENT / INDUSTRIAL	650	E	VALLEY	YES
MK	19	STONE STRUCTURE / WATCHTOWER	855	N	HILL TOP	?
MK	20	T7 AD38	880	E	VALLEY	NIL
AD	1	TOMB / SANCTUARY	520	W	RIDGE	NIL
AD	2	PITCH SITE	620	SE	HILL SIDE	NIL
AD	3	DAD AND DAVE	500	W	VALLEY	NIL
AD	4	MANDRA	480	SE	VALLEY	YES
AD	5	PITCH EXTRACTION SITE T15 AD69	580	N	HILL SIDE	NIL
AD	6	STONE STRUCTURE	630	W	HILL SIDE	YES
AD	7	MANDRA (BYZ) T16 AD64/51	610	W	RIDGE	YES
AD	8	SQUARE STRUCURE T16 AD59 Q	740	W	HILL SIDE	YES
AD	9	SQUARE STRUCTURE T16 AD59	800	WSW	HILL SIDE	YES
AD	10	SQUARE STRUCTURE T16 AD62	800	WSW	RIDGE	YES
AD	11	INDUSTRIAL/KILN T15 AD69	580	SE	HILL SIDE	YES
AD	12	SETTLEMENT @ BISHOP'S LEAP T15 AD66	520	N	HILL SIDE	YES
AD	13	SETTLEMENT / INDUSTRIAL? T15 ENCLAVE	600	N	HILL SIDE	YES
AD	14	INDUSTRIAL - KILN T15 AD65	580	E	HILL SIDE	YES
AD	15	CIRCULAR STONE ARRANGEMENT T16 AD44 Q	700	W	HILL SIDE	YES
AD	16	SETTLEMENT - MANDRA FORD	560	W	VALLEY	YES
AD	17	CIRCULAR STONE ARRANGEMENT AD43	700	W	RIDGE	YES
AD	18	SQUARE STONE STRUCTURE AD43	540	W	RIDGE	YES
AD	19	SQUARE STONE STRUCTURE - MANDRA AD45	530	W	RIDGE	YES
AD	20	MANDRA AD64	590	W	RIDGE	YES
AD	21	KILN (PITCH?) AD45	650	W	RIDGE	YES
AD	22	MANDRA (EARLY MOD?)	650	W	VALLEY	YES
AD	23	SETTLEMENT ENCLAVE	620	E	HILL SIDE	NIL
AD	24	MANDRA - (EARLY MOD) AD45	580	W	RIDGE	YES
AD	25	VILLAGE / INDUSTRIAL? T18 AD19	540	E	HILL SIDE	NO
AD	26	SETTLEMENT / MANDRA T18 AD16	480	S	HILL TOP	NIL
AD	27	SQUARE STONE STRUCTURE T9 AD	890	E	HILL SIDE	YES

Appendix 3: Archaeological Database

Pottery	Geology	Kiln Present	Distance from Water		
YES	FULUVIAL TCE	NIL	50	130	127
YES	FULUVIAL TCE	NIL	30	127	48
YES	DIABASE	NIL	500	65	
NIL	DIABASE	NIL	500	65	13
YES	DIABASE	NIL	ADJACENT	65	
YES	DIABASE	NIL	ADJACENT	65	58
YES		NIL	ADJACENT	65	67
YES	BASAL	YES	ADJACENT	65	58
YES	DIABASE	NIL	500	65	58
YES	DIABASE	NIL	ADJACENT	65	74
YES	DIABASE	NIL	800	65	5
YES	DIABASE	NIL	50	65	74
NIL	DIABASE	NIL	700	74	66
YES	DIABASE	NIL	100	65	67
YES	DIABASE	NIL	50	17	67
YES	DIABASE	NIL	200	92	5
NIL	DIABASE	NIL	ADJACENT	65	19
YES	DIABASE	NIL	500	65	74
YES	DIABASE	NIL	700	65	17
YES	PILLOW LAVA	YES	ADJACENT	65	17
NIL	DIABASE	NIL	1000	65	48
NIL	DIABASE	YES	30	65	74
YES	PILLOW LAVA	NIL	100	65	15
YES	PILLOW LAVA	YES	75	65	15
YES	FULUVIAL TCE	YES	25	65	15
YES	FULUVIAL TCE	NIL	ADJACENT	72	49
YES	DIABASE	NIL	70	65	48
YES	DIABASE	NIL	1000	65	15
YES	DIABASE	NIL	800	65	15
YES	DIABASE	NIL	800	65	17
YES	DIABASE	NIL	200	65	67
TES	DIABASE	NIL	200	65	15
YES	DIABASE	YES	25	65	15
YES	DIABASE	NIL	30	65	127
YES	DIABASE	YES	100	65	48
NIL	DIABASE	YES	40	65	
NIL	FULUVIAL TCE	NIL	300	65	15
YES	FULUVIAL TCE	NIL	30	48	125
NIL	PILLOW LAVA	NIL	400	65	112
?	DIABASE	NIL	600	65	15
YES	DIABASE	NIL	300	65	125
YES	DIABASE	NIL	50	65	5
NIL	PILLOW LAVA	YES	200	48	17
YES	FULUVIAL TCE	NIL	40	17	48
YES	PILLOW LAVA	NIL	100	48	125
NIL	PILLOW LAVA	NIL	70	65	67
YES	PILLOW LAVA	YES	50	65	17
YES	PILLOW LAVA	YES	25	65	48
NIL	DIABASE	NIL	900	65	17

Appendix 3: Archaeological Database

Associated Flora Species										
125	5	15								
48	67									
17	126	127	67	19						
82	58	17	128	129	130	68	131	48	132	34
129										
74	67	15								
58										
17	58	8	15	70	5	48				
58	65	32	41	90	81					
17	15	125								
8	74	15	6	65						
125										
15										
3	15	5								
19	15	5								
1	47	55	5	101	65	58	15			
15										
1	47	67	78	15	5	87	81	12	19	20
72										
5	106	90	55	8	125					
5	112	125								
133	112	115	65	67	48	17	127			
67	15	125	55	6	57	112				
94	72	5	112							
94	72	5	112	125						
15	5	125								
5	112	125	72							
125	5									
67	125	61								
5	70	125								
67	1	127	17	130	133	125				
125	112									
127										
5	15									
5	11									
112	15	69	90							
72	15	67	17	55	68	81	47	125		
15	65	125								
67	111	15	125	112						
5	112	127	48	125	5					
48	125	5	112							
133	135	15	4	6	111					
109										
67	15	5	63	112	11	125				

Appendix 4: Forest Compartment Description Form

Front Page. Forest Compartment Description Form.
Numbers shown correlate to database entries.

OPERATION SPINOSUM

J Ellis Burnet School of History, Philosophy & Politics and School of Biological Sciences Macquarie University, Sydney

Forest Compartment Description Data.

Transect no		**Quadrant no**	
Date		**Time**	
Visibility			
excellent 1	good 2	moderate 3	poor 4
Latitude		Longitude	
Forest		**Compartment**	
Access			

Geology and Landforms

Rock Types			
U.Pillow 1	diabase 2	serpentine 3	
L.Pillow 4	gabbro 5	sandstone 6	limestone 7
pyroxenite 8	periodites 9	other (specify) 10	

Topography			
valley 1	hillside 2	hilltop 3	ridge 4
plateau 5	scarp 6		

Slope			
flat 1	slight 2		moderate 3
steep 4	very steep 5		

Elevation

Aspect

Climatic features
Wind direction

Catastrophic weather	yes no		
cloud cover 1	sunshine 2	rain 3	snow 4

Appendix 4: Forest Compartment Description Form

Second Page. Forest Compartment Description Form.
Numbers shown correlate to database entries.

Soil
red loam | brown loam | terra rossa | sand
podsolic | clay | chalk | other

Soil fertility
good 1 | moderate 2 | poor 3

Soil depth
deep 1 | moderate 2 | shallow 3 | denuded 4

Soil moisture
excellent 1 | good 2 | moderate 3 | poor 4

Soil temperature | **Soil pH**

Soil Structure
blocky 1 | granular 2 | prismatic 3 | platy 4

Drainage pattern/water catchment
river 1 | creek 2 | gully 3 | depression 4
none 5

Name:

Snow fall | Snow depth | snow melt | isolated patches

Riparian buffer vegetation
present T | absent F | light 1 | dense 2

Potential vegetation
Forest type
natural 1 | plantation 2 | agricultural invasion 3 | secondary 4
artificial regeneration 4

Site class
A class 1 | B+ 2 | B- 3 | C 4

Seed bed condition
good 1 | moderate 2 | poor 3 | ash 4

Understory vegetation
dense 1 | moderate 2 | sparse 3 | denuded 4

Ground litter depth
deep 1 | moderate 2 | slight 3 | none 4

Stand composition
pure conifer 1 | pure hardwood 2
mixed conifer 3 | mixed hardwoods 4
conifer/hardwoods 5

Species identified

Appendix 4: Forest Compartment Description Form

Third Page. Forest Compartment Description Form.
Numbers shown correlate to database entries.

Ratio of tolerant species

<25% 1	50% 2	>75% 3

Dominant age

sappling 1	pole 2	early mature 3
mature 4	senescent 5	uneven aged 6

Stand structure

simple 1	complex 2	

Canopy cover

<30% 1	50% 2	>70% 3

Crown condition

supressed 1	released 2	mature 3

Flora associations

Olea europea 1	Quercus coccifera 2	Pteris 3
Cistus 4	Quercus alnifolia 5	other 6

Notable injuries

wild fire	grazing	resin tapping
erosion	clear felling	insects
disease	pollution	storm damage
Date		

Felling sequence

Management techniques

production 1	tourist area 2	scientific 3
other 4		

Stage of ecological succession

Stage I	Stage II	Stage III
Stage IV		

Site disturbance

mechanical reafforestation		
habitation	current	abandoned
wells	tombs	terracing
aquaducts/pipes	mines	other
high tension transmission lines		

Anthropogenic evidence

pottery	garbage	masonary
industrial waste	landscape modification	slag
hunting debris	other	

Biotic associations

mouflon	birds	hares
feral cat	feral dog	fox
other		

Appendix 4: Forest Compartment Description Form

Fourth Page. Forest Compartment Description Form.

Numbers shown correlate to database entries.

Quadrant Map of Vegetation Cover.

Key

Documentation

Appendix 5: Archaeological Site Form

First Page. Archaeological Site Forms.

Operation Spinosum

J.Ellis Burnet School of History,Philosophy & Politics and School of Biological Sciences,Macquarie University,Sydney

Archaeological Site Form

MAP NAME	EDITION	SCALE	REFERENCE	FOREST TRANSECT QUADRANT COMP.NO.

Latitude | Longitude | Elevation

Date | Time

Village | Toponym
Alt Toponyms

Site Type | Plan/sketch attached? YES/NO
How many?

Aerial photo refs. | Photographs taken? YES/NO
How many attached?

Access to the site (refer to permanent features, give best approach to site)

Other sites in Locality? YES/NO

Reference sources/Documentation

Checklist:
surface visibility
damage/disturbance/
threat to site | Condition of site:

Recommendations for management & protection

Department of Antiquities notified | Forestry Department notified

Appendix 5: Archaeological Site Form

Second Page. Archaeological Site Forms.

SITE POSITION & ENVIRONMENT					
Landform valley	hillside	hilltop	ridge	plateau	scarp
Site aspect					
Slope flat	slight	moderate	steep	very steep	
Description:					
Geology		Landuse			
Distance from water		Source			
Resource zone					
Vegetation type					
Faunal resources noted					
Other exploitable resources known					

PLAN OF SITE

Appendix 6: Flora Identification Numbers

Taxa	Database Reference Number.
Acer obtrusifolium	1
Arabis purpurea	2
Arbutus andrachne	3
Arisarum vulgare	4
Asphodelus microcarpus	5
Asparagus acutifolius	6
Astralagus lusitanicus	7
Rhamnus palaestina	8
Scirpus lacustris	9
Cladonia cf. convoluta	10
Carlina coryrubosa	11
Cedrus brevifolia	12
Ceterach offiinarum	13
Cistus salviaefolius	14
Cistus villosus creticus	15
Clematis cirrhosa	16
Crataegus azarolus	17
Valeriana italica	18
Cupressus sempervirens	19
Cyclamen cyprium	20
Inula viscosa	21
Crupina crupinastrum	22
Crepis reuteriana	23
Umbilicus rupestris	24
Spring ephemerals	25
Equisetum ramosissimum	26
Ptilostemon chamaepeuce	27
Hypnum cupressiforme	28
Selaginella denticulata	29
Vicia sp.	30
Phaonalon rupestre	31
Galium aparine	32
Geranium sp.	33
Calycotome villosa	34
Bromus sp	35
	36
	37
	38
	39
	40
Lonicera etrusca	41
Lonicera etrusca	42
Juniperus oxycedrus	43
Peltigera cf. canina	44
Lamium moschatum	45
Limodorum abortivrm	46
Laurus nobilis	47
Lavandula stoechas	48
Evernia prunastri	49
Lithospenum diffusum	50
Mentha Longifolia	51
Veronica cymbalaria	52
Muscari parviflorum	53
Agaricus campestris	54
Myrtus communis	55
	56
Teucrium creticum	57
Olea europa	58
Papaver somniferum	59
Xanthora parietina	60

Appendix 6: Flora Identification Numbers

Orchis anatolica	61
Phagnalon graecum	62
Sedum cyprium	63
Paxillus involutus	64
Pinus brutia	65
Pistacia lentiscus	66
Pistacia terebinthus	67
Plantanus orientalis	68
Cheilanthes pteridioides	69
Poterium spinostrum	70
Romulea tempskayana	71
Vicia cassia	72
Styrax officinalis	73
Quercus alnifolia	74
Quercus coccifera	75
Geranium purpureum	76
Phragmites communis	77
Rhus coriaria	78
Sinapis arvensis	79
Biscutella didyma	80
Rubus sanctus	81
Salvia cypria	82
Scilla shyacinthoides	83
Asplenium	84
	85
Cladonia sp.	86
Smilax aspersa	87
Sorghum	88
	89
Pleurochaete squarrosa	90
	91
Thymus integer	92
Geastrum triplex	93
Aymenocarpos circinnatus	94
Trifolium stellatum	95
	96
Cladonia furcata	97
Valeriana italica	98
	99
Zosima absinthifoia	100
Allium trifoliatum	101
Styrax officinalis	102
Achillies sp.	103
Fumana arabiea	104
Cistus salviifolius	105
Trifolium campestre	106
	107
Gagea juliae	108
Zizyphus spina-christi	109
Sedum microstachyum	110
Capparis spinosa	111
Allium ampeloprasum	112
Crocus hartmannianus	113
Muscari inconstricitum	114
Micromeria myrtifolia	115
Helichryysum italicum	116
Prasium majus	117
Origanum majorana	118
Rubia tennifolia	119
Smyrnium connatum	120
Micromeria chionistrae	121

Appendix 6: Flora Identification Numbers

Lathynis aphaca	122
Muscari incoustrictum	123
Cephalorrhynchus cyprius	124
Avena barbata	125
Parmelia tiliacea	126
Alnus orientalis	
Orobanche cypria	
Anthemis plutonia	
Sinapis arvensis	
Dianthus strictus	
Vitis vinifera	

Appendix 7: Makheras Logbook

Q1 T1 MK102

The quadrat lies from a rise to a steep gully line along the south west. The terrain has been badly scored by goat and sheep tracks rendering the top soil highly unstable. The larger trees in the quadrat, that is, those with a CBH>56cm were predominantly suppressed and crown damaged, only a few would suit commercial harvesting. The goat damage in the environment, particularly on saplings and sub-strata species, is extreme with lower branches denuded and dead, and ring-barking frequent. No seedling regeneration was observed. A good sub-strata of *Pistacia terebinthus* is present though depleted by goat grazing. No hunting debris or garbage was present. One lizard was observed but no small mammals. The soil was eroded through goat passage, fire, and probably easily eroded during winter. *Pinus brutia* needles and small twigs were "chufted" along narrow ridges caused by the goat tracks.

Q2 T1 MK103

The quadrat was laid out in relationship to boundary cairn 27. A large *Pinus brutia* dominated the quadrat which also displayed sapling growth. The sub-strata had been damaged by goats in the past, (that is 2-3 years ago), *Pistacia terebinthus* being the species most impacted upon. There was good grass cover over the quadrat and a heavy cover of *Cistus* and *Lavandula Stoechas*. Some cleavers and wild Asparagus were noted. *Quercus alnifolia* seedlings were observed under the *Pinus brutia* canopy cover. The aspect (westerly) gave the quadrat considerable shelter. An abandoned farm lay below the quadrat to the west. Terracing grew grapes, almonds, but now the *Pinus brutia* is invading the
abandoned land. A fire break runs along the ridge line above the quadrat.

Q3 T1 MK107

The quadrat was laid out so as to include features of the compartment. Some previous felling had occurred and the area severely burnt by wildfire in 1958. Subsequently, artificial regeneration had been practiced with one gradoli passing through the quadrat. *Pinus brutia* saplings from the reforestation project appear stunted with deformed trunks. Thinnings for the "Christmas Tree" market were evident.

The area of natural regeneration carried a thin coverage of *Quercus alnifolia* and *Pistacia terebinthus* with *Sumach* seedlings growing vigorously on the edge of the fire trail.

Small rodent burrows were evident and a snake skin was collected. The ground cover was rich in species diversity though in places the stony ground remained bare. *Lavandula Stoechas* co-dominated with the *Cistus* with *Thymus* also evident. Where soil had collected in depression there was a marked increase in fertility. One seedling *Pinus brutia* was observed growing on the fire trail.

Sheep and goat pellets were observed on the fire break at the top of the ridge.

Q4 T1 MK108

The area of the quadrat lying N-S on the transect line straddles a small, currently dry creek with rich and dense riparian cover. The area has been contoured mechanically for re-afforestation following the wildfire of 1958. Some of the larger sapling are cone bearing and some thinnings have occurred. A large number of new saplings are growing on the disturbed gradoli. Slopes are covered by *Cistus* with only the occasional *Lavandula stoechas* - mostly on the gradoli. The soil is deep and loose - easily eroded. *Pinus brutia* middens are evident. Small rodent burrows and a larger abandoned burrow near the creek line were also observed. A dense regeneration of *Olea europea* along the creek line is evident with a small birds nest found in one.

The quadrat demonstrates the capacity for the forest to recover under good silvicultural management. The *Quercus alnifolia* is the dominant species on the Southern facing slope though not as numerous as in MK107. Rodent burrows are also concentrated on the southern facing slope though the *Pinus brutia* middens occur on the northern facing slope.

A small snake was observed just to the NW edge of the quadrat. Some burnt and rotten stumps were noted but not plotted.

Q5 T1 MK109

The quadrat was laid out so as to obtain an accurate location with regard to the boundary cairn 91. The topography was very steep with evidence of the earlier (1958) wildfire present. Only a few mature trees, in an open, scattered formation were in the vicinity. The undergrowth was sparse with a few *Quercus alnifolia* seedlings under the canopy cover of the remaining trees. The ground cover of *Cistus* was fairly dense with *Thymus* and *Gorse* also present. Some fallen trees were rotting. The direction of fall was SW. Some rodent burrows were sighted. The aspect was southerly to SW and very exposed. There was, from the evidence of *Galium aparine* and other moisture indicators, the possibility of better regeneration if some broadleaf species were reintroduced.

Q6 T1 MK135

Quadrat laid out in a ridge lying on the N-S transect. The outer line was an outcrop of sheeted dyke with numerous habitat locations. With the exception of one stunted sapling all the trees were mature. No seedlings were observed. The undergrowth was tussocky *Cistus* with a heavy litter accumulation. Trees had been felled in the quadrat on at least 2 different occasions. The area surrounding the fellings was severely depleted. A winter spring or seep was situated on the southern area of the quadrat. Small, isolated shrubs of *Quercus alnifolia* were recorded.

Dieback in the *Pistacia terebinthus* was also observed but without explanation. Some *Myrtus communis* was growing downslope from the seep. A Chukar scrap and droppings were noted near the sheeted dyke area. A few large *Pinus brutia* middens made by the rodents were also noted and photographed.

Q7 T1 MK143

Quadrat was laid out above creek line on steep slope with a northerly aspect. Some rock outcrops carried Lichen. The area had been previously clear felled - within quadrat 9 stumps were noted and one small pole tree (stem) shattered by larger tree fall. Very rich substrata exists with good *Cistus* cover between canopy patches. The area is very sheltered and a range of moisture indicator plants and one Orchid were recorded.

Appendix 7: Makheras Logbook

Seed middens occurred outside scraps and these not previously been observed.

Q8 T1 MK142 28.10.94

Q9 T1 MK138 28.10.94

Q10 T1 MK139 28.10.94

Q11 T1 MK140 28.10.94

Q12 T2 MK101

Quadrat laid out above a newly graded road line on a badly eroded slope with heavily scored goat tracks. Ground cover and sub-strata are denuded to a large degree. One *Pistacia terebinthus* destroyed by grazing. Two felling sequences were noted - the first apparently clear felling, the second, more recent selection felling. The ground is unstable due to goat tracks and loss of vegetation. Some grass occurs on the slope and knolls. No regeneration is evident.

Q13 T2 MK100

Quadrat on SE aspect.

The area is very badly damaged through grazing, heavily felled, and no natural regeneration of dominant canopy or sub-strata species is evident. Rare occurrence of orchid species noted. Most sub-strata *Pistacia terebinthus* has been destroyed by goats. The ground cover is dominated by *Lavandula Stoechas* with some *Cistus* sp. Hunting debris light. Some seed mounds with lizards in association were observed. Fallen trees contain termites.

The soil is generally unstable with loose surface stones. Goat hairs adhering to *Pinus* bark indicate illegal trespass.

Q14 T2 MK99

Quadrat laid out from East to West across a steep ravine with a NNE aspect. Rock screes had stabilised to a large degree. The area was planted with *Pinus brutia* and felling has taken place. No natural regeneration was evident. Rocks were heavily mossed.

Many cartridge wads were collected although no sign of game species was observed. Several bottles were collected from the quadrat - one with insects inside. The area is in need of immediate remedial afforestation with the introduction of broadleaf species necessary. Standing trees are stem deformed due to lack of light.

Q15 T2 MK98

Quadrat laid out N-S crossing a creek line and an abandoned roadway just north of its junction with the Prophet Elias road. Plantation trees in early maturity on Western side of creek. Some felling evident. Trees on the Eastern side seem less vigorous with only a very few seedlings from natural regeneration observed, however, a good scattering of *Quercus alnifolia* occurs in the quadrat although a seed tree was not evident. Some abandoned rodent burrows but no hunting debris. Some early riparian cover over the creek line was developing - mainly *Myrtus communis* and *Olea europea*. *Pistacia terebinthus* growing under the *Pinus brutia*.

Q16 T2 MK104

Quadrat laid out on a spur jutting westward from the peak. It has the appearance of former plantation and terracing, however there are wide gaps within the stand and very open scrub on the Northern side possibly due to wildfire. Good distribution of *Crataegus azarolus* as a major component of the substrata was observed. The ground is well covered and grassed. Although very open to the West the soil seems stable and capable of moisture retention due to the depth of litter. Birds nests fallen from trees were collected - generally a very tranquil spot worth revisiting.

Q17 T2 MK105

Quadrat laid out along ridge to the West of road line on further side of a gully. Old plantation stand with some stunted saplings. No seedlings in the understorey. Good ground cover of *Cistus* and grasses with thickets of *Quercus alnifolia* were evident. The Southern portion of the quadrat is open though no signs of past felling is evident. The Northern portion has a dense canopy cover.

A Viper was encountered in a bush (very small) also numerous small burrows and scraps were noted.

The *Quercus alnifolia* has stabilised the gully line.

Q18 T2 MK106

Quadrat laid out with some difficulty on a steep slope just below the ridge line leading to the summit. Extensive rock falls onto the scree running through the quadrat has broken and severely damaged a dense *Quercus alnifolia* thicket. Extensive wildlife is associated with the area which is of high elevation. A hare scrap was noted just South of the quadrat and a bird's nest was located in a nearby *Olea europea*.

The scattered *Pinus brutia* may be remnants of a former plantation - some trees seem to be regularly spaced however the trees are in generally poor condition and no regeneration was observed. Many seedlings from the *Quercus alnifolia* seemed to had died at the two leaf stage. Apart from the dense band of *Quercus alnifolia* which has stabilised the lower scree, the quadrat was open.

Q19 T2 MK111

Quadrat situated North of Prophetis Elias Monastery crossing the ridge line. Trees are part of a plantation which has not survived on the Southern slope. Growth on the ridgeline and the Western aspect appears quite vigorous. Garbage pits containing large quantities of tins, many of which have eroded out or may have been scavenged by foxes, were evident. Good regeneration of *Quercus alnifolia* whilst no *Pinus brutia* regeneration observed. Good ground cover on Eastern (ridgeline) and Northern aspects.

Appendix 7: Makheras Logbook

Rubbish collected from the pits was recorded and sampled, goats teeth sampled and an insect casing or cocoon. A mule path crossed the SE corner of quadrat, and it was later found to be part of the Walking Path to Kionia.

Birdsong was pleasant with numerous species present.

Q20 T2 MK132

Quadrat laid out N-S across an Easterly facing ridge (looking toward boundary cairn 91). Formerly a plantation stand it has been heavily felled with little regeneration. No seedlings were apparent. Several small clumps of *Quercus alnifolia* were recorded. Moss occurs on the NE facing slope. The ridge has an iron ore / sulphur carrying seam - a sample of which was collected. Several burrows were evident although many were abandoned. No wildlife was sighted in the immediate area.

Sunday afternoon war zone with hunters on both sides affecting the bird life.

Q21 T2 MK133

Quadrat laid out below ridge line - former plantation heavily felled. No natural regeneration of *Pinus brutia* was evident although few thickets of *Quercus alnifolia* were present. There was a good ground cover of *Cistus* but little *Lavandula Stoechas*, and no grasses. Some *Pistacia terebinthus* grew along Northern edge of quadrat. The site is in good condition - but trees showing very little increment potential. Exposed ground generally red/orange in colour.

Q22 T2 MK145

Quadrat dominated by large rock outcrop dropping to a cliff face in the NE corner. Below NE - SW sector there is a gully and associated scree which has stabilized. The Northern end carries a dense thicket of *Quercus alnifolia*. The open (and higher) Southern end has been clear felled and carries *Cistus*. Only a few trees remain in the quadrat with a few saplings on the South Eastern section of the rock outcrop. No regeneration was observed in the *Cistus* patch. The rock outcrop provided an excellent habitat area for wildlife and a partial skeleton was recovered from below a high hollow in the cliff face. Five different moss types were evident.

Q23 T2 MK144

Quadrat laid out on rocky slope with an Easterly aspect. The former production stand is now heavily felled. Good regeneration of *Pinus brutia* seedlings and saplings and *Quercus alnifolia* are present. *Pistacia terebinthus* carrying a heavy gall infection. Many small burrows and middens occur on the otherwise exposed areas. Active bird life was observed. One fox casting was noted. Some charcoal scatter, but little other evidence of wildfire other than lack of sub-strata species. No garbage or hunting debris recorded although side is close to the road line.

Q24 T2 MK141

Quadrat laid out with boundary cairn 142 as Southern corner. Slight depression tending NNW. Original plantation had experienced some felling, however, the undergrowth severely damaged by wildfire.

Good regeneration of *Quercus alnifolia* and ground covers were evident with young Buckthorn germinating. *Arbutus andrachne* was coppicing from roots with only 30cm maximum growth. Young scattered *Pinus Brutia* saplings occurred in an area of open canopy. Dense growth was observed in the sheltered gully line.

Area was grazed and used as goat dump for dead animals. *Olea europea* and *Crataegus azarolus* show extensive browse. Old goat tracks cross the area. One well grown *Quercus alnifolia* was present with others much smaller and a few seedlings. Toadstools and Mushrooms, veterinary medicine and syringe were recovered from the quadrat.

Q25 T3 MK95

Quadrat laid out below road line on a steep slope (spur) straddling a gully line. The stand was even aged plantation, heavily grazed, sub-strata and ground cover to the extent that the few *Pistacia terebinthus* had been completely destroyed. Some scattered *Salvia* have had the flower heads eaten. A lone *Cistus* was observed in a patch where a tree has been felled creating a canopy gap - otherwise the ground is very sparse.

Garbage and industrial waste, dumped from the roadway, litter the gully line. Feathers from a bird were noted but not sampled due to their poor condition. Fresh goat droppings were observed. Some spindly *Pinus brutia* saplings occur within the stand but are of very poor quality.

Q26 T3 MK94

Quadrat laid out on hillside above an abandoned road. It is again an old plantation which has been felled. Very good regeneration of *Pistacia terebinthus* and *Quercus alnifolia* was observed with a good ground cover of *Cistus* and some grassy patches. Trees show vigorous growth though some storm damage was noted. Some *Pinus brutia* saplings have established but are spindly and suppressed.

An adjacent area had a warren of Fox earths and a scatter of goat bones in a nearby gully.

Q27 T3 MK93

Quadrat laid out across a saddle in a former plantation now heavily felled with no regrowth or regeneration observed. One senescent tree was present together with a few deformed *Pinus brutia* saplings. The area was crossed by old goat tracks and fresh goat droppings were observed. *Pistacia terebinthus* in the sub-strata had its lower branches destroyed by grazing.

One bone was collected and described, probably brought into the area by a fox from the carcass dump on the nearby road. On the higher ground of the quadrat, that is across the saddle - the ground cover and the litter depth was minimal. However, on both the Northern and Southern slopes of the quadrat there was a noticeable improvement.

Q28 T3 MK76

Appendix 7: Makheras Logbook

Quadrat laid out on steep slope below a ridge line. A plantation of *Pinus brutia* had been clear felled, however, original *Olea europea* remain - 15 in total. Some are growing closely together and initially looked like a thicket of wild *Olea europea*. Shoots from the base of the *Olea europea*, most of which have lost their central core, have reverted to the wild form, but where the trees have retained shoots on the old wood they change the appearance of the cultivated form.

The plantation is on stabilised scree now moss covered. A quantity of felling debris or fallen timber contributes to the litter level. The open areas of the quadrat carry a dense cover of *Cistus*.

A Chukar's nest was located and broken egg shell was collected. Numerous small burrows and middens were sighted.

Q29 T3 MK114

Quadrat laid out across a small ridge running from the road line along a spur to the SE. The soil along the ridge had been disturbed 2-3 years ago with good seedling regeneration. The older trees were in a planted formation; some of the NE slope were deformed through wind damage.

Very little litter was present along the ridge line though some accumulation has occurred under the mature trees on the Eastern slope.

Q30 T3 MK115

Quadrat laid out along spur trending South with a steep slope to the West where a road dissected on a SE to NW line. Seedlings had established in the soft soil of a very steep embankment. The area had been heavily felled with immature trees retained. Very little regeneration had occurred. Several scrape-like forms were observed and may be attempts at soil disturbance, as seedlings were present.

A snake skin was collected from within the quadrat and a rock outcrop provided a habitat area.

Q31 T3 MK116

Quadrat laid out across a small creek with a water flow at the time of recording. Steep to very steep scarps heading down to the water line. The southern facing slope was drier and more open while the northern facing slope featured moss-covered rocks and ferns. The creek floor was dominated by *Rubus sanctus* and reeds with many species of mints. One large *Platanus orientalis* had fallen into the quadrat and was sending up vertical limbs. Twigs from a *Quercus coccifera* were collected for identification. Only two mature *Pinus brutia* remain in the quadrat. Others, particularly on the Southern facing slope having been clear felled.

An egg shell fragment observed and collected while animal pathways were also discerned as running parallel to the creek bed.

Q32 T3 MK119 06.11.94

Quadrat laid out on steep slope trending NNW-SSE. The old plantation *Pinus brutia* showed remarkable regeneration and an uneven aged stand is developing without, as yet, any senescent trees. Sapling and pole trees appear vigorous.

A hare run appears to cross the quadrat and two separate areas of spoor were noted.

Very few sub-strata trees grow in the quadrat - only one *Crataegus azarolus* and the *Quercus alnifolia* appear only as seedlings or small shrubs. Good ground cover of *Cistus* with *Thymus* on the rocky areas. Some *Lavandula stoechas* but not as heavy as in many other areas.

Q33 T3 MK129

Quadrat laid out on a steep spur below road line trending North. The majority of quadrat lies on the sunny Eastern side of the spur. A very cold wind was blowing from the snow of the previous evening.

The plantation has been heavily felled with some larger saplings and pole trees retained. A good sub-strata of *Quercus alnifolia* and several mature *Arbutus andrachne* exist which show signs of previous coppice. Ground litter depth is of variable deep - very deep depending on felling debris. Many small stumps from Christmas tree harvesting. The quadrat carried in excess of 25 stems.

Q34 T3 MK130

Quadrat on Southern side follows ridgeline as noted for MK146, however the ridgeline is wider and affords considerable shelter. A good quality stand of plantation timber exists and has been partially felled. A dense mat of *Pinus Brutia* needle litter exists, with scattered *Cistus*. However, immediately below ridge line to the North there is a dramatic decline with no *Pinus brutia* evident - only a few stunted *Quercus alnifolia* and *Arbutus andrachne* growing on the exposed rocky slopes and unstable rock outcrops. Almost no ground cover on this area. Dominant factors may be the extreme wind conditions. The slope to the South-East (Outside forest) supports a dynamic stand of uneven-aged *Pinus brutia* which appear to have been an extension of the original plantation.

Q35 T3 MK146

Southern quadrat section runs East to West along a narrow ridge between boundary cairn 168 and 169. Animal tracks follow the ridgeline. The quadrat has been formerly severely damaged by wildfire and possible grazing. Only one *Pinus brutia* of a CBH > 56cm was recorded. Some few scattered seedlings occur just below the ridgeline. The quadrat was dominated by dense thickets of *Quercus alnifolia* and *Arbutus andrachne*. The *Arbutus andrachne* has formed thick spreading clumps up to 10M in diameter. The bases show former coppicing and wildfire damage and breaks into the bark may indicate browsing. A large rock outcrop in the NE corner of the quadrat may provide some habitat shelter. The quadrat is crossed by animal tracks which contour around the very steep hillside. Adjacent on the Southern facing slope lies abandoned agricultural terracing with a completely different character.

Q36 T4 MK88

Quadrat laid out on steep slope above river line and below ridge line. A small gully with terracing below the quadrat was observed. The original plantation has been thinly felled and has produced saplings and pole sized trees. Some older trees are deformed probably by wind damage. The dominant sub-strata is *Olea europea* with long narrow leaves. *Crataegus azarolus* is also present but rare. Spring ephemerals are germinating, toadstools were also noted. No *Lavandula stoechas* grows within the quadrat though it is observed in an adjacent clearing. The area is crossed heavily by old goat tracks with fresh goat droppings. The *Olea europea* have been eaten back. Also, the *Pistacia terebinthus* and *Crataegus azarolus* show browse damage. Samples of the *Olea europea* varieties were collected as botanical specimens.

Q37 T4 MK87

Quadrat laid out on a gentle to moderate slope to cover an area of recent regeneration below a stand of older plantation. Trees have been seeded into gradoli and heavy thinnings (felled to waste) have occurred. Most saplings have responded and appear vigorous, a few have suppressed crowns. None of the seedlings have a Cbh > 54cms. Goat bones located within the quadrat and one scrap or old burrow noted.

Q38 T4 MK86

Quadrat laid out on hillside and short spur below the picnic area. Pit plantings (failed) and contour terracing support a young plantation of *Pinus brutia*. Stumps of older, mature trees are evident in some areas. Good growth is exhibited particularly on the northern side of the quadrat. Scattered *Crataegus azarolus*, *Pistacia terebinthus*, and *Cistus* in the more open areas. Trees have been thinned though no fell to waste was noted in the quadrat. Some burrows and two bullets were recovered.

Open areas were well grassed. Some evidence of browse on the hawthorn. Scattered bones with sharp splintered ends recovered.

Q39 T4 MK90

Quadrat lad out over small gorge and slopes above the creek line. An old plantation area with the better trees felled. Some pit planting has been attempted but failed. Water holes exist along the bottom of a small gorge. The canopy was open and good cover of *Cistus* is carried, except on very barren slopes where the gullies joined. No *Pinus brutia* regeneration was noticed. The area appears sheltered. Some old fox castings were sampled and an egg collected, possibly fallen from a nearby tree. No evidence of rodents was recorded.

Q40 T4 MK91

Quadrat laid out over two spurs and a gully with a southern aspect. It is an open area in an old plantation. Remnants of old wild *Olea europea* - one with browse scars evident - lie on the western slope. Regrowth from the base of one was noticed however others appear to have been uprooted. A sample of vine was taken - this species has not been observed previously. Evidence of hares, that is, droppings were abundant and a nest-like structure at the base of a *Pinus brutia* was noted. Some bird life present but not in great numbers.

Q41 T4 MK81

Quadrat laid out on a slope with NNE aspect. Very few trees exist in the stand and only a scatter of seedlings were present. No evidence of wildfire or felling so it is presumed that the plantation did not establish well on the stony ground. Soil moisture appears high but soil horizon is very shallow. A light layer of *Pinus brutia* needles has developed and *Cistus* seedlings are abundant along the lower level of the quadrat. A goat head was recovered from the quadrat.

Q42 T4 MK74

Quadrat laid out on steep slope high along the ridge line trending N-S along the transect bearing. The dominant standing species were *Crataegus azarolus* with two *Pinus brutia*, one old mature tree exhibits some signs of wind damage. A very dense cover of forbs, the main species asphodel with other species also represented. No *Pinus brutia* seedlings, some young *Crataegus azarolus*, with wild *Olea europea* occurring outside the quadrat but in the immediate vicinity. The soil is very shallow and rocky though some moss was encountered.

Q43 T4 MK73

The quadrat is approx. 100 m due North from MK118 - and a great contrast in species and habitats is apparent. The southern section of the quadrat is dominated by a sharp rock outcrop with a narrow passage. The quadrat lies on a natural terrace above a steep slope. One old mature *Pinus brutia* and 3 seedlings of various ages are present. The ground cover is dominated by *Cistus* with areas of moss and fine grasses. The rock outcrop has weathered with plants, some ephemerals, rununculi and *Lavandula stoechas* growing in the cracks. Patches of thick moss also occur. The area seems sheltered from the prevailing wind.

Q44 T4 MK118

Quadrat laid out below a mule path and incorporates a stand of trees in excess of 91 years old. Excellent regeneration of *Quercus alnifolia* in the thick ground litter and of *Pinus brutia* seedlings occur in areas previously felled and where the ground had been slightly disturbed.

A black rat was disturbed while the CBH measurement was being carried out. Notes on its characteristics were taken and the specimen photographed. The soil was grey in appearance - may be decomposing gabbro. A midden associated with rat habitat was recorded.

Some selection fellings have taken place, date of harvest currently unknown. Two trees in quadrat with CBHs in excess of 1.91M are yet to be measured.

Q45 T4 MK119

Quadrat laid out below E-W ridge line marking Southern edge of compartment and above, to the SE of a series of agricultural terraces supporting ancient *Olea europea* trees and wild seedling *Olea europea*. All the mountain *Pinus brutia* have been clear felled leaving the area open & barren apart from *Cistus* cover and clumps of *Quercus alnifolia*.

Rodent burrows located under the *Quercus alnifolia* thickets while the scree has stabilised and carries moss rocks and the occasional ferns and cleavers. Some scattered, rare, thorny burnet present. Fox like droppings noted and one sample was collected.

Appendix 7: Makheras Logbook

Q46 T4 MK120

Quadrat laid out on South facing slope of former plantation now clear felled except for bent stems. Some evidence of charcoal associated with rodent burrows in stumps below ground level. Debris from felling still evident but decomposing. The hillside is dotted with rodent burrows and runs under a thick cover of *Cistus*. Not a great variety in the flora - many of the *Quercus alnifolia* look old - some *Pistacia terebinthus* exhibit old stem die back. Just outside the quadrat to the South there is an *Acer obtusifolium* with CBH 77cm. Old wood (fallen) associated with the clump which seems to be a feature of the mixed thickets within the quadrat.

Q47 T4 MK121

Quadrat laid out on a steep slope heavily disturbed and eroded after clear felling of most commercial trees. A stand of *Quercus alnifolia* has been damaged by rock fall. Screes are unstable. A good regeneration of hardwood species in stable areas has occurred in areas with litter cover. *Galium aparine*, vines & ferns indicate a high soil moisture content. Several examples of the unidentified *Quercus alnifolia* species recorded. Some thorn, *Lavandula stoechas* and *Myrtus communis* present. Geological samples taken from scree as some intrusions seemed evident.

The quadrat represents an extreme example of environmental degradation following ill-advised forestry practice.

Some evidence of localised rodent activity, and a fox casting was found immediately outside the quadrat.

Q48 T4 MK127

Quadrat laid out over a creek floor at the Northern age of the compartment so as to encompass a rich riverine area. The southern bank of the creek rose into a steep cliff formation covered by mosses and some *Quercus alnifolia*. The creek line was dominated by ancient, coppiced *Laurus noblis* and *Platanus orientalis* and wild, old spreading *Myrtus communis*. The prickly vine and *Clematis cirrhusea*, and brambles climb into the sub-strata upper branches making a dense mat. Ferns dot the mossy rock areas. A rat run was noted. Only a few of the *Acer obtusifolia* carry seed, however, regeneration appears to be dynamic. Many *Laurus noblis* trees at both "pole" and sapling stage were noted. The first cyclamen of season was noted also.

Q49 T5 MK13.5

Quadrat laid out in SE section of the compartment below the summit of Skordokephalos. The forest stand appeared natural with two old trees deformed by forking at ground level. A few saplings and pole trees were associated with those in the quadrat and a 1 year old seedling was noticed. A large number of young *Crataegus azarolus* were in the quadrat and a few wild *Olea europea* and *Quercus alnifolia*. *Pistacia terebinthus* was also present. *Lavandula stoechas* was rare. Most of the burrows appeared to be under the shelter of trees and shrubs.

Q50 T5 MK84

The quadrat may have been plantation but there is insufficient evidence. Heavy fellings and fell to waste have occurred with a thick regeneration of wild *Olea europea* and thorn with *Crataegus azarolus*, *Clematis cirrhusea* and prickly vine create a very dense ground cover and sub-strata. Old runs - either fox or hare - cross the quadrat but currently appear unused. Rodent burrows occur under trees. A thick accumulation of litter protects the slope which has only one small area of scree. There are no *Quercus alnifolia* in the quadrat or immediate area adjacent.

Q51 T5 MK83

Quadrat laid out on slope during rain shower. Originally plantation -clear felled with a sapling and pole *Cupresses Sempervirenes* stand gaining dominance. Scattered hawthorn also present while there was no occurrence of *Quercus alnifolia*. One midden under a *Cypress* was recorded and no other faunal activity noticed although tin cans littering the area may have been dug out by foxes. The area seems to have been pit planted possibly for the *Cupresses Sempervirenes*, however, most pits are barren.

Q52 T5 MK82

Quadrat laid out below summit of "Mushroom Hill", a steep conical mountain with a ridge to the East. The area is exposed and carries a few *Pinus brutia* both in senescent and storm damaged condition. Both saplings recorded are deformed. Numerous wild *Olea europea* and *Crataegus azarolus* occur within the quadrat. The slope is steep to very steep with moss covered stones. No cleavers or wild vine are evident. Very little faunal activity was recorded though two vertical snake like cavities were recorded. A large number of mushrooms were gathered from one spot within the quadrat under a *Quercus alnifolia*.

Q53 T5 MK79

Quadrat laid out on a steep slope just below the ridge line and above a forest road from Mandra tou Kambiou. The area was a former plantation with heavy fellings occurring and some sapling regrowth.

Abundant evidence for faunal activity. Some areas of ground were bare and denuded of soil, whilst other areas, particularly under good canopy cover, carried a moderate litter level. The tops of several young trees showed storm damage.

Q54 T5 MK77

Possible plantation severely storm damaged with deformed trees and several wind-blown. Quadrat laid out South of the ridge line. The slope was steep with only a thin soil cover, some grasses and litter occurred under the canopy. Every *Quercus alnifolia* had at least one midden site. The hare droppings were old. An old fox den was located on transect outside quadrat just below the mule path.

Q55 T5 MK72 Quadrat laid out below bend in road on transect line according to compartment map. Original plantation with some fellings and considerable storm damage evident with several fallen trees and many deformed crowns. Some sapling and pole regeneration. Many young *Quercus alnifolia* and two well-grown *Pistacia terebinthus* are present in the stand.

Appendix 7: Makheras Logbook

Q56 T5 MK71

Quadrat laid out from a steep spur trending East in alignment with MK72. The terrain is very steep with frequent rock outcrops. The forest cover is uneven aged with some scattered *Pinus brutia* saplings. The dominant regrowth appears to be *Quercus alnifolia* with one clump of *Crataegus azarolus* and *Acer obtusifolium* growing in the shelter of a *Pistacia terebinthus* thicket. Middens occur under many of the *Quercus alnifolia*. Some ferns are growing in rock crevices and *Clematis cirrhusea* and *Galium aparine* occur under denser canopy cover. A small brown snake about 7cm in length was also observed. It disappeared into the *Pinus brutia* litter after first being sighted on a scree slope. Robins were observed in the lower branches of a *Pistacia terebinthus* though many bird calls were not identified.

Q57 T5 MK68

Quadrat was laid out on a very steep, partially stabilised scree. It was not measured out with a tape due to the hazardous nature of the terrain. Loose, moss covered rocks and mobile scree dominated the area with a few mature *Pinus brutia*. No *Pinus brutia* regeneration was noted. Several clumps of *Quercus alnifolia* provided shelter and stabilisation to the slope. Scattered *Cistus* occurred under the canopy breaks. Very little birdsong - no middens or evidence of rodent activity was recorded.

Q58 T5 MK66

Quadrat laid out across spur jutting out onto river line. The spur ends with a rock outcrop before a steep cliff face dropping to the gorge. An old track crosses the quadrat and appears to descend to an alluvial area at the junction of a creek with the river. One pottery handle fragment was collected from the track. No other pieces were noted. The area has been heavily disturbed by hunters - hides have been built from green branches and rocks. The hunting debris is dense. A good scatter of *Crataegus azarolus* exists with many small saplings. Several large middens were located on the southern aspect of the quadrat while hides - one inside the quadrat and one just outside were built on the northern edge of the ridge line. 83 Cartridges were collected from the quadrat.

Q59 T5 MK59

Quadrat laid out from the ridge line down a western slope through an old plantation, partially felled and felled to waste, and exhibits storm damage. Some marginal *Pinus brutia* regeneration whilst the *Quercus alnifolia* has very strong seedling cover. Thickets of *Quercus alnifolia* dominate the quadrat with *Arbutus andrachne* thickets also co-dominant.

A fox casting, burrows and hunting debris were recorded. Mushrooms were collected for analysis as a food source supply as many had been scratched out of the ground.

Q60 T5 MK124

Quadrat laid out below a spur line dividing MK124 & MK59. The terrain was a steep slope with stabilised scree and with an almost total canopy cover of *Quercus alnifolia* with *Arbutus andrachne*. The *Arbutus andrachne* had been coppiced by axe and had suffered die back with new growth from the base. Many of the stems of the *Quercus alnifolia* carried lichen. A good cover of litter was observed. Many *Quercus alnifolia* seedlings had established and one seedling *Arbutus andrachne* was noted. Two *Arbutus andrachne* middens were recorded along the eastern edge of the quadrat. A few small scrapes and burrows were also recorded.

Q61 T5 MK125

Quadrat laid out on the northern, exposed slope over very steep terrain just below ridge line marking forest boundary. Local boundary cairns not sighted apart from one approx. 300m to the east along cliff face. Strong gusts of cold wind were experienced - snow patches lay on the ground. One snow covered stone had a bird imprint but no other activity was noticed. Insect casing and hare dropping were noted. The climatic conditions made recording difficult - canopy plan is not very accurate - only 2 *Pinus brutia* grew in the quadrat which was otherwise dominated by *Quercus alnifolia* (uneven aged) and an occasional *Acer obtusifolium*. One *Arbutus andrachne* was seen. *Quercus alnifolia* of the older age group all suffered storm damage - one had died and was rotting *in situ*. The quadrat appeared devoid of faunal activity.

Q63 T6 MK9

Quadrat laid out N-S along a bearing on a steep hillside. The terrain was partially stabilized and partially a very mobile, scree. Stabilized and semi-stabilized areas carried a heavy moss cover - several varieties of which were evident.

Only three *Pinus brutia* were standing. A heavy cover of *Quercus alnifolia* to the centre and western sectors of the quadrat.

Rat activity observed and recorded. A dog or fox skull was collected from an area to the east of the quadrat.

NOTE: No quadrat was put into MK8 as almost the entire area is currently being quarried.

Q64 T6 MK10

Quadrat laid out over small ridge sloping to the South West. A small cluster of *Pinus brutia* occurred to the Northern sector of the quadrat above old terracing . All trees show suppressed crowns. Small saplings of the *Olea europea* and *Pistacia terebinthus* were present. A Chukar was disturbed in laying out the quadrat. Some hunting debris occurred and was recorded.

Q65 T6 MK11

Quadrat laid out across a narrow gully giving a North and South aspect. The ground between scattered *Cistus* was moss covered. Some small *Crataegus azarolus* and wild *Olea europea* dominated the quadrat. A birds nest was observed, recorded and photographed in *Crataegus azarolus* in the creek line. Some seedling growth of annual species was evident.

Q66 T6 MK16

Quadrat laid out over ridge line trending west. A Rock outcrop in the centre of the quadrat appeared as an archaeological ruin after pottery was sighted during the laying of the quadrat. An archaeological report has been made. Several *Olea europea* trees may have been planted in association with the structure which afforded excellent views both up and down the valley.

Appendix 7: Makheras Logbook

Q67 T6 MK15

Quadrat laid out over an area mechanically pit planted (unsuccessfully). The sub-strata was dominated by *Crataegus azarolus* and was rich in bird life. A rodent burrow was discovered under a *Cistus* bush. Ground cover was generally patchy with grasses growing between the sparsely placed *Cistus*.

Q68 T6 MK14

Quadrat laid out with the western section lying along the transect line. The character of the area was one of former terraces now heavily disturbed by clear felling and pit planting with success rate about 25%. The terrain was stable with spoil from pits carrying grass cover. Some older *Crataegus azarolus* had been damaged by mechanical reforestation. Saplings from *Crataegus azarolus* were evident.

Faunal evidence included a newly killed, gnawed bone and small amount of fur recovered from an area of rodent burrows and runs. Bird call was local.

Q69 T6 MK25

Quadrat laid out along spur trending east and with a slope to the north. Originally a plantation across old terracing - possibly agricultural, the are now carries some remnant *Olea europea* with wild seedlings evident. Mature trees on the northern aspect have been clear felled. No regeneration has occurred. Some trees exhibit storm damage. The area was well grassed though the gully line seems unstable with large rocks from the road line above in a semi mobile condition. Bird call was local. Rodent activity was recorded and some evidence of hare activity. No hunting debris was sighted.

Q70 T6 MK24

Quadrat laid out on gentle to moderate slope. Old plantation *Pinus brutia* dominated although there were extensive canopy gaps where felling had occurred. A good regeneration of sub-strata species was noticeable and is dominated by *Crataegus azarolus*. Seedlings of *Quercus alnifolia* are occurring under larger trees in sheltered areas. An old road way crosses the southern section of the quadrat and is well grassed. Two terraces, possibly abandoned agricultural, were also crossed. The upper terrace has disintegrated more than the lower where several lengths showed a height of 2-3 feet.

An "elite" tree was recorded in the quadrat and photographed. Rodent activity present and no hunting debris was noted. No bird call was heard in the locality though recent droppings were noted.

Q71 T6 MK23

Quadrat laid out on very steep, moss covered slope of stabilised scree. Very few old trees with no regeneration apart from one seedling *Quercus alnifolia*. Wild *Olea europea* occur and exhibit storm damage. Although the aspect is the one most preferred, the rocky nature of the slope and very little soil put the quadrat into the 'C' class.

Rodent activity recorded, a few birds, and no hunting debris.

Q72 T6 MK22

Quadrat laid out along a spur below terracing with scattered *Pinus brutia* seedlings, *Crataegus azarolus* and *Cupresses sempervirenes*. The area was burnt by the wildfire of 1986 and the regrowth on the area of the quadrat has been natural. Larger *Olea europea* show fire scars on dead branches but have regenerated from the base. Seedling *Olea europea* also occur. Many sapling *Crataegus azarolus* occur on the quadrat and appear vigorous. A few *Pistacia terebinthus* also exist and it is difficult to assess whether they have been coppiced from the root stock or not. One *Pinus brutia* sapling is growing on the top of a rock outcrop.

From the *Pinus brutia* cones lying in the quadrat it appears that the burning occurred before the cones had reached maturity. Some blackened stumps *Pinus brutia* are evident. The rodent activity is present together with evidence of fox and ideal conditions exist for reptiles. The open aspect has a cover of *Cistus* with some bare soil. Many areas have been heavily disturbed by rodents digging for small bulbs. Very little bird activity was evident.

Q73 T6 MK29

Quadrat laid out on the edge of an old plantation and crossed onto a barren rocky scree. The treed area was stable with a good depth of litter. Some *Quercus alnifolia* seedlings had failed and *Pinus brutia* saplings were spindly due to lack of light. One fallen tree was recorded. Bird feathers collected. The quadrat is not typical of the compartment as a whole.

Q74 T6 MK30

Quadrat laid out from along ridge line marking the edge of the compartment and above an old mule path marked on the maps but now severely eroded. The terrain is steep to very steep - semi stable scree with areas of mobile scree. Moss covers rocks on the more stable areas. The quadrat appears to have been planted, though very few trees remain. The area may have been burnt 30 - 50 years ago but there is no evidence of wild fire apart from the erosion. The sub-strata is developing well but the lack of canopy species will inhibit the long term rehabilitation of the slope. Bird call was local but very little faunal activity was evident. No hunting debris was recorded.

Q75 T6 MK63

Quadrat laid out on a SE sloping hillside. An old plantation with some old *Quercus alnifolia* coppicing with both axe marks and sawn pieces. Good regeneration of *Quercus alnifolia*. Die back noticeable in the *Cistus*. Good evidence for faunal activity and birds were in a reasonable quantity.

The quadrat is in marked contrast to MK30 which lies approx. 40 m to the north. The quadrat lies below a well made path leading from Mandra tou Kambiou to Kionia. The quadrat exhibits a great potential for an internal dynamic under better silvicultural practice.

Appendix 7: Makheras Logbook

Q76 T6 MK64

Quadrat aligned between the base tree and marked by a cairn as terrain was very steep, rocky and dangerous. A variety of sub-strata species were evident - mostly very old specimens with lichen growing on low trunks and lower branches. Two *Pinus brutia* were providing a needle litter cover which was assisting in the stabilisation of the slope. The area was photographed prior to the commencement of the survey. A sample of lichen was taken. Large Chukar like droppings were observed and some rodent activity was recorded.

Q77 T6 MK61

Quadrat laid out well below ridge line and incorporates some well grown plantation stock with old *Quercus alnifolia* with good *Quercus alnifolia* regeneration. The *Pinus Brutia* exhibit a distinct yellow colour on needles - very little regeneration is apparent although many of the older *Pinus brutia* have attained a good height. Some stems are deformed through crown suppression. Good cover of litter to the south of the quadrat while the northern sector is more open, barren and rocky. Some older *Crataegus azarolus* are in stag condition.

Q78 T6 MK57

Quadrat laid out on the west of a spur trending north and separating forest compartments 57 & 56. The terrain is steep with a fine, semi stabilised scree with areas of mobile scree. The quadrat is crossed S->N by a gully line which contains a small spring. The dominant species is *Quercus alnifolia* which shows signs of axe coppicing. Occasional *Arbutus andrachne* and *Acer obtusifolium* are also present. One mature *Pinus brutia* and one developed sapling are examples of the original plantation. Under the canopy cover a level of leaf litter has developed. Some *Cistus* are present on the higher, more open areas of the quadrat. Thick moss occurs in dense canopy areas and below the spring. Bird call local and one large midden was recorded.

Q79 T6 MK56

Quadrat laid out on a very steep slope and is dominated by a dense, old *Quercus alnifolia* thicket, and a series of forestry terraces now badly eroded. The terracing has been colonised by many *sumach* trees which, though badly storm damaged, are rehabilitating the area. Only two *Pinus brutia* were located in the quadrat. A bird of prey, possibly a falcon, was observed above the quadrat. No rodent activity was noticed.

Just below the quadrat, approximately 3m to the north, a small spring was located. Two pieces of red coarseware body sherds were located: one in the spring basin and one just outside. Neither were diagnostic and were replaced *in situ*. The *Quercus alnifolia* thicket had been coppiced for many years and stumps of different ages and diameters were noticed.

Q80 T6 MK55

Quadrat laid out on a southern slope below peak of the "Double Nose." The area is windswept and exposed - storm noise inhibited the faunal survey - particularly for bird call.

The original *Pinus brutia* plantation has one old fallen tree now mostly decayed. One seedling *Pinus brutia* was observed. Heavy lichen is carried on almost all the *Quercus alnifolia* and *Pistacia terebinthus*. Soil below humus layer is lighter brown and a sticky consistency. Open areas carried *Cistus* and *salvia*. Some *Cistus* appeared very old with thick stems. Some rat activity was recorded with a large midden just outside quadrat.

No bird call was heard.

Q81 T7 MK1

Quadrat laid out on steep slope and within an area of remnant plantation not burnt by the 1986 fire. A few mature trees were standing and a few stunted saplings. Several thickets of *Quercus alnifolia* were present some showing axe coppicing. A few young *Pistacia terebinthus* were also present and a group of *Crataegus azarolus* on the lower slope.

The flora was species rich with a new orchid specie observed, although the flower was not open. Old goat tracks crossed the quadrat and were concentrated in the lower sectors. Hunting debris was noted and collected. Hare droppings were also recorded.

Q82 T7 MK2

Quadrat laid out on a steep slope and transected by an old forest road. The are had been burnt with some standing trees showing fire scars. Sub-strata trees, particularly the *Pistacia terebinthus* exhibit post-fire regrowth. Some faunal activity noted and bird call local.

The quadrat was largely open and exposed with soil depth and humus concentrated under the standing trees. Open areas carry a thin cover of grass. A geranium species was observed in flower. The area below the road was largely of unstable scree - a result of the road line. Trees and shrubs exhibited storm or rock-fall damage. The upper area of quadrat appeared stable.

Q83 T7 MK3

Quadrat laid out on recent gradoli below MK.A.12, some old terrace and wall still evident. Several *Olea europea* trees destroyed by fire were standing. Saplings were growing on two gradoli - pottery protruding from cuttings and eroding down slope. Fox evidence was recorded.

An adjacent archaeological site was measured and recorded as MK.A.13. One large piece of pithos was located in quadrat.

Q84 T7 MK4

Quadrat laid out from the gully line where fire had coppiced the *Platanus orientalis* and *Alnus orientalis*. A thick cover of *Rubus sanctus* grew along the creek line and a clump of bamboo was evident. Old terracing was partially obscured by the new gradoli. Pithoi and tile fragments occurred in a scatter. Older trees were burnt and stumps remained. Some coppicing of *Olea europea* was evident.

Q85 T7 MK20

Appendix 7: Makheras Logbook

Quadrat laid out below old road line and above gorge. A small stream was flowing outside the western sector of the quadrat. The area had been planted with *Pinus brutia* - one seedling had grown spontaneously. An old *Olea europea* tree with cutting across its base had re-sprouted. One *Pinus brutia* had fallen as a stag following storm damage. The base tree was lifting from the ground.

Much rodent activity was evident and several large middens were recorded. Local bird call was noted.

An old path crosses the northern sector of the quadrat. Much storm damage was noted outside the quadrat on the eastern side.

Q86 T7 MK21

Quadrat laid out below forest road and across a series of forest terraces established after the fire of June, 1986. *Pinus brutia* seedlings appear well established with some scattered *Cupresses Sempervirenes* and *Cedrus brevifolia.* Seedlings have been thinned. Older *Pistacia terebinthus* show coppice marks (axed) and fire damage. Older trees were felled post fire and a few stumps remain. No hunting debris noted. Faunal activity was recorded. Pottery density indicates an archaeological site (MK.A.8). *Pistacia terebinthus* carried heavy quantity of galls. Old trees damaged by the fire and subsequently felled carried annual ring lines of up to 110 years (counted by S.Swiny)

Q87 T7 MK38

Quadrat laid out on a transect line above a gorge with a flowing creek. Two standing trees remain from the original plantation: one tree showing burn marks across a saw line while some old senescent branches also carry burn marks. *Quercus alnifolia* and *Pistacia terebinthus* which occur along the gorge have also coppiced due to burning. The quadrat area has been re-afforested using gradoli and sown to *Pinus brutia*, *Cupresses Sempervirenes* and *Cedrus brevifolia* (possibly planted). Growth for both the *Pinus brutia* and *Cupresses Sempervirenes* appears vigorous. The *Quercus alnifolia* has, to a varying degree, a deeply serrated edge. A flat stone near the bast tree shows plough marks. Old artificial terracing remnants - mostly bulldozed indicate past agricultural activity. A sample of baked earth - possibly due to intense wild fire heat was collected. Slag like substance was noted but not collected.

Q88 T7 MK36

Quadrat laid out below a spur line trending NNE. Old plantation trees standing all showing burn marks from the 1986 fire. The extent of the fire and burn marks on the western face of trees extending up in excess of 12 feet and between 3-4 feet on eastern face. Trees higher up the slope are more greatly affected. Several *Quercus alnifolia* were coppiced by fire and have dead branches still standing. The slope is very steep with retaining walls for old agricultural terraces evident and these are built up to approximately 3 feet high. *Acer obtusifolium* is also coppiced. The eastern side of the quadrat is dominated by trees with some unstable scree developing. Good cover of *Cistus* is carried on open ground.

Q89 T7 MK34

Quadrat laid out over a rocky slope above loop in the roadway around the spur. A few scattered, stunted *Pinus brutia* to the west of the quadrat have produced a needle layer and this helps to stabilise the lower slope. The upper slope is bare and exposed with unstable scree. Some old to very old *Crataegus azarolus* present and coppiced *Quercus alnifolia* with old axe marks. A few *Pistacia terebinthus* with lichen growth. There was no noticeable regrowth of any species. Bird call was local.

Q90 T7 MK62

Quadrat laid out on moderate slope below the road line in a slight depression. The area was Dominated by an old *Quercus alnifolia* thicket which shows signs of axe and saw coppicing. The standing *Pinus brutia* have lichen adhering to their bark and the *Quercus alnifolia* also carry heavy coatings of lichen. Two small *Pistacia terebinthus* were noted and numerous small and seedling *Acer obtusifolium.* Many small and seedling *Quercus alnifolia* were also observed. The small *Pinus brutia* may be old. Either undeveloped trees or seedlings occurring on the edge of the *Quercus alnifolia* thicket. Apart from the small area of open canopy which supported a *Cistus* ground cover the forest floor was not covered. All the trees measured showed little capacity for further increment.

Q91 T7 MK52

Quadrat laid out on sunny spur on the transect line from the north. The old plantation stand with a few saplings in very poor condition and evidence of several fallen saplings. An *Olea europea* lay just outside the quadrat and old footpath ran along western sector. Local bird call recorded and no hunting debris. All *Pinus brutia* appear to have a small amount of increment potential -those occurring in the middle of the quadrat more so than others.

Q92 T7 MK51

Quadrat was laid out in sight of boundary cairns on a moderate south facing slope. Area shows signs of old terracing and plantings with some restricted *Cupresses sempervirenes.* The standing *Pinus brutia* have a distinct yellow to their needles perhaps indicating some trace element deficiency. The *Quercus alnifolia* clumps are also growth stunted and carry a heavy lichen cover on trunks and branches. Rat and hare activity was recorded.

Q93 T8 MK5

Quadrat laid out below old agricultural terracing. The area was burnt in 1986 and shrubs were coppicing from root stock. Pinus saplings grew on the gradoli. Bird life was local.

Q94 T8 MK41

Quadrat laid out below the spur and above the Ped River gorge. The slope was steep, very rocky with some scree patches. Storm damaged *Quercus alnifolia* was stunted. The area was crossed by telegraph power lines supplying the village of Lazania. The slope appeared very damp with moss covered rocks on the north facing. Many ephemeral plants growing between the rocks. Dandelion type plants were being eaten by rodents.

Appendix 7: Makheras Logbook

Q95 T8 MK40

Quadrat laid out between two spurs trending east with an old terraced gully line between them. The terraces were eroded but remain stable. An old *Olea europea* grove lay below the quadrat with some *Olea europea* trees above. The plantation trees remaining have growth defects due to insufficient light. Hare runs lay along the eastern sector of the area with droppings associated. Bird call was local. The area was sheltered with early morning sun.

Quercus alnifolia was coppiced. Wild garlic was growing in clumps on the old terrace lines.

Q96 T8 MK49

Quadrat was laid out on a spur trending west above Ped River line. It was on open exposed rock with no standing *Pinus brutia* though there was evidence of an old fallen stag tree now termite ridden. Stunted *Quercus alnifolia* and *Acer obtusifolia* with scattered *Olea europea* composed the sub-strata. An unusual petal-shaped sedum noted growing on the rock and may be very rare.

In the adjacent area some plantation *Pinus brutia* were evident in sheltered pockets. Otherwise the terrain was steep, very rocky, barren and exposed.

Q97 T8 MK46

Quadrat laid out on a steep NE facing slope - scree largely stabilised by *Pinus brutia* needle litter and various mosses. The trees are largely storm damaged but still show signs of possible increment. Bird call was local and rodent or large animal activity was recorded. There was considerable *Quercus alnifolia* seedling regeneration evident.

Just outside quadrat a tree (*Pinus brutia*) shows signs of resin tapping. No *Pinus brutia* regeneration was observed though some stems may represent stunted pole growth.

Q98 T8 MK45

Quadrat laid out across a narrow ridge line above a small stand of plantation trees to the west. One very storm stunted tree was positioned on the ridge line with small stand of mixed *Pistacia terebinthus* and *Quercus alnifolia* on the eastern aspect. Some rodent activity under *Quercus alnifolia* on eastern aspect. Bird call was local.

Appendix 8: Adelphi Logbook

T13 AD85 08:02 08.04.95

Quadrant laid out across gully line on a moderate slope. Gully with evidence of old retaining terracing - no riparian cover.

Trees in standard plantation form - two biggest felled (stumps wider diameter than standing trees. Much stem deformity in sapplings. Heavy litter in gully from felling. Area open with no sub-strata apart from two seedling hawthorn. No asphodels - some short grasses. Good seedling regeneration, however, very poor quality standing timber. Bird call local. Bones collected. Botannical samples collected for identification.

(Assistants: MJM Given, MC Kean & Jenny (K9))

T13 AD86 10:24 08.04.95

Quadrant laid out over creek line with gully confluence - slope moderate. Clump of myrtle and buckthorn in thicket and junction of creek lineswith heavy carry of clemetis and prickly thorn. Creek now dry though line remains moist. Coal tits nesting in quadrant in burrow into bank behind fallen branch. Close observation of habitat made by MJMG - photographed and activity videoed by MC Kean. Recent fellings in quadrant revealed heavy insect infestation in sapwood rendering the timber uncommercial. Insect lava may be food source for tits.

Gully lines have small, single layer of stone retaining walling to prevent erosion - of old workmanship. Area rich floristically though very open.

T13 AD90 11:57 08.04.95

Quadrant laid out on moderate slope with depression line running central to stand composed of several mature stems with numerous sapplings and a few pole stems. Mature trees show little potential for increment. Some turning of litter under pines - one midden recorded - processional caterpillar nest recorded and videoed. Overall impression of stand OK. Dry, no wind recorded. Cistus white flowering variety. Spoar from some animal collected - red growth observed under cistus.

T13 AD89 13:10 08.04.95

Quadrant laid out below road line just to east of ridge - slope steep, two gully lines which converge below quadrant. Scatter of standing trees all with deformed stems (dancing). Purple cistus scattered. One clump of cleavers with small terebinth and hawthorn on gully line. Capers sprouting. Bird call local.

T14 AD83 12:10 13.04.95

Quadrant laid out on steep slope to east of transect line. Area open with two mature trees (both with resin scarring at the base). Pit plantings generally failed with pits dry and exposed. Two hawthorn in flower. Pole trees in good condition.
(Assistant:MCKean)

T14 AD82 15:51 08.04.95

Quadrant laid out above road line and reaching almost to the ridge line - slope steep, ground open, bare with surface rocks. Good cover of onion weed, flowers and cistus. One wild olive / feral olive with birds nest. All standing stems - including spllings - with deformed trunks. \big midden under olive. Patches of heavy litter from fellings - all commercial timber removed. Bird call local. Hunting debris recovered.
(Assistants:MJMGiven, MCKean, Jenny(K9))

T14 AD75 14:10 13.04.95

Quadrant laid out on moderate south facing slope below fire trail. All commercial stems removed - some pit planting, mostly failed.

One small hawthorn otherwise area open, dry with scarse, scattered cistus and grasses. No faunal activity in quadrant noted.

T14 AD76 09:20 14.04.95

Quadrant laid out on slope between ridgeline and clearing above gully. All standing stock young pole, plantation pines dbh>56 cm. Contour gully lines had been setablished at time of planting and have grassed over. Stand dense with some fell to waste trees. Ground cover sparse -little cistus and some dried grasses between plantation rows. Some storm damage to tops of trees. Branches low to ground creating fire hazard.

T14 AD78 10:58 14.04.95

Quadrant laid out on open sunny spur. SOme early mature trees standing with pole and sapplings in "shelterwood" plantation form. Good regeneration of younger trees though some stem deformity noted. Ground disturbed through felling and most regeneration occuring in these areas. Pit plantings (failed) immediately outside quadrant. No sub-strata species.

T14 AD93 12:53 14.04.95

Quadrant laid out along spur - area open with good stand of mature trees and sapplings of different ages. Some felling evidence but not too heavy. No Sub-strata or species in quadrant other than Pinus brutia. Grass cover in open areas with scattered cistus. Almost all stems without deformity.

T14 AD100 13:10 17.04.95

Quadrant laid out on steep north facing slope. Dense plantation of sapplings with some pole trees. No sub-strata. Dense layer of needles with some fell-to waste. Many lower limbs dead and potential wild-fire hazard. Stems straight - one or two stag sapplings.

Appendix 8: Adelphi Logbook

T14 AD103 10:26 19.04.95

Quadrant laid out on very steep south facing slope just below the peak. Good standing timber - some trees marked for felling - some old fellings evident. Good regeneration with un-even aged stand developing. Some stems bent due to storm damage. Good cover of litter with branches from fellings in patches.

T14 AD102 11:53 19.04.95

Quadrant laid out on very steep exposed slope with northern aspect. Several mature trees standing with numerous poles and sapplings established. Heavily storm damaged with 9 fallen or broken trees. (mostly sapplings.)

T14 AD101 13:50 19.04.95

Quadrant laid out over hillside with a diagonal gully line. Good stand of mature timber with some pole trees - a few (not many) sapplings. Base tree struck by lightnint. (Photographed.) Some fellings have occured - some trees fallen through storm damage.

T14 AD74 10:18 20.04.95

Quadrant laid out over gully lines with both north and south facing aspects. Northern aspect with moss and lichen growth, southern slope dry and stoney. Standing trees clumped with rare seedling regeneration - some sapplings mostly with deformed trunks - a few straight. Open areas carry thin grasses and a few cistus - mostly on the northern aspect.

T15 AD77 13:13 20.04.95

Quadrant laid out on steep north facing slope and above a gorge line. Several old mature trees dominated the quadrant. These trees were either deformed or resin tapping scarred. Some sappling had developed. Ground cover mainly spring ephemerals with mosses. Area very dry on top though moisture appeared trapped beneath leaf litter.

T14 AD104 17:18 22.04.95

Quadrant laid out on steep slope - semi-stable scree, moss covered with many cavities. Several old mature trees standing withsome poles. Older trees tend to have minor trunk deformity - younger trees much straighter. Some seedling terebinthus. Ground cover thick with onion weed.

T14 AD105 18:11 22.04.95

Quadrant laid out on steep slope - some scree patches. Open areas carrying thin grass cover, some cistus (pink & white), ORCHES (seedling), onion weed also past flowering. Many small seedlings (this years germination) but very few older seedlings. Two olea europa seedlongs (small shrubs).

T14 AD107 08:59 23.04.95

Quadrant laid out over very steep slope from a ridgeline trending west. Two mature trees standing with a few sapplings and poles. Rock outcrops frequent. Some fallow trees outside quadrant. Not enough trees to discern plantation form.

T14 AD106 11:00 23.04.95

Quadrant laid out over rocky slope extending from a spur trending west. Clumps of senescent trees - deformed trunks, little girth and small height. Some sub-strata of olea europa, terebinthus & Quercus coccifera evident. Open areas between rock outcrops carrying grasses, scattered throny burnet & cistus (pink)

T15 AD81 14:45 23.04.95

Quadrant laid out between two ridge lines with small gully intersecting. Only one mature tree - several pole and some older, small girth and deformed limbs. A very few small sapplings, most with deformed stems. Terrain stoney with open areas carrying dry grasses.

T15 AD79 16:03 23.04.95

Quadrant laid out on moderate SW slope. Trees predominantly old mature or senescent with very few sapplings. Cistus mainly in areas of broken canopy cover. Few small buckthorn bushes - no sub-strata species. Open areas carrying thin dry grasses.

T15 AD73 09:20 26.04.95

Quadrant laid out over a series of small gully lines joining deeper gully to the east. Good stand of mature trees - some felling has occured but majority of original trees remain. Felling debris thick in patches. No sub-strata and very little ground cover apart from occasional cistus. One thorny burnet noted. Eastern gully line with some myrtle and a white flowered LABOUTAe. A few sappling and several good pole trees

T15 AD95 11:25 26.04.95

Quadrant laid out over moderate spur between two gully lines. Nice stand of mature trees in plantation lines with some inter-dispersed natural vegetation. Some deformity to young stems. Open areascarry thin, dry grasses. Semi-shaded areas with scattered cistus. No sub-strata. Some fellings have occured.

T15 AD94 13:40 26.04.95

Quadrant laid out from ridge line trending south with steep slope to above gully line to the east. Several mature trees standing along ridge line with some stunted and deformed sapplings above gully. Gground, open, bare and very rocky. Some scattered cistus (pink), thyme (rare) and occasional asphodel.

Appendix 8: Adelphi Logbook

T15 AD96 08:54 02.05.95

Quadrant laid out on moderate slope above gully line. Good stand of mature trees with sapplings in various growth stages. No new seedling germination noted. Area with old contour and pit planting (failed). Some site disturbance due to recent fellings. Heavy litter in patches. Most standing trees with straight HOLE.

T15 AD66 10:17 04.05.95

Quadrant laid out from spur line down steep slope. Old path crossed quadrant in the eastern sector with a branch path cutting across southern sector. Heavy litter patches from old fellings - some fell-to-waste. Standing timber straight pole, some good quality sapplings.

T15 AD68 12:42 04.05.95

Quadrant laid out from ridge line on edge of compartment. Slope very steep with rock outcrops and scree areas. Very dry with withered grasses and scattered cistus. Standing trees in generally good condition. Some stem and crown storm damaged. Some sapplings - dominantly bent stems. Patches of litter from fellings.

T15 AD67 13:52 04.05.95

Quadrant laid out on steep slope, stand of young mature trees - older ones with some bark burning to about 1m on northern side. Area very dry. Patches of dense litter from heavy felling. Scattered cistus - some very young pines and sapplings.

T15 AD69 11:00 05.05.95

Quadrant laid out over very steep gully line using base line N->S. Southern facing slope steep rock outcrops dropping sharply to gully line. Northern facing slope open - dry carrying thin grasses and a few pine sapplings. Area heavily cleared.

T15 AD70 12:30 05.05.95

Quadrant laid out from spur line down steep north-facing slope. Old goat tracksstabilised and moss-covered. Good stand of trees though previously heavily felled. No sub-strata. Scattered cistus. No seedlings observed. Some trees have reached maximum increment with girths <1m. Crowns generally well developed.

T15 AD65 10:08 08.05.95

Quadrant laid out on dry slope below abandoned roadline, SOme standing trees-generally good condition but no capacity for further increment. Most <1m girth. Area heavily felled. Some sappling regeneration in open areas with scattered cistus. Some litter debth under mature trees, most areas denuded apart from felling debris patches. Areas of disturbance around previous felling sites - no regeneration - minor erosion. No sub-strata. Annual species have dried and seeded.

T15 AD57 12:34 08.05.95

Quadrant laid out across ridgeline with very steep slope either side. Very good stand of trees - some burning at base of older tree - some lighning struck. Old fellings evident. Very little regeneration as trees are close together. Cistus scattered. Orchids in one patch sampled and photographed.

NOTE:Area photographed from roadway.

T16 AD45 09.05.95

Quadrant laid out between gully lines on steep slope. Good standing timber with poles and sapplings. Deep pine needle litter under older trees. Mature trees exhibit scarring on the base in north-facing direction. Scarring appears square in shape with sap excretion. One tree is lightning scarred.

T16 AD46 09:01 10.05.95

Quadrant laid out over very steep south facing slope. 2/3 of the quadrant covered by unstable scree. Small ridgeline carries good quality mature trees and some poles. Limited number of seedlings. Deep to moderate needle litter under trees, bare and denuded areas of scree adjacent.

T16 AD64 10:45 10.05.95

Quadrant laid out over wide, gentle spur between two deep gully lines. Area previously heavily felled with few mature trees standing. Several with girths <1m show little signs of further increment. Some seedling regeneration in canopy gap. Good stand of young pole trees in northern sector of quadrant. Some older trees show burning on bark at base.

T16 AD61 10.05.95

Quadrant laid out over steep south facing slope above gully line. Good standing trees along spur with some unstable scree in gully lines. Very little natural regeneration. Some scarring on base of trees.

T16 AD56 09:26 11.05.95

Quadrant was laid out using an E-W base line only as terrain was very steep and unstable. Good stand of early mature trees with some poles -limited number of sapplings and very little natural regeneration. Girth of trees was not measured due to safety factors. Ground carried a light cover of pine needle litter but was very unstable beneath this layer. One Quercus alnifolia acorn was located in the quadrant near some bird droppings. No Quercus alnifolia located in this area. Acorn buried.

T16 AD55 10:30 11.05.95

Quadrant laid out over very steep, dry open slope. Very stoney ground carrying only thin grasses and a few sapplings. The quadrant lies above a small COPSE of very good young pole trees, otherwise the compartment has only scattered mature treea with evidence of clear felling. Widespread erosion. The compartment has scattered Quercus alnifolia sighted for the first time in the Adelphi and sumach in some gully lines. The quadrant has been clear felled with very limited natural regeneration.

T16 AD59 12:25 11.05.95

Quadrant laid out from spur line down northern facing slope. Area clear felled with only very few standing trees and very little regeneration. Thick covering of cistus. Habitation indicated by presence of wild garlic.

T16 AD62 14:10 11.05.95

Quadrant laid out over steep northern facing slope. Clear felling has occured with few trees over the whole compartment. Terrain stoney with moss and lichen - now dried - on the rocks. Some wild asparagus. Little regenration and a heavy cover of litter from former fellings. Good cover of cistus.

Appendix 9: Makheras Fox Report

Makheras Forest: Fox Reports: Michael Given

MK 142 [T1, Q8] N of Pano Teratsotos. Alt 510m. 28/10/94
Fox castings. [Collected]. 3 * 1, 2.5 * 1, 3.5 * 1.
5 mandible fragments (from 2 lower mandibles, 1 individual), & 2 lower incisors. Size about same as in MK 139, though rather smashed up. Small Rattus rattus.
1 vertebra: width 5mm.
2 long bones (but not leg): 15mm.
1 long bone: 6mm.
12 unidentified bone fragments.

MK 138 [T1, Q9] Pano Teratsatos, steep NW slope, nr gully. Alt 705m. 28/10/94
Fox castings. White, 2-3cm long, bullet shaped. Feathers inside. Also several along ridge path above MK 142.

Boundary of compartments 138 & 139 [T1] Pano Teratsotos, high ridge. 28/10/94
Various castings, producing:
Upper incisor of rat (rather smaller than Jerry).
5 chunky bone fragments (larger than rat).
3 flat bone fragments: cranium? scapula?
9 unidentified bone fragments.
1 claw.

MK 139A [T1, Q10] Pano Teratsotos, high ridge. Alt 830m. 28/10/94
[MK 139A (ii)] Fox casting. In centre of quadrant, on top of ridge. 5 * 2cm. Very twisted, hard, whitish, not much hair visible from outside. Absolutely thick with hair inside. Mandibles found together.
2 mandibles (found together in casting); 1 in 2 pieces. Noticeably smaller than Jerry, though morphology identical. Cheek tooth row 6mm, ie borderline rat/mouse.
Cranium fragment, 7 * 8mm.
12 unidentified fragments.
1 long bone: 11mm, with one joint missing.

MK 140 [T1, Q11] Pano Teratsotos, S of summit. Alt 850m. 28/10/94
Fox casting. [Collected], under 3-4m Pinus brutia sapling. NB large vertebra found separately nearby. 6.5 * 2, 3.5 * 1.5. Twisted, many bone fragments showing, & coarse hair.
13 fragments of shrapnelled long bones. Diameter of one shaft 3mm.
7 feather quills. Length 10-20mm. Thickest 3mm.
3 fragments ?cranium.
2 long bones: 4mm, 10mm.
12 unidentified bone fragments.

MK 111 [T2, Q19] Prosefkis, spur above Prophitis Elias. Alt 830m. 30/10/94
Fox castings on paths. Not under trees. [Collected].
[MK 111 (ii) a] 5 * 1cm; 3 * 1cm. Whitish, twisted, tapering to tail.
5 feathers: 12-19mm. More smaller feathers not kept.
5 bone fragments, of large long bones.
[MK 111 (ii) b] 5.5 * 1.7; 4 * 2. Rounded ends. Not twisted.
7 whitish feathers: 12-13mm.
1 whitish feather: 6mm. Many more small feathers [Not kept]
1 unidentified fragment. No bones.
[MK 111 (ii) c] 1 complete: 5 * 1.5cm, + 3 fragments. Light grey/brown.
6 small bone fragments: including cranium?
[MK 111 (ii) d] 2 small pellets. Not twisted. Fairly fine consistency. Some fur, but not much. Very gritty.
4 chunky fragments from large bones, with much honeycomb structure.
6 fragments from long bones.
1 unidentified bone fragment.

MK 144 [T2, Q23] High spur & steep S slope opposite Moutti Mandra tou Klosmatou. Alt 760m. 4/11/94
3 fox castings. On track going down spur. Moist, dark, fresh.

[Collected]. 4 * 1.8, 3 * 1.5, 3 * 1.5cm. All bones in one 3 * 1.5 casting.
4 long bones: 22-27mm (Smaller than Jerry).
2 half long bones: 9-11mm (belong together?)
8 long bones: 10-12mm.
7 long bones: 4-6mm.
1 joint from long bone.
2 stones.

MK 94 [T3, Q26] Vromonera, gully bottom. Alt 660m. 2/11/94
100m up gully to NW: fox dens. In south-facing bank above stream bed & old track, and on slope on top of bank. c.10 holes, including 1 double. 2 footprints. Light cobwebs over many holes. Sheep or goat bones in vicinity. [1 slide, 2 prints].
CASTING near dens: 6 * 2cm, twisted, lightish brown; full of fur (thick, solid whitish, a little grey brown: rat?); no bones in it.
Outer corner of road 30m SE of quadrant, where met by old track. 10m wide scatter of ?goat bones, including 4 skulls. Fully-grown, but immature: teeth not worn, `wisdom' teeth not through yet. Dry & white, ie. old. Probably dumped at corner of road, then scavenged & scattered by foxes (cf. further up road towards MK 94: goat carcasses of varying ages dumped from road). [1 skull & lower jaw collected]

MK 130 [T3, Q34] High ridge, steep slope, Kambi tou Arayi. Alt 890m. 7/11/94
Fox casting by vague track in pine needles just below ridge. [Collected] 3.5 * 1.8cm, and fragment. Dark; rounded & bulbous. Chock full of 100+ small light brown seeds; 1 stem; coarse white ?hair; no bones.

MK 146 [T3, Q35] High ridge NE of Moutti tou Pelekanou. Alt 915m. 7/11/94
Fox castings along (very open) ridge path. 5 together, 1 (the smoother one) separately. [Collected]:
7 pieces, all white, fairly old: 5 * 1.5, 4.5 * 2, 3.5 * 1.5, fragments. Slightly twisted. Very hard. Yellow inside (as were bones).
CONTENTS: 18 unidentified bone fragments. Large chunks; too big to be rat. Biggest 14 * 11 * 10mm.
1 medium-sized casting: soft, dark, not yellow, clearly a different batch; no bones.
NB Fox castings common on all high ridge paths walked on 7/11/94 (MK 129, 130, 146)

MK 90 [T4, Q39] Just above gully, Mandra Kambiou. Alt 700m. 9/11/94
Fox castings. 4 small, in open space between cistus plants; not particularly a track. 2+ 2+ 1 on faint track along top of little gully running through quadrant. Single in open beside cistus, no track. [All collected].
[MK 90 (i)] 3 middle-sized castings. Bones mostly in one. Seeds in all.
11 seeds: black to dark red-brown; globular to heart-shaped; 1.5-3mm.
4 bone fragments: ?incisors of large non-rat.
1 long bone: 7mm.
2 V-shaped bone fragments: 9mm.
8 unidentified bone fragments.
[MK 90 (iii) a] 4.5 * 2, 3 * 1.5, 3.5 * 1.5. White. Blunt, not twisted.
15 unidentified bone fragments.
[MK 90 (iii) b] 7 * 1.5, twisted & very tapered, whitish. 3.5 * 1.5cm, tapered.
2 round black seeds: 3mm.
1 incisor with rectangular cross section. Probably not rat. 10mm.
4 bone fragments: like incisor; claw-shaped.
2 bone chunks.
12 unidentified bone fragments.

MK 74 [T4, Q42] High bare ridge & W slope, Arkolioudhia. Alt 810m. 10/11/94
2 small old fox castings in crevice in rock on ridge path.
[Collected: MK 74 (i)]. 3.5 * 1.5, 2.5 * 1.3. Whitish.
CONTENTS: 1 long bone, 6mm. 1 unidentified fragment.

Appendix 9: Makheras Fox Report

MK 73 [T4, Q43] N-facing slope just below ridge. Alt 960m. 10/11/94
Small fox casting fallen just below path near top of quadrant.
[Collected: MK 73 (i)]. 2cm fragment.
CONTENTS: 5 small fragments.

MK 79 [T5, Q53] E slope. Alt 1140m? 18/11/94
1 fox casting. [Collected]. Flat, very hairy: almost all light grey/brown fur. 7 * 2cm.
CONTENTS: grey fur. 3 claws: one 2.5mm, two 4mm. 1 brown fragment.

MK 59 [T5, Q59] Ridge & N-facing slope. Alt 1120m. 4/12/94
MUSHROOMS. 4 areas of underground cream-gilled mushrooms uprooted, partially ripped & bitten. Possible animals involved:
Fox. Digging too large to be rat, too small to be people (besides, mushrooms not removed). Bite marks 2-3cm. Scratch on edge of mushroom with 3 claws (slightly dirty); fox? NB fox casting found nearby [collected]
CASTINGS. 2 medium, 1 small. All similar, as if found together. Each thick with light grey fur; some slightly gritty patches. Contents:
Upper mandible with incisor. Rattus rattus, but smaller; sub-adult?
4 cranium fragments (1 with part of orbit rim).
1 scapula.
1 ?incisor.
3 long bones: 8mm, 7+mm, 5mm.
3 unidentified fragments.

Appendix 10: Fox Casting Analysis: Dr. Sebastian Payne

Material collected by Julia Ellis Burnet in Cyprus

A: Larger polythene bags:

Label: Near T6
MK 8 or 9
Skull
(and, in another colour) = Q63

Content: Vulpes skull and tooth fragments
Comment: adult, fairly large

Label: T5
MK 125 (iv)
4.12.94
BONES

Content: Canis series of 12 hind thoracic and lumbar vertebrae and sacrum
- three caudal vertebrae and rib, probably belonging to above
Comment: adult, fairly large

Label: none.
loose label in box may be from this bag:
T4
MK 87
9/11/94
Bones

Content: Ovis left side of skull, including frontal, orbit and maxilla, and dP4-M3
?Ovis left tibia - unfused distal epiphysis
?Ovis left calcaneum (traction epiphysis unfused) and lateral malleolus
Ovis/Capra left naviculo-cuboid

Comment: teeth suggest animal in second year; note horn appears to have been removed by human action; other parts consistent with same animal, fairly large for age.

B: Ziplock bags:

Label: T16 AD44
9 MAY '95
BONES

Content: Sus right proximal ulna and radius, both unfused
Comment: probably same animal, fairly large for age

Label: T6
MK94
Jan 8 95
Bone - cut

Content: Ovis/Capra right distal humerus, fusing
Comment: Appears to have had part of distal epiphysis chopped off. fairly young animal (probably second half of first year; fairly large for age; may be goat rather than sheep, but not sure.

Label: T2
MK III (IV)
30 OCT 95
CAPRID JAW?

Content: Ovis/Capra right maxilla fragment with M1-3
Ovis/Capra left maxilla fragment with P4,M1-3

Appendix 10: Fox Casting Analysis: Dr. Sebastian Payne

Ovis/Capra premolar
- fragments

Comment right and left maxilla both with heavily worn teeth - probably same elderly animal; other fragments consistent.

Label: T4 MK 88 (ii)
9 NOV 1994
BONE FRAGS

Content: cf Ovis/Capra axis vertebra fragment

Label: AD81 T14
Bone & egg
24/4/95
JEB

Content: bone unidentified fragments
eggshell 3 fragments, thin white, fairly large egg

Label: AD A IV
4.12.95
BONE

Content: Vulpes frontal, right and left mandibles, loose mandibular and maxillary teeth, atlas, axis and two cervical vertebrae (central epiphyses unfused), scapula, right humerus (distally fused, proximally unfused), ulna (proximally fused, distally unfused), radius (proximally fused, distally unfused), two metacarpals.

Comment: probably from same subadult fox

Label: T17
AD 32
19 DEC 95

Content: Capra left calcaneum (traction epiphysis unfused), astragalus and naviculo-cuboid
- fragments probably of distal tibia unfused epiphysis

Comment: probably from same subadult goat - fairly large for age

---

Label: T15 AD95
2 MAY 1995
BONES:FOX?

Content: Vulpes right and left mandibles, loose teeth, fourteen assorted cervical, thorcic, lumbar and caudal vertebrae, fragments of at least six ribs, sternal element, right and left scapulae, right and left humeri (proximally and distally fused), ulna, radius, four metacarpals, right and left femora (proximally and distally fused), tibia (proximally and distally fused), calcaneum, two metatarsals

Comment: probably all from same fairly small adult fox

Label: T15 AD95
26 APRIL 95
BONES

Content: cf. Vulpes seven cervical and thoracic vertebrae (all with unfused central epiphyses), ten ribs, one sternal element

Comment: all appear to be consistent with being from same subadult fox

---

Label: T5
MK 13A
16/11/94
Bones on trans.

Appendix 10: Fox Casting Analysis: Dr. Sebastian Payne

Content: assorted head bones, including lower and upper dentaries, from fairly large fish

Comment: sorry - can't identify more closely because of lack of experience and reference material

---

Label: ADELPHI
T15 ADA IV
BONE-RATTUS
6 DEC 1995

Content: Rattus rattus skull, mandibles, axis, right and left hidlimbs (most bones, still in articulation)

Comment: probably same animal, subadult

Label: MK A IX
8 JAN 95
MANDRA TOU JERAMIAS
BONE-RATTUS?

Content: Rattus rattus skull, vertebral column and ribs (still in articulation), mandibles, upper right and left forelimbs (still in articulation), upper right and left hindlimbs (still in articulation)

Comment: probably same animal, adult

Label: T2
MK 145 (ii0
4/11/94
backbone, rat?

Content: cf. Rattus rattus groups of articulated vertebrae, one with sacrum and right innominate

---

Label: T14 AD 79
17 APRIL 95
BONE FRAG

Content: Sus second phalanx, with fusing epiphysis

Comment: subadult pig

Label: T16 AD44
9 MAY 95
BONE FRAGS

Content: reptile front of skull, and left mandible

Comment: sorry - can't identify more closely because of lack of experience and reference material

---

Label: T6
MK 14
6/1/95
(and, in blue on bag) Q68

Content: lagomorph scapula fragment

Comment: fairly large - either large hare or large domestic rabbit

---

Label: MK A VIII
MANDRA TO KAMBIOU
1 Jan 1995

Content: lagomorph one cervical and one thoracic vertebrae

Appendix 10: Fox Casting Analysis: Dr. Sebastian Payne

Label: T2 Just
outside MK 104
Bird bones
[?]MMG

Content: small bird, cf. Turdus sp.one humerus; synostosed vertebrae; right and left femora, tibio-tarsi and tarso-metatarsi; phalanges

C: Ziplock bags containing envelopes of bones from fox scats

Label: ADELPHI
T18 AD25
27 DECEMBER 1995
Bone material from fox casting

Content: frs; incl. cf. Rattus rattus ti dist unfused, phal 3 and caudal vert. epiphysis

---

Label: T15 AD 69
AD A XI
4 DECEMBER 1995
Bone material extracted from fox casting

Content: frs; incl. ?lagomorph phal 3 and ? femur prox ep. unfused

---

Label: T19 AD10
10 JANUARY 1996
Bone material extracted from fox castings

Content: frs; none id.

Label: T4 MK 89
MK A V
9 NOVEMBER 1995
Bone material from fox casting

Content: frs. incl: small bird beak; vertebra

Label: ADELPHI
T15 AD A 1
1 DECEMBER 1995
Bone material (and shell) from fox casting

Content: frs. incl. ?Rattus rattus metatarsal, distally unfused, and articulating tarsal
?bird ?eggshell fragment

Label: ADELPHI
T15 AD A IV
5 DECEMBER 1995
Bone material from fox casting

Content: frs. incl: Rattus rattus frontal fr.; atlas vertebra; astragalus

---

Label: T14-T15 AD75 (RIDGELINE)
Bone material extracted from five fox castings
5 DEC 1995

ontent: frs. incl: Rattus rattus mandible fr. with M1, lower incisor, distal humerus, proximal radius, ulna fr., distal tibia, calcaneum - all appear adult.
juv. lagomorph upper first incisor
bird carpal
insect frs.

Appendix 10: Fox Casting Analysis: Dr. Sebastian Payne

Label: ADELPHI
T15 ENCLAVE AD A XIII
6 DECEMBER 1995
Bone material from fox casting

Content: unid frs: not sure whether bone

Label: T16 AD 44
9 MAY 1995
Bone material extracted from fox casting

Content: frs. incl: Rattus rattus mandible with M1, M2; lower and upper incisors; isolated molar; proximal radius; proximal ulna

---

Label: T15 AD 57
8 MAY 1995
Bone material extracted from fox casting

Content: unid frs; one could be caudal vertebra, possibly of rat ...

Label: ADELPHI
T18 AD23
27 DECEMBER 1995
Bone material from fox castings

Content: frs. incl: juv. lagomorph maxilla fr.; isolated tooth; phal 3
Rattus rattus two right mandibles; frontal; maxilla fragments; upper and lower incisors; scapula; proximal humerus epiphysis; distal femur
small bird wishbone

Comment: minimum of two rats

Label: ADELPHI
T16 AD40
6 JULY 1995
Bone material from fox casting

Content: frs. incl: Mus musculus right maxilla; left and right mndibles; innominate, femur; left and right tibiae

---

Label: T16 AD42
9 MAY 1995
Bone extracted from fox casting

Content: frs, incl: ? avian frs
also fragments of feather quill

Label: T15 AD70
5 MAY 1995
Bone material extracted from fox casting

Content: frs. incl: ?Rattus rattus proximal humerus

Label: AD93
Bone material from 8 fox castings

Content: frs. incl: Rattus cf. rattus innominate; right and left proximal femora; distal femur; vertebral fragments etc.
insect fragments

Appendix 10: Fox Casting Analysis: Dr. Sebastian Payne

D. Small plastic boxes

Label: T2 MK142
4 NOV 94
BONES

Content: Hemiechinus auritus left mandible; 12 assorted vertebrae; sacrum; ribs; loose teeth; right innominate

Comment: all consistent with being from same individual

Label: AD 106 T14
Bone
24/4/95
[?]M/Mb

Content: ?Gallus gallus distal femur - in sevaral pieces

E: Ziplock bags containing envelopes of bones from fox scats

Label: T1 MK140
28.10.94
Bones extracted from fox casting

Content: fragments include: cf. Lepus europaeus distal humerus
bird quill base fragments

Label: T1 MK142
28.10.94
Bones extracted from fox casting

Content: fragments include: Rattus rattus right maxilla with M12; right mandible with M12; incisor; ulna; vertebra

---

Label: T2 MK144(ii)
4.11.94
Bones extracted from fox casting

Content: fragments include: Rattus cf. rattus right and left ilia, femora, tibiae; astragalus; metapodia and phalanges

---

Label: T3 MK146(ii)
7.11.94
Bones extracted from fox casting

Content: fragments include: medium mammal vertebral fragment

Comment: probably lumbar vertebrae; possibly sheep, goat, pig etc.

Label: T7 MK1(i)
5 FEBRUARY 1995
Bones extracted from fox casting

Content: fragments include: Rattus rattus left maxilla with M2; left upper M1; left mandible with M123; left lower M1; M2; four upper incisors; lower incisor; proximal humerus epiphysis; distal tibia unfused; calcaneum unfused largerunidentified fragments

Comment: Two individuals - one adult, the other juvenile; also fragments of something larger.

Appendix 10: Fox Casting Analysis: Dr. Sebastian Payne

Label: T14 AD101
19 APRIL 1995
Bones extracted from fox casting

Content: fragments include small rodentatlas vertebra; caudal vertebra; calcaneum; astragalus

Comment: smaller than rat - possibly Mus, Acomys etc.

Label: T14 AD82
8 APRIL 1995
Bones extracted from fox casting

Content: fragments include: Rattus rattus right maxilla with M123; two broken uopper incisors; right mandible with M12; lower incisor and fragment; M3; ulna
Mus musculus upper left maxilla with M2; premaxilla with incisor

---

Label: T6 MK63
8 JANUARY 1995
Bones extracted from fox casting

Content: fragments include: large rodent caudal vertebrae; fragments of other vertebrae etc.

Comment: could be rat

Label: T7 MK2
5 FEBRUARY 1995
Bones extracted from fox casting

Content: fragments include: Rattus rattus frontal; right maxilla with M1; right maxilla with M23; right mandible with M123; left lower M1 crown; caudal vertebrae; very juvenile humerus shaft;
larger unidentified fragments

Comment: minimum of two rats, one very young (lower M1 crown, humerus shaft), the other older. Also fragments of something larger.

Label: T3 MK95(ii)
2 NOVEMBER 94
Bones extracted from pellet or fox casting

Content: fragments include Rattus rattus tooth and jaw fragments; distal humerus; proximal ulna; calcaneum

---

Label: T15 AD79
24 APRIL 95
Bones extracted from fox casting

Content: fragments include: Sus scrofa lumbar/sacral vertebra fragments, some butchered; tarsal; lateral second phalanx; proximal sesamoid
Comment: presumably scavenged from refuse

Label: T7 MK1(ii)
5 FEB 95
Bones extracted from fox casting

Content: fragments include: immature bird phalange fragments; distal tarsometataral fragment

Comment: could be from foot of fowl

Label: T5 MK59
4.12.94
Bones extracted from fox casting

Appendix 10: Fox Casting Analysis: Dr. Sebastian Payne

Content: fragments include Rattus cf. rattus upper incisor; skull fragments; caudal vertebrae; scapula; phalanx

---

Label: ADELPHI
T15 AD73
26 APRIL 1995
Fox casting: bone extracted

Content: small bird, cf. Turdus sp. right and left proximal coracoids; humeri, proximal radii; ulnae, carpometacarpals;
smaller bird upper and lower beak fragments; furcula and sternum fragment probably also from smaller bird
Mus musculus three right maxillae; three left maxillae; four right mandibles; four left mandibles; assorted postcranial including seven humeri and five femora

Comment: at least four mice and two birds

Label: ADELPHI
AD bc 442
Fox castings - extracted
bone fragments
feathers
lizard skin?

Content: fragments include: bird quill shaft fragments; claw sheath
?bird bone fragments; skin (?from foot)

Comment: fairly large immature bird - could be fowl

Label: T5 MK
27.11.94
HAIR/FUR & CLAWS
FROM FOX CASTING

Content: fragments include ?Rattus rattus claw and claw sheath
mammal fur

Label: T4 MK73(i)
10.11.94
Bones extracted from fox casting

Content: nothing identified

Label: T4 MK74(i)
10.11.94
Bones extracted from fox casting

Content: corroded bird phalanx

Label: T4 MK90(i)
9.11.94
Bones and botanical specimens
extracted from fox casting/pellets?

Content: fragments include bird two mandible fragments; claw sheath
seeds Vicia, Lathyrus and indet. (kindly id. by Dr. D de Moulins)

Comment: minimum of two birds!

Label: T4 MK90(iii)a
9.11.94
Bones extracted from fox casting

Appendix 10: Fox Casting Analysis: Dr. Sebastian Payne

Content: fragments include bird distal phalanx

Label: T3 MK130
7.11.94
Seeds and botanical material extracted from fox casting

Content: includes: 11 fig seeds (kindly id. by Dr. D de Moulins)

Label: T4 MK90(iii)b
9.11.94
Bones and seeds? extracted from fox casting

Content: fragments include ?juvenile lagomorph lower incisor germ; molar germ fragments; ?claw fragments
?Rattus rattus skull fragments; sacrum fragment; ?ulna fragment

---
Label: T2 MKIII(ii)a
30.10.94
Bones extracted from fox casting

Content: nothing identified except ?feather fragments

Label: T2 MKIII(ii)b
30.10.94
Bone fragment and feathers extracted from fox casting

Content: fragments include bird feather fragments

Label: T2 MKIII(ii)c
30.10.94
Bones extracted from fox casting

Content: nothing identified

Label: T2 MKIII
(ii)d
30.10.94
Bones extracted from fox casting
Content: nothing identified

Comment: larger mammal - possibly sheep/pig size

Label: T1 138/139 BOUNDARY
28.10.94
Bones from fox castings (various)

Content: fragments include cf. Rattus rattus upper incisor; proximal femur; thrid phalanx and claw sheath

unid fragments of larger vertebrate

Label: T1 MK139(ii)
28.10.94
Bones extracted from fox casting

Content: fragments include: Rattus rattus left maxilla with M12; left mandible with M12; ulna

Appendix 11: Pottery Register

MK A 4 1 RIM OPEN BOWL INTERIOR GLAZE MANDRA TOU KLOSMATOU
MK A 4 2 OPEN BOWL INTERIOR GLAZE 2 PIECES JOINED
MK A 4 3 RIM FRAGMENT OPEN BOWL INTERIOR GLAZE SEE REPORT
MK A 4 4 RIM OPEN BOWL INTERIOR EXTERIOR 2 TONE GREEN GLAZE
MK A 4 5 BODY SHERD OPEN BOWL INTERIOR GREEN GLAZE
MK A 4 6 BODY SHERD OPEN BOWL INTERIOR YELLOW GLAZE
MK A 4 7 BODY SHERD OPEN BOWL INTERIOR CREAM GLAZE
MK A 4 8 HANDLE FRAGMENT CLOSED VESSEL
MK A 4 9 BODY SHERD OPEN BOWL INTERIOR GREEN GLAZE SEE REPORT
MK A 4 10 RIM PLATE? BLUE AND WHITE GLAZE, RED TRIM MODERN
MK A 5 1 1 BODY SHERD PLAINWARE RAISED LINE DECORATION
MK A 5 1 2 SPOUT FRAGMENT? PLAINWARE MANDRA TOU KAMBIOU
MK A 5 1 3 RIM INCISED DECORATION - GLAZE DECORATION?
MK A 5 1 4 AMPHORA HANDLE FRAGMENT
MK A 5 1 5 BODY SHERD PLAINWARE INCISED DECORATION
MK A 5 1 6 PORCELAIN ELECTRICAL INSULATOR FRAGMENT MODERN
MK A 5 1 7 PORCELAIN ELECTRICAL INSULATOR FRAGMENT MODERN
MK A 5 1 8 PORCELAIN ELECTRICAL INSULATOR FRAGMENT MODERN
MK A 5 1 9 BASE FRAGMENT
MK A 66 HANDLE ATTACHED TO RIM T5
MK A 5 1 11 3 BODY SHERD (JOINED) LOWER NECK / SHOULDER
MK A 5 1 12 HANDLE FRAGMENT PLAIN WARE
MK A 5 1 13 RIM FRAGMENT PLATE MOTTLED PLAINWARE
MK A 5 1 14 RIM FRAGMENT CLOSED VESSEL
MK A 5 1 15 BODY SHERD COIL RIBBING
MK A 5 1 16 BODY SHERD COMB AND PIERCED DECORATION
MK A 5 1 17 HANDLE FRAGMENT CARSE WARE, RED SLIP?
MK A 5 1 18 BODY SHERD CLOSED VESSEL RAISED DECORATION?
MK A 5 1 19 RIM AND HANDLE FRAGMENT OPEN BOWL
MK A 5 1 20 RIM AND SPOUT FRAGMENT COOKING POT?
MK A 5 1 21 SHOULDER SHERD OPEN VESSEL BROWN GLAZE
MK A 5 1 22 BODY SHERD CLOSED VESSEL RIB DECORATION
MK A 5 1 23 BASE CLOSED VESSEL WHEEL MADE FLOWER POT
MK A 5 1 24 HANDLE FRAGMENT PLAINWARE
MK A 5 1 25 RIM FRAGMENT CLOSED VESSEL COARSE WARE
MK A 5 1 26 HANDLE FRAGMENT COARSE WARE
MK A 5 1 27 RIM FRAGMENT OPEN VESSEL GREEN GLAZE
MK A 5 1 28 BODY SHERD OPEN VESSEL CLEAR INTERIOR GLAZE
MK A 5 1 29 BODY SHERD COOKING POT?
MK A 5 1 30 RIM SHERD COFFEE CUP MODERN
MK A 5 1 31 RIM SHERD PLATE FLORAL PATTERN MODERN
MK A 5 1 32 BODY SHERD OPEN VESSEL YELLOW INTERIOR GLAZE
MK A 5 1 33 TILE / BRICK FRAGMENT RIBBED
MK A 5 1 34 NECK SHERD CLOSED VESSEL BROWN EXTERIOR GLAZE
MK A 5 1 35 RIM PLATE GREEN INTERIOR GLAZE
MK A 5 1 36 RIM CLOSED VESSEL COARSE WARE
MK A 5 1 37 RIM CLOSED VESSEL COARSE WARE MORTAR?
MK A 5 1 38 BODY SHERD OPEN VESSEL BROWN INTERIOR GLAZE
MK A 5 1 39 BASE FRAGMENT CLOSED VESSEL COARSE WARE
MK A 5 1 40 RIM FRAGMENT CLOSED VESSEL COARSE WARE
MK A 5 1 41 BODY SHERD COMB PATTERN
MK A 5 1 42 BASE OR RIM FRAGMENT COARSE WARE
MK A 5 1 43 BODY SHERD OPEN VESSEL INTERIOR/EXTERIOR WHITE GLAZE
MK A 5 1 44 BODY SHERD OPEN VESSEL INTERIOR/EXTERIOR WHITE AND GREEN GLAZE
MK A 5 1 45 BODY SHERD OPEN VESSEL INTERIOR GREEN GLAZE
MK A 5 1 46 BODY SHERD PLATE? THICK CREAM GLAZE ONE SIDE
MK A 5 1 47 BODY SHERD CLOSED VESSEL YELLOW/CREAM SLIP
MK A 5 2 1 RIM FRAGMENT CLOSED VESSEL ROMAN?
MK A 5 2 2 RIM FRAGMENT CLOSED VESSEL RED EXTERIOR SLIP

Appendix 11: Pottery Register

MK A 5 2 3 RIM FRAGMENT PLATE PLAINWARE
MK A 5 2 4 BODY SHERD RIBBED DECORATION
MK A 5 2 5 RIM AND HANDLE FRAGMENT OPEN SHALLOW DISH BROWN INTERIOR GLAZE
MK A 5 2 6 RIM OF NECK OPEN VESSEL INTERIOR YELLOW GLAZE
MK A 5 2 7 RIM OPEN VESSEL INTERIOR / EXTERIOR GREEN GLAZE
MK A 5 2 8 NECK SHERD OPEN VESSEL GREEN INTERIOR GLAZE
MK A 5 2 9 RIM FRAGMENT TWO TONED GREEN INTERIOR GLAZE
MK A 5 2 10 HANDLE FRAGMENT?
MK A 5 2 11 BODY SHERD EXTERIOR RED SLIP
MK A 5 2 12 BODY SHERD GREEN INTERIOR GLAZE
MK A 5 2 13 BODY SHERD GREEN INTERIOR GLAZE
MK A 5 2 14 NECK SHERD GREEN INTERIOR GLAZE, STRIPED EXTERIOR
MK A 5 2 15 RIM FRAGMENT INTERIOR/EXTERIOR GREEN GLAZE
MK A 5 2 16 NECK FRAGMENT INTERIOR GREEN GLAZE
MK A 5 2 17 BODY FRAGMENT INTERIOR GREEN GLAZE
MK A 5 2 18 RIM FRAGMENT OPEN VESSEL INTERIOR BROWN GLAZE
MK A 5 2 19 RIM FRAGMENT OPEN VESSEK CREAM EXTERIOR GLAZE
MK A 5 2 20 BODY SHERD OPEN VESSEL GREEN INTERIOR GLAZE
MK A 5 2 21 BODY SHERD OPEN VESSEL GREEN INTERIOR GLAZE
MK A 5 2 22 BODY SHERD OPEN VESSEL GREEN INTERIOR GLAZE / EXTERNAL STRIPS
MK A 5 2 23 BODY SHERD OPEN VESSEL BLUE INTERIOR GLAZE
MK A 5 2 24 BODY SHERD OPEN VESSEL BLUE AND GREEN INTERIOR GLAZE
MK A 5 2 25 BODY SHERD GREEN INTERIOR GLAZE
MK A 5 2 26 BODY SHERD OPEN VESSEL BROWN INTERIOR GLAZE
MK A 5 2 27 BODY SHERD GREEN INTERIOR GLAZE
MK A 5 2 28 BODY SHERD PLATE BIRD PATTERN BROWN ON WHITE MODERN
MK A 5 2 29 BASE FRAGMENT PLATE WHITE GLAZEMODERN
MK A 5 2 30 BODY SHERD PLATE LEAF AND ROSE DESIGN MODERN
MK A 5 2 31 BODY SHERD PLATE LEAF AND ROSE BUD DESIGN MODERN
MK A 5 2 a GLASS FRAGMENT BASE
MK A 8 1 4 PIECES JOINED OPEN PEDESTALLED BOWL CREAM / GREEN GLAZE SEE REPORT, BISHOPS RETREAT
MK A 8 2 BASE FRAGMENT OPEN BOWL YELLOW GLAZE / BROWN LINE DECORATION SEE REPORT
MK A 8 3 BASE FRAGMENT OPEN BOWL CREAM/YELLOW/BROWN DECORATION SEE REPORT
MK A 8 4 BODY SHERD OPEN BOWL CREAM/LIMEBROWN DECORATION SEE REPORT
MK A 8 5 BASE FRAGMENT OPEN BOWL CREAM INTERIOR GLAZE
MK A 8 6 BASE FRAGMENT OPEN BOWL BROWN INTERIOR GLAZE
MK A 8 7 BODY SHERD OPEN BOWL CREAM/GREEN INTERIOR DECORATION NOTE POSSIBLE WORN JOIN WITH MK(A)VIII:(1)
MK A 8 8 RIM FRAGMENT OPEN BOWL CREAM INTERIOR/EXTERIOR GLAZE
MK A 8 9 BODY SHERD LAMP FRAGMENT?
MK A 8 10 BODY SHERD OPEN BOWL CREAM/GREEN/BROWN DECORATION SEE REPORT
MK A 8 11 RIM FRAGMENT WITH LID HANDLE OPEN BOWL BROWN INTERIOR GLAZE
MK A 8 12 HANDLE FRAGMENT
MK A 8 13 BASE FRAGMENT OPEN BOWL?
MK A 8 14 RIM FRAGMENT OPEN BOWL RED INTERIOR SLIP?
MK A 8 15 BODY SHERD CLOSED VESSEL
MK A 8 16 RIM FRAGMENT OPEN BOWL CREAM SLIP SEE REPORT
MK A 8 17 BODY SHERD WITH HANDLE ATTACHMENT CLOSED VESSEL
MK A 8 18 RIM AND HANDLE FRAGMENT
MK A 8 19 BODY SHERD OPEN BOWL GREEN INTERIOR - CREAM/BROWN EXTERIOR GLAZE
MK A 8 20 RIM FRAGMENT BROWN INTERIOR GLAZE
MK A 8 21 RIM FRAGMENT BROWN INTERIOR GLAZE

Appendix 11: Pottery Register

MK	A	8		22	BASE FRAGMENT OPEN BOWL BROWN INTERIOR SLIP
MK	A	8		23	RIM FRAGMENT CLOSED VESSEL
MK	A	8		24	HANDLE FRAGMENT?
MK	A	8		25	RIM FRAGMENT CREAM/BROWN INTERIOR GLAZE 3 PIECES JOINED
MK	A	8		26	BODY SHERD OPEN BOWL RED INTERIOR SLIP/BROWN EXTERIOR GLAZE SEE REPORT
MK	A	8		27	BDY SHERD OPEN BOWL GREEN/BROWN INTERIOR DECORATION SEE REPORT
MK	A	8		28	RIM FRAGMENT CLOSED VESSEL
MK	A	8		29	HANDLE FRAGMENT COOKING POT?
MK	A	8		30	BASE FRAGMENT CLOSED VESSEL
MK	A	8		31	BODY SHERD CLSOED VESSEL
MK	A	8		32	BODY SHERD CLOSED VESSEL
MK	A	8		33	BODY SHERD - BROKEN RIM? FRAGMENT
MK	A	8		34	BODY SHERD CLOSED VESSEL
MK	A	8		35	BODY SHERD CREAM/BROWN GLAZE, EXTERIOR CR.
MK	A	8		36	BODY SHERD CLOSED VESSEL
MK	A	8		37	BODY SHERD FORMER INTERIOR/EXTERIOR GLAZE ERODED
MK	A	8		CS.	CHIPPED STONE CHIPPED STONE
MK	A	8		38	RIM FRAGMENT CLOSED VESSEL
MK	A	8		39	RIM WITH LIP HANDLE? FRAGMENT PUNCHED DECORATION
MK	A	8		40	RIM LARGE OPEN VESSEL
MK	A	8		41	FLAT RIM BOWL
MK	A	8		42	BASE MEDIUM BOWL
MK	A	8		43	FLAT RIM SMALL-MEDIUM BOWL
MK	A	8		44	STRIP HANDLE FRAGMENT AND RIM BOWL
MK	A	8		45	STRIP HANDLE AND RIM BOWL
MK	A	8		46	BODY SHERD VERY GRITTY SURFACE
MK	A	8		47	BODY SHERD
MK	A	8		48	BODY SHERD
MK	A	8		49	HANDLE FRAGMENT AND RIM FRAGMENT JOINED
MK	A	8		50	RIM FRAGMENT OPEN OR CLOSED VESSEL
MK	A	8		51	RIM FRAGMENT OPEN VESSEL
MK	A	8		52	FLAT RIM FRAGMENT
MK	A	8		53	BODY SHERD
MK	A	8		54	THIN BODY SHERD
MK	A	8		55	BODY SHERD
MK	A	8		56	FLAT RIM FRAGMENT
MK	A	8		57	RIM FRAGMENT CLOSED VESSEL
MK	A	8		58	BODY SHERD
MK	A	8		59	FRAGMENT COOKING POINT
MK	A	8		60	TILE
MK	A	5	2	33	HANDLE FRAGMENT AND BODY FRAGMENT
MK	A	5	2	34	HANDLE FRAGMENT AND BODY FRAGMENT
MK	A	5	2	35	HANDLE FRAGMENT
MK	A	5	2	36	HANDLE FRAGMENT
MK	A	5	2	37	RIM FRAGMENT LARGE OPEN VESSEL
MK	A	5	2	38	RIM FRAGMENT LARGE OPEN VESSEL
MK	A	5	2	39	RIM FRAGMENT CLOSED VESSEL
MK	A	5	2	40	RIM FRAGMENT CLOSED VESSEL
MK	A	5	2	41	BASE FRAGMENT
MK	A	5	3	1	RIM FRAGMENT
MK	A	5	3	2	RIM FRAGMENT CLOSED VESSEL
MK	A	5	3	3	BASE FRAGMENT VESSEL
MK	A	5	3	4	TILE FRAGMENT
MK	A	5	3	5	2 INTERIOR RIM FRAGMENTS PIPE (FRAGMENT) YELLOW GLAZED INTERIOR
MK	A	5	3	6	PIPE FRAGMENT YELLOW GLAZED
MK	A	5	3	7	PIPE FRAGMENT YELLOW GLAZED
MK	A	5	3	8	PIPE FRAGMENT YELLOW GLAZED TERMINAL JOINT

Appendix 11: Pottery Register

MK	A	5	3		9	PIPE FRAGMENT YELLOW GLAZED
MK	A	5	3		10	PIPE FRAGMENT YELLOW GLAZED, RESIDUAL CEMENT (REAR) ACCIDENTAL BLACK SPOT ADDED
MK	A	5	3		11	PIPE FRAGMENT YELLOW GLAZED
MK	A	5	3		12	PIPE FRAGMENT YELLOW GLAZED
MK	A	5	3		13	PIPE FRAGMENT YELLOW GLAZED
MK	A	5	3		14	FRAGMENT
MK	A	5	3		15	FLOOR TILE?
MK	A	5	3		16	PIPE?
MK	A	5	3		17	PIPE FRAGMENT GREEN GLAZE
MK	A	5	3		18	PIPE FRAGMENT GREEN GLAZE
MK	A	5	3		19	PIPE FRAGMENT GREEN GLAZE
MK	A	5	3		20	PIPE FRAGMENT GREEN GLAZE - RESIDUAL CEMENT
MK	A	5	3		21	PIPE FRAGMENT GREEN GLAZE
MK	A	5	3		22	PIPE FRAGMENT GREEN GLAZE - RESIDUAL CEMENT
MK	A	5	3		23	PIPE FRAGMENT GREEN GLAZE
MK	A	5	3		24	PIPE FRAGMENT GREEN GLAZE
MK	A	5	3		25	PIPE FRAGMENT GREEN GLAZE
MK	A	5	3		26	PIPE FRAGMENT GREEN GLAZE
MK	A	5	3		27	PIPE FRAGMENT GREEN GLAZE
MK	A	5	3		28	PIPE FRAGMENTS PRE-JOINED WITH CEMENT AS FOUND
MK	A	14			1	CUP FROM BASE TO RIM CUP 6 PIECES JOINED, FORT SITE
MK	A	14			2	RIM FRAGMENT GREEN AND BROWN GLAZE
MK	A	14			3	BASE FRAGMENT GREEN INTERIOR GLAZE
MK	A	14			4	BODY SHERD CLOSED VESSEL PLAINWARE
MK	A	18	8	A.1	1	BODY SHERD IMPRESSED "WHEEL" DECORATION PIE CRUST RIM?
MK	A	18	8	A.1	2	BODY SHERD IMPRESSED "WHEEL" DECORATION PITHOS?
MK	A	18	8	A.1	3	BODY SHERD IMPRESSED "WHEEL" DECORATION PITHOS?
MK	A	18	2	A	1	BASE FRAGMENT OPEN BOWL INTERIOR GLAZE WITH DECORATION
MK	A	18	2	A	2	BASE FRAGMENT OPEN BOWL OLIVE GREEN INTERIOR GLAZE
MK	A	18	2	A	3	BODY SHERD OPEN VESSEL CREAM INTERIOR GLAZE
MK	A	18	2	A	4	BODY SHERD OPEN VESSEL YELLOW INTERIOR GLAZE 2 PIECES JOINED
MK	A	18	2	A	5	RIM FRAGMENT OPEN VESSEL COARSE WARE 2 PIECES JOIONED
MK	A	18	2	A	6	BODY SHERD OPEN VESSEL GREEN/CREAM INTERIOR DECORATION
MK	A	18	2	A	7	HANDLE FRAGMENT PLAIN WARE
MK	A	18	2	A	8	BODY SHERD OPEN BOWL GREEN/CREAM INTERIOR GLAZE
MK	A	18	2	A	9	3 BODY SHERDS OPEN VESSEL BROWN/CREAM/GREEN INTERIOR GLAZE NOT JOINED
MK	A	18	2	A	10	CHIPPED STONE CHERT. THREASHING SHED FLINT
MK	A	18	3	A	1	BODY SHERD CLOSED VESSEL INCISED EXTERIOR DECORATION
MK	A	18	3	A	2	BODY SHERD CLOSED VESSEL IMPRESSED LINE AND PUNCH DECORATION
MK	A	18	3	A	3	BODY FRAGMENT OPEN BOWL CREAM/GREEN INTERIOR GLAZE
MK	A	18	3	A	4	BODY SHERD OPEN VESSEL CREAM/YELLOW/BROWN INTERIOR DECORATION
MK	A	18	3	A	5	BASE FRAGMENT OPEN VESSEL PLAINWARE
MK	A	18	3	A	6	BODY SHERD OPEN VESSEL CREAM/GREEN INTERIOR GLAZE
MK	A	18	3	A	7	BODY SHERD OPEN VESSEL MOTTLED GREEN INTERIOR GLAZE
MK	A	18	3	A	8	BODY SHERD OPEN VESSEL CREAM/GREEN INTERIOR DECORATION
MK	A	18	3	A	9	RIM SHERD PLATE LIGHT/DARK GREEN INTERIOR + EXTERIOR DECORATION
MK	A	18	3	A	10	BODY SHERD CREAM/BROWN STRIPED INTERIOR DECORATION
MK	A	18	3	A	11	BODY SHERD OPEN VESSEL CREAM/GREEN WAVE DECORATION
MK	A	18	3	A	12	BODY SHERD OPEN VESSEL? GREEN INTERIOR/EXTERIOR GLAZE 2 PIECES JOINED
MK	A	18	3	A	13	BODY SHERD OPEN VESSEL CREAM INTERIOR GLAZE
MK	A	18	3	A	14	BODY SHERD DARK GREEN EXTERIOR/LIGHT GREEN INTERIOR GLAZE 2 PIECE JOINED
MK	A	18	2	A	11	BODY SHERD COARSE WARE DARK BROWN EXTERIOR SLIP
MK	A	18	2	A	12	BODY SHERD COARSE WARE IMPRESSED EXTERIOR DESIGN

Appendix 11: Pottery Register

MK	A	18	2	A	13	BODY SHERD COARSE WARE BROWN INTERIOR, ORANGE EXTERIOR SLIP
MK	A	18	2	A	14	BODY SHERD COARSE WARE BROWN SLIP? LINE DECORATION
MK	A	18	5	A	1	BODY SHERD CLOSED VESSEL PINK FABRIC
MK	A	18	5	A	2	RIM FRAGMENT RIBBED EXTERIOR DESIGN ROMAN?
MK	A	18	5	A	3	BODY SHERD PLAINWARE RIBBED EXTERIOR DESIGN
MK	A	18	3	A	15	BASE FRAGMENT OPEN BOWL BROWN/CREAM INTERIOR GLAZE
MK	A	18	3	A	16	RIM FRAGMENT PITHOS
MK	A	18	3	A	17	BODY SHERD PITHOS - WAVE AND PUNCHED IMPRESSED DESIGN
MK	A	18	4	A	1	BODY SHERD PITHO - DECORATED BAND BELOW RIM
MK	A	18	4	A	2	BODY SHERD PITHOS
MK	A	18	6	B	1	BASE FRAGMENT
MK	A	18	6	B	2	BASE FRAGMENT PLATE? GROVE DECORATION?
MK	A	18	6	B	3	RIM FRAGMENT
MK	A	18	6	B	4	RIM FRAGMENT LINE DESIGN AROUND RIM
MK	A	18	6	B	5	RIM FRAGMENT PLATE?
MK	A	18	6	B	6	BODY SHERD PLATE? WIDE GROVE
MK	A	18	8	B	1	HANDLE FRAGMENT
MK	A	18	9	B	1	RIM FRAGMENT SHALLOW BOWL VERY WORN IMPRESSED DECORATION
MK	A	18	9	B	2	RIM FRAGMENT
MK	A	18	9	B	3	HANDLE FRAGMENT
MK	A	18	9	B	4	BODY SHERD CLOSED VESSEL
MK	A	18	9	B	5	BASE FRAGMENT
MK	A	18	9	B	6	RIM FRAGMENT CUP? FINE FABRIC
MK	A	18	9	B	7	RIM/SPOUT FRAGMENT STRAP DESIGN
MK	A	18	9	B	7.5	BASE FRAGMENT? BOWL?
MK	A	18	9	B	8	BASE OR RIM FRAGMENT
MK	A	18	9	B	9	RIM FRAGMENT CUP OR SMALL BOWL
MK	A	18	9	B	10	RIM FRAGMENT JUGLET OR FLASK
MK	A	18	9	B	11	BODY SHERD GROVE DESIGN
MK	A	18	9	B	12	BODY SHERD HANDLE ATTACHMENT
MK	A	18	9	B	13	BASE FRAGMENT
MK	A	18	9	B	14	BODY SHERD WITH SMALL PORTION OF RIM OR LIP
MK	A	18	9	B	15	BODY SHERD PITHOS? RAISED LINE DECORATION
MK	A	18	9	B	16	HANDLE FRAGMENT
MK	A	18	9	B	17	RIM OR BASE FRAGMENT IMPRESSED LINE DECORATION ON 2 SIDES
MK	A	18	9	B	18	NECK FRAGMENT LARGE CLOSED VESSEL
MK	A	18	9	B	19	RIM FRAGMENT BROWN INTERIOR/EXTERIOR SLIP
MK	A	18	9	B	20	RIM FRAGMENT PLATE?
MK	A	18	9	B	21	RIM FRAGMENT FINE WARE
MK	A	18	9	B	22	RIM FRAGMENT BLACK MICA IN FABRIC
MK	A	18	9	B	23	BODY SHERD PLATE? RAISED EDGE PORTION
MK	A	18	9	B	24	RIM FRAGMENT WHITE FABRIC
MK	A	18	9	B	25	LAMP FRAGMENT
MK	A	18	9	B	26	HANDLE FRAGMENT
MK	A	18	9	B	27	HANDLE FRAGMENT
MK	A	18	9	B	28	RIM FRAGMENT PITHOS
MK	A	18	9	B	29	RIM AND BODY SHERD PITHOS, WIDE GROVE DECORATION
MK	A	18	9	B	30	RIM SHERD INTERIOR PROJECTING RIDGE
MK	A	3			1	RIM FRAGMENT DARK GREY INTERNAL GLAZE, BROWN DECORATION PROPHETIS ELIAS
MK	A	3			2	BASE FRAGMENT OPEN BOWL LIGHT AND DARK GREEN GLAZE DECORATION
MK	A	3			3	RIM FRAGMENT OPEN BOWL BROWN INTERIOR GLAZE
MK	A	3			4	HANDLE FRAGMENT
MK	A	3			5	BODY SHERD COARSEWRE, RIDGE DECORATION
MK	A	3			6	HANDLE FRAGMENT
MK	A	3			7	BODY SHERD CLOSED VESSEL RIB DECORATION
MK	A	3			8	MODIFIED PITHOS SHERD

Appendix 11: Pottery Register

MK A 3 9 BODY SHERD COOKING POT?
MK A 3 10 BODY SHERD KORNOS WARE? IMPRINT DESIGN
MK A 3 11 NECK FRAGMENTS 2 JOINED
MK A 3 12 RIM FRAGMENT BRWON SLIP?
MK A 3 13 BODY SHERD WIDE GROOVE DESIGN
MK A 3 14 BODY SHERD INCISED LINE DECORATION
MK A 3 15 BODY SHERD RED INTERIOR GLAZE, CARBON DEPOSIT
MK A 3 16 BODY SHERD INCISED LINE DECORATION
MK A 9 1 BASE FRAGMENT OPEN BOWL LIGHT GREEN INTERIOR GLAZE AYIOS ONOUPHRIOUS
MK A 9 2 BODY SHERD WHITE INTERIOR/EXTERIOR GLAZE, YELLOW INTERIOR DECORATION
MK A 9 3 BASE FRAGMENT
MK A 5 4 1 HANDLE FRAGMENT WIDE STRAP MANDRA TOU KAMBIOU FIELDS
MK A 5 4 2 BODY SHERD lARGE COARSEWARE VESSEL WITH HANDLE BASE
MK A 5 4 3 HANDLE FRAGMENT
MK A 5 4 4 BODY SHERD HANDLE BASE
MK A 5 4 5 BODY SHERD RAIZED/RIB DECORATION, BLUE CORE
MK A 5 4 6 BODY SHERD HANDLE BASE SLIGHTLY CURVED
MK A 5 4 7 BASE FRAGMENT WHITE/LIGHT PINK FABRIC
MK A 5 4 8 HANDLE FRAGMENT BROWN SLIP/GLAZE? INCISED DECORATION

MK A 5 4 9 HANDLE FRAGMENT WIDE STRAP, DEEP CURVE
MK A 5 4 10 HANDLE FRAGMENT
MK A 5 4 11 HANDLE FRAGMENT PINK FABRIC, WHITE SLIP
MK A 5 4 12 HANDLE FRAGMENT SMALL
MK A 5 4 13 HANDLE / RIM FRAGMENT
MK A 5 4 14 RIM? FRAGMENT CLOSED VESSEL PLAIN WARE
MK A 5 4 15 BODY SHERD WHITE INTERIOR GLAZE
MK A 5 4 16 HANDLE FRAGMENT
MK A 5 4 17 HANDLE FRAGMENT
MK A 5 4 18 HANDLE FRAGMENT FLARING TO BASE
MK A 5 4 19 RIM FRAGMENT CLOSED VESSEL
MK A 5 4 20 BODY SHERD INCISED RIB DECORATION
MK A 5 4 21 RIM / HANDLE FRAGMENT FINE RED FABRIC / BLUE CORE
MK A 5 4 22 RIM FRAGMENT
MK A 5 4 23 BODY SHERD BROWN/GREEN EXTERIOR GLAZE, WHITE INTERIOR GLAZE WITH DECORATION
MK A 5 4 24 RIM / HANDLE FRAGMENT
MK A 5 4 25 HANDLE FRAGMENT COARSE WARE
MK A 5 4 26 BASE FRAGMENT WHITE FABRIC WITH PINK SLIP
MK A 5 4 27 HANDLE FRAGMENT WHITE EXTERNAL/INTERIOR GLAZE INCISED PATTERN ON HANDLE SURFACE
MK A 5 4 28 RIM FRAGMENT
MK A 5 4 29 BODY SHERD CREAM/BROWN/YELLOW INTERIOR GLAZE
MK A 5 4 30 HANDLE FRAGMENT INCISED LINE DESIGN
MK A 5 4 31 RIM FRAGMENT CLOSED VESSEL
MK A 5 4 32 BODY SHERD WITH "NDB" HANDLE
MK A 5 4 33 HANDLE FRAGMENT AND BASE
MK A 5 4 34 RIM FRAGMENT WHITE FABRIC RED SNE INCLUSIONS
MK A 5 4 35 HANDLE FRAGMENT
MK A 5 4 36 RIM FRAGMENT OPEN BOWL
MK A 5 4 37 RIM FRAGMENT WHITE FABRIC
MK A 5 4 38 RIM FRAGMENT OPEN BOWL
MK A 5 4 39 RIM FRAGMENT OUTER LIP ONLY
MK A 5 4 40 BASE FRAGMENT WHITE EXTERNAL SLIP
MK A 5 4 41 LAMP FRAGMENT?
MK A 5 4 42 RIM / HANDLE FRAGMENT WIDE SHALLOW BOWL
MK A 5 4 43 HANDLE FRAGMENT
MK A 5 4 44 RIM FRAGMENT SHALLOW BOWL

Appendix 11: Pottery Register

MK A 5 4 45 RIM FRAGMENT LINE DECORATION BELOW RIM
MK A 5 4 46 RIM FRAGMENT NECK OF FLASK SOME BROWN GLAZE
MK A 5 4 47 RIM FRAGMENT SOFT PINK FABRIC
MK A 5 4 48 HANDLE FRAGMENT WITH BASE
MK A 5 4 49 HANDLE FRAGMENT
MK A 5 4 50 RIM FRAGMENT OPEN BOWL FINE WARE
MK A 5 4 51 RIM FRAGMENT
MK A 5 4 52 RIM FRAGMENT EXTERNAL CARBON COATING
MK A 5 4 53 BODY SHERD WHITE INTERNAL GLAZE
MK A 5 4 54 BODY SHERD INCISED LINE DECORATION
MK A 5 4 55 BODY SHERD PINK FABRIC WHITE SLIP
MK A 5 4 56 BODY SHERD SOFT PINK FABRIC
MK A 5 4 57 RIM FRAGMENT
MK A 5 4 58 RIM FRAGMENT
MK A 5 4 59 RIM FRAGMENT WHITE FABRIC RED SAND INCLUSIONS?
MK A 5 4 60 RIM FRAGMENT SHALLOW BOWL HELLENISTIC?
MK A 5 4 61 BASE FRAGMENT
MK A 5 4 62 RIM FRAGMENT / HANDLE
MK A 5 4 63 HANDLE FRAGMENT
MK A 5 4 64 BASE FRAGMENT
MK A 5 4 65 BODY SHERD WOTH HANDLE BASE?
MK A 5 4 66 RIM SHERD WHITE INTERIOR/EXTERIOR GLAZE
MK A 5 4 67 RIM SHERD BROWN PAINT LINE DECORATION
MK A 5 4 68 BASE FRAGMENT
MK A 6 1 BODY SHERD PITHOS RAISED DECORATION MANDRA TOU
PETRI
MK A 6 2 TILE FRAGMENT BYZANTINE
MK A 6 3 BODY SHERD PITHOS RAISED DECORATION, FAINT LINES
MK A 6 4 RIM FRAGMENT INTERNAL 'PIE CRUST' PATTERN AND CARBON
RESIDUE
MK A 6 5 TILE FRAGMENT BYZANTINE
MK A 6 6 TILE FRAGMENT BYZANTINE
MK A 6 7 HANDLE FRAGMENT COARSE WARE, BLACK MICA FABRIC
MK A 6 8 RIM FRAGMENT PITHOS, BLUE CORE
MK A 6 9 BODY SHERD BLUE CORE
MK A 6 10 BODY SHERD INCISED WAVY LINE AND PUNCH DECORATION
MK A 6 11 HANDLE / RIM FRAGMENT
MK A 6 12 BODY FRAGMENT PITHOS, RAISED LINE DECORATION
MK A 11 1 1 BODY SHHERD SLIGHT "COB" PATTERN MANDRA TOU
JEREMIAS
MK A 11 1 2 RIM SHERD SHALLOW DISH?
MK A 11 1 3 BODY SHERD
MK A 11 1 4 BODY SHERD SAME VESSEL AS ...iii?
MK A 11 1 5 BODY SHERD WHITE/GREEN INTERIOR GLAZE
MK A 11 1 6 BODY SHERD "COMB" DECORATION
MK A 11 1 7 RIM FRAGMENT SHALLOW OPEN BOWL
MK A 11 1 8 RIM FRAGMENT BLUE CORE
MK A 11 1 9 BASE FRAGMENT? "COMB" DECORATION
MK A 11 1 10 BODY SHERD "COMB" DECORATION DECORATION
MK A 11 1 11 BODY SHERD COOKING POT?
MK A 11 1 12 BODY SHERD
MK A 11 1 13 BODY SHERD "COMB" DECORATION DECORATION
MK A 11 1 14 BODY SHERD "COMB" DECORATION DECORATION
MK A 11 1 15 BODY SHERD "COMB" DECORATION DECORATION
MK A 11 1 16 RIM FRAGMENT CLOSED VESSEL POSSIBLE JOIN WITH ... xiv
MK A 11 1 17 RIM FRAGMENT PITHOS
MK A 11 2 1 BODY SHERD IMPRESSED SQUARE DESIGN
MK A 11 3 1 BODY SHERD PITHOS 2 PIECES JOINED
MK A 11 3 2 RIM FRAGMENT CUP?
MK A 11 3 3 RIM FRAGMENT PITHOS
MK A 11 4 1 BODY SHERD PITHOS, PRESSED "COIL"? DECORATION

Appendix 11: Pottery Register

MK A 11 4 2 RIM FRAGMENT SHALLOW BOWL
MK A 11 4 3 BODY SHERD BASE? FRAGMENT
MK A 11 4 4 HANDLE FRAGMENT
MK A 11 4 5 RIM FRAGMENT WIDE IN-CURVING RIM WITH PIE CRUST DECORATION
MK A 11 4 6 BODY SHERD VESSEL SHOULDER OR NECK? PUNCH AND LINE DECORATION
MK A 11 4 7 BODY SHERD CURVED - NO DECORATION
MK A 11 4 8 RIM FRAGMENT FLASK?
MK A 11 4 9 BODY SHERD WITH HANDLE BASE
MK A 11 4 10 BODY SHERD LARGE VESSEL WAVY "COMB" PATTERN
MK A 7 1 BASE FRAGMENTS OPEN TROUGH 2 PIECES JOINED, LOOM WEIGHT SITE
MK A 7 2 BASE FRAGMENTS OPEN TROUGH
MK A 7 3 RIM FRAGMENT / TILE? WITH RAISED RIB
MK A 7 4 BASE FRAGMENTS OPEN TROUGH
MK A 7 5 BASE FRAGMENTS OPEN TROUGH
MK A 7 6 RIM FRAGMENT PITHOS
MK A 7 7 BASE FRAGMENTS OPEN TROUGH 2 PIECES JOINED
MK A 7 8 RIM FRAGMENT LARGE CLOSED VESSEL
MK A 7 10 RIM FRAGMENT OPEN BOWL OR PLATE
MK A 7 11 BODY SHERD EXTERNAL LINE DECORATION
MK A 7 12 HANDLE FRAGMENT
MK A 7 13 BASE FRAGMENT OPEN BOWL?
MK A 7 14 RIM FRAGMENT
MK A 7 15 HANDLE FRAGMENT
MK A 7 16 BASE FRAGMENT RIBBED PATTERN, FINE WARE
MK A 7 17 BODY SHERD EXTERNAL YELLOW SLIP
MK A 7 18 BODY SHERD WHITE FABRIC, INTERNAL BROWN SLIP
MK A 7 19 BODY SHERD WHITE FABRIC, EXTERNAL BROWN SLIP
MK A 7 20 RIM FRAGMENT CLSOED VESSEL
MK A 7 21 3 RIM FRAGMENTS VERY SMALL, FINE WARE
MK A 7 22 LOOM WEIGHT
MK A 7 23 LOOM WEIGHT WITH USE WEAR?
MK A 7 24 LOOM WEIGHT
MK A 138-139 BODY SHERD IMPRESSED DECORATION "SPRING OF DARKNESS"
MK A 130 BODY SHERD CLOSED VESSEL PLAIN WARE STEEP SPUR
MK A 131 4 BODY SHERDS, 2 JOINING ADJACENT TO TEST QUADRANT
MK A 73-118 3 BODY SHERDS ABOVE MULE TRACK NEAR RIDGE GAP
MK A 66 HANDLE FRAGMENT TRACK CROSSING QUADRANT
MK A 16 3 BODY SHERDS "COMB" PATTERN, JOINED
MK A C44-45 BODY SHERD VERY WORN
MK A 30 VESSEL NECK (10 PIECES), HANDLE (2 PIECES) BASE FRAGMENT (2 PIECES) ALL JOINED
MK A 73-75 RIM FRAGMENT 2 PIECES, JOINED
AD A 1 1 HANDLE FRAGMENT WHITE FABTIC
AD A 1 2 RIM FRAGMENT 8 PIECES JOINED
AD A 1 3 RIM FRAGMENT WHITE FABRIC
AD A 1 4 RIM FRAGMENT SHALLOW BOWL HELLENISTIC
AD A 1 5 HANDLE FRAGMENT
AD A 1 6 RIM FRAGMENT
AD A 1 7 RIM FRAGMENT
AD A 1 8 RIM FRAGMENT
AD A 1 9 RIM FRAGMENT
AD A 1 BAG BODY SHERDS 290 PIECES BULLDOZED, UNDIAGNOSTIC
AD A 2 1 RIM FRAGMENT PITHOS PITCH ADHERING TO INTERNAL SURFACE
AD A 2 2 BODY FRAGMENT PITHOS PITCH ADHERING TO OUTER SURFACE
AD A 2 3 BODY SHERD PITHOS
AD A 2 4 BODY SHERD PITHOS BLACK
AD A 2 5 BODY SHERD PITHOS BLACK

Appendix 11: Pottery Register

AD A 2 BAG BODY SHERDS PITHOS SIX PIECES
AD A 4 1 RIM FRAGMENT PITHOS
AD A 4 2 BODY SHERD FINE PARALLEL LINE MARKINGS
AD A 4 3 BODY SHERD PLAIN WARE
AD A 5 1 BODY SHERD PITHOS GREY FABRIC, EXTERNAL LINE PATTERN
AD A 5 2 BODY SHERD
AD A 5 3 BODY SHERD
AD A 5 4 BODY SHERD THIN
AD A 5 5 COARSE WARE
AD A AD66 T15 BODY SHERD CIRCULAR IMPRESSED PATTERNING + LINES FROM GULLY LINE ABOVE SPRING, 2 PIECES JOINED
AD A AD70 T15 BODY SHERD VERY WORN EXTERIOR LINE DECORATION ROAD CUTTING POTTERY
AD A 7 1 BODY SHERD PLATE WHITE INTERIOR / EXTERIOR GLAZE
AD A 8 1 HANDLE CLOSED VESSEL
AD A 1 10 HANDLE FRAGMENT WITH RIM
AD A 2 1 RIM FRAGMENT PLATE INTERNAL RED SLIP
PA A 2 (AD) 2 RIM FRAGMENT EXTERNAL LINE DECORATION, RED EXTERNAL SLIP
PA A 2 (AD) 3 BODY SHERD WITH HANDLE BASE
PA A 2 (AD) 4 BODY SHERD WHITE INTERIOR / EXTERIOR SLIP
PA A 2 (AD) 5 BODY SHERD WITH HANDLE BASE
PA A 2 (AD) 6 RIM FRAGMENT CLOSED VESSEL COARSE WARE
PA A 2 (AD) 7 HANDLE BASE BROWN EXTERIOR SLIP
PA A 2 (AD) 8 RIM FRAGMENT PLATE
PA A 2 (AD) 9 RIM FRAGMENT CLOSED VESSEL
PA A 2 (AD) 10 RIM FRAGMENT
PA A 2 (AD) 11 HANDLE FRAGMENT RIB DESIGN
PA A 2 (AD) 12 HANDLE FRAGMENT
PA A 2 (AD) 13 BODY SHERD WHITE EXTERNAL SLIP
PA A 2 (AD) 14 RIM FRAGMENT
PA A 2 (AD) 15 BASE FRAGMENT
PA A 2 (AD) 16 RIM FRAGMENT WHITE EXTERIOR SLIP
PA A 2 (AD) 17 RIM FRAGMENT
PA A 2 (AD) 18 HANDLE FRAGMENT
PA A 2 (AD) 19 HANDLE FRAGMENT
PA A 2 (AD) 20 HANDLE FRAGMENT
PA A 2 (AD) 21 BODY SHERD EXTERNAL INCISED DESIGN
PA A 2 (AD) 22 BODY SHERD BROAD RIB EXTERNAL PATTERN
PA A 2 (AD) 23 RIM FRAGMENT
PA A 2 (AD) 24 RIM FRAGMENT PLATE? INCISED TRIANGLE DESIGN
PA A 2 (AD) 25 BODY SHERD NECK FRAGMENT
PA A 2 (AD) 26 BASE FRAGMENT BLACK INTERIOR GLAZE
PA A 2 (AD) 27 HANDLE FRAGMENT
PA A 2 (AD) 28 HANDLE FRAGMENT
PA A 2 (AD) 29 HANDLE FRAGMENT?
PA A 2 (AD) 30 BASE FRAGMENT FINEWARE?
PA A 2 (AD) 31 RIM FRAGMENT LARGE CLOSED VESSELBROWN EXTERNAL SLIP
PA A 2 (AD) 32 RIM FRAGMENT WIDE GROOVE EXTERNAL IMPRESSION
PA A 2 (AD) 33 RIM FRAGMENT
PA A 2 (AD) 34 RIM FRAGMENT
PA A 2 (AD) 35 BASE FRAGMENT
PA A 2 (AD) 36 BODY SHERD BROWN EXTERNAL SLIP - WIDE BANDS
PA A 2 (AD) 37 RIM FRAGMENT
PA A 2 (AD) 38 BODY SHERD BROWN INTERNAL SLIP, RAISED DECORATION
PA A 2 (AD) 39 RIM FRAGMENT FINEWARE
PA A 2 (AD) 40 BASE FRAGMENT?
PA A 2 (AD) 41 BODY FRAGMENT PALTE? RED INTERIOR SLIP
PA A 2 (AD) 42 RIM FRAGMENT RAISED EXTERIOR LIP
PA A 2 (AD) 43 BASE OR RIM FRAGMENT
PA A 2 (AD) 44 BODY SHERD STRAP DECORATION

Appendix 11: Pottery Register

PA A 2 (AD) 45 BODY SHERD RAISED DECORATION
PA A 2 (AD) 46 BODY SHERD WHITE EXTERIOR? SLIP
PA A 2 (AD) 47 BODY SHERD BROWN EXTERIOR SLIP
PA A 2 (AD) 48 BODY SHERD WHITE EXTERIOR SLIP
PA A 2 (AD) 49 BASE FRAGMENT PLATE OR OPEN BOWL RED INTERIOR
PA A 2 (AD) 50 BODY SHERD BROWN INTERIOR / EXTERIOR PAINT
PA A 2 (AD) 51 RIM FRAGMENT
PA A 2 (AD) 52 RIM FRAGMENT
PA A 2 (AD) 53 RIM FRAGMENT
PA A 2 (AD) 54 BASE FRAGMENT PLATE
PA A 2 (AD) 55 BODY SHERD WHITE EXTERIOR SLIP
PA A 2 (AD) 56 RIM FRAGMENT
PA A 2 (AD) 57 BODY SHERD WHITE EXTERNAL SLIP
PA A 2 (AD) 58 BODY SHERD FINEWARE
PA A 2 (AD) 59 BODY SHERD WHITE EXTERNAL SLIP
PA A 2 (AD) 60 RIM OR BASE FRAGMENT
PA A 2 (AD) 61 BODY SHERD RED EXTERNAL GLAZE
AD A 13 K 1 PIE CRUST RIM LARGE OPEN BOWL
AD A 13 K 2 RIM FRAGMENT BOWL OR PLATE
AD A 13 K 3 BODY SHERD WHITE STRAP DECORATION
AD A 13 K 4 RIM FRAGMENT BROWN EXTERNAL SLIP
AD A 13 K 5 RIM FRAGMENT INCISED DECORATION
AD A 13 K 6 RIM FRAGMENT INCISED SYMBOL |;(|);+;| (;=NEW LINE)
AD A 13 K 7 HANDLE FRAGMENT
AD A 13 K 8 RIM FRAGMENT LARGE OPEN VESSEL
AD A 13 K 9 BODY SHERD WHITE EXTERIOR SLIP
AD A 13 K 10 BASE FRAGMENT
AD A 13 K 11 BASE OR RIM FRAGMENT
AD A 13 K 12 RIM FRAGMENT LARGE VESSEL BROWN EXTERIOR SLIP
AD A 13 K 13 HANDLE BASE WHITE EXTERIOR SLIP
AD A 13 K 14 RIM FRAGMENT
AD A 13 K 15 RIM FRAGMENT SAME VESSEL AS (viii)?
AD A 13 K 16 MODIFIED SHERD INCISED CROSS
AD A 13 K 17 RIM FRAGMENT PLATE FINEWARE
AD A 13 K 18 BODY SHERD WHITE EXTERIOR SLIP
AD A 13 K 19 RIM FRAGMENT
AD A 13 K 20 BASE FRAGMENT FINEWARE
AD A 13 K 21 HANDLE FRAGMENT WITH RIM
AD A 13 K 22 BODY SHERD LINE PATTERN
AD A 13 K 23 RIM FRAGMENT
AD A 13 K 24 RIM FRAGMENT?
AD A 13 K 25 BODY SHERD CLOSED VESSEL? WHITE FABRIC
AD A 13 K 26 BODY SHERD FAINT RAISED DESIGN
AD A 13 K 27 BODY SHERD BROWN EXTERIOR SLIP
AD A 13 K 28 RIM FRAGMENT
AD A 13 K 29 RIM FRAGMENT PITHOS? GREY INTERIOR
AD A 13 K 30 RIM FRAGMENT PITHOS
AD A 13 K 31 BODY SHERD INTERNAL RESIDUE?
AD A 13 K 32 BODY SHERD FAINT LINE PATTERN
AD A 13 K 33 BODY SHERD INTERESTING FABRIC CORES
AD A 13 K 34 BASE FRAGMENT
AD A 13 K 35 BASE FRAGMENT
AD A 13 K 36 HANDLE FRAGMENT
AD A 13 K 37 BASE FRAGMENT
AD A 13 K 38 TILE FRAGMENT
AD A 13 K 39 RIM FRAGMENT COARSEWARE
AD A 13 K 40 BODY SHERD WIDE STRAP DESIGN
AD A 13 K 41 RIM FRAGMENT
AD A 13 K 42 RIM FRAGMENT DARK BROWN SLIP
AD A 13 K 43 HANDLE FRAGMENT
AD A 13 K 44 RIM FRAGMENT

Appendix 11: Pottery Register

AD	A	13		K	45	HANDLE FRAGMENT		
AD	A	13		K	46	HANDLE FRAGMENT		
AD	A	13		K	47	RIM AND BASE	OPEN TROUGH?	2 PIECES JOINED
AD	A	13		K	48	RIM FRAGMENT	LARGE COARSEWARE VESSEL	
AD	A	13		K	49	RIM FRAGMENT?		
AD	A	13		K	50	HANDLE FRAGMENT		
AD	A	13		K	51	RIM FRAGMENT		
AD	A	13		K	52	RIM FRAGMENT		
AD	A	13		K	53	BODY SHERD	STRAP DESIGN	
AD	A	13		K	54	RIM FRAGMENT		
AD	A	13		K	55	BODY SHERD	FAINT INCISED STRAP DESIGN	
AD	A	13		K	56	BODY SHERD FINEWARE?	STRAP DESIGN	
AD	A	13		K	57	BODY SHERD	PITHOS, BLUE CORE, 4 PIECES JOINED	
AD	A	12		Bishop's Leap	1	BODY SHERD HANDLE BASE		
AD	A	12		Bishop's Leap	2	HANDLE FRAGMENT		
AD	A	12		Bishop's Leap	3	HANDLE FRAGMENT		
AD	A	12		Bishop's Leap	4	HANDLE FRAGMENT		
AD	A	12		Bishop's Leap	5	BODY SHERD	VESSEL NECK	
AD	A	12		Bishop's Leap	6	BODY SHERD	RAISED DESIGN	
AD	A	12		Bishop's Leap	7	RIM FRAGMENT		
AD	A	12		Bishop's Leap	8	BODY SHERD	BROWN EXTERNAL SLIP	
AD	A	12		Bishop's Leap	9	RIM FRAGMENT		
AD	A	12		Bishop's Leap	10	BODY SHERD	HANDLE BASE	
AD	A	12		Bishop's Leap	11	BODY SHERD	WHITE EXTERNAL SLIP	
AD	A	12		Bishop's Leap	12	HANDLE FRAGMENT		
AD	A	12		Bishop's Leap	13	BODY SHERD	WHITE EXTERNAL SLIP	
AD	A	12		Bishop's Leap	14	RIM FRAGMENT		
AD	A	12		Bishop's Leap	15	BODY SHERD	BROWN EXTERIOR PAINT?	
AD	A	12		Bishop's Leap	16	BASE FRAGMENT	FINEWARE	
AD	A	12		Bishop's Leap	17	RIM FRAGMENT	BROWN EXTERIOR SLIP?	
AD	A	12		Bishop's Leap	18	BODY SHERD FINEWARE		
AD	A	12		Bishop's Leap	19	BODY SHERD	WHITE FABRIC, FINE INCISED	
AD	A	12		Bishop's Leap	20	BODY SHERD	STRAP DESIGN	
AD	A	12		Bishop's Leap	21	HANDLE FRAGMENT		
AD	A	12		Bishop's Leap	22	RIM FRAGMENT	PITHOS	
AD	A	12		Bishop's Leap	23	RIM FRAGMENT	PITHOS	
AD	A	12		Bishop's Leap	24	RIM OR TILE FRAGMENT		
AD	A	12		Bishop's Leap	25	TILE FRAGMENT		
AD	A	12		Bishop's Leap	26	TILE FRAGMENT		
AD	A	16	2	Ford		FORD SITE	SESTTLEMENT / MANDRA?	
AD	A	16	2	Ford 1		RIM FRAGMENT		
AD	A	16	2	Ford 2		BODY SHERD CLOSED VESSEL	NECK	
AD	A	16	2	Ford 3		HANDLE FRAGMENT		
AD	A	16	2	Ford 4		BODY SHERD	WHITE FABRIC, INTERNAL GROOVE	
AD	A	16	2	Ford 5		BODY SHERD OPEN BOWL	WHITE EXTERIOR / GREEN INTERIOR GLAZE	
AD	A	16	2	Ford 6		BASE FRAGMENT	PLATE?	
AD	A	16	2	Ford 7		BODY SHERD	FABRIC IMPERFECTION	
AD	A	16	2	Ford 8		HANDLE FRAGMENT		
AD	A	16	2	Ford 9		BODY SHERD CLOSED VESSEL	INTERNAL GROOVES	
AD	A	16	2	Ford 10		BODY SHERD	CURVED WALL?	
AD	A	16	2	Ford 11		RIM FRAGMENT	CLOSED VESSEL	
AD	A	16	2	Ford 12		RIM FRAGMENT	PITHOS	
AD	A	16	2	Ford 13		RIM FRAGMENT	PLATE	
AD	A	16	2	Ford 14		BODY SHERD	BROWN EXTERNAL SLIP	
AD	A	16	2	Ford 15		BODY SHERD	CURVED	
AD	A	16	2	Ford 16		BODY SHERD	WHITE INTERNAL / EXTERNAL SLIP	
AD	A	16	3	Ford		PITHOS FIELD ABOVE FORD		
AD	A	16	3	Ford 1		HANDLE FRAGMENT		
AD	A	16	3	Ford 2		RIM FRAGMENT	PITHOS	
AD	A	16	3	Ford 3		BASE FRAGMENT		

Appendix 11: Pottery Register

AD A 16 3 Ford 4 RIM FRAGMENT CLOSED VESSEL
AD A 16 3 Ford 5 BODY SHERD GROOVE DESIGN
AD A 16 3 Ford 6 BODY SHERD INTERNAL GROOVE COIL BUILT?
AD A 16 3 Ford 7 BODY SHERD COOKING POT
PA A 1 3a Panagia Bridge
PA A 1 3a Panagia Bridge 1 BODY SHERD OPEN BOWL WHITE EXTERNAL GLAZE, GREEN AND WHITE INTERNAL GLAZE DESIGN
PA A 1 3a Panagia Bridge 2 RIM FRAGMENT OPEN VESSEL COARSEWARE
PA A 1 3a Panagia Bridge 3 HANDLE FRAGMENT
PA A 1 3a Panagia Bridge 4 HANDLE FRAGMENT RED SLIP?
PA A 1 3a Panagia Bridge 5 HANDLE FRAGMENT
PA A 1 3a Panagia Bridge 6 BODY SHERD FINEWARE EXTERNAL PATTERN
PA A 1 3a Panagia Bridge 7 HANDLE FRAGMENT SLIGHT INCISED EDGE
PA A 1 3a Panagia Bridge 8 HANDLE FRAGMENT? MICA IN SLIP OR WORN GLAZE
PA A 1 3a Panagia Bridge 9 BASE FRAGMET? OPEN BOWL INTERIOR GLAZE DESIGN
PA A 1 3b Cultivated potato field
PA A 1 3b Cultivated tomato field 1 RIM FRAGMENT OPEN BOWL FINWARE
PA A 1 3b Cultivated tomato field 2 BASE FRAGMENT RED EXTERNAL SLIP
PA A 1 3b Cultivated tomato field 3 BODY SHERD EXTERNAL INCISED DESIGN
PA A 1 3b Cultivated tomato field 4 BODY SHERD WHITE EXTERNAL SLIP
PA A 1 3b Cultivated tomato field 5 RIM FRAGMENT OPEN BOWL WHITE INTERNAL/EXTERNAL SLIP
PA A 1 3b Cultivated tomato field 6 RIM FRAGMENT FISH PLATE
PA A 1 3b Cultivated tomato field 7 BODY SHERD
PA A 1 3b Cultivated tomato field 8 BODY SHERD FINEWARE ORANGE INTERNAL SLIP

PA A 1 3b Cultivated tomato field 9 RIM FRAGMENT OPEN VESSEL COARSEWARE
PA A 1 3b Cultivated tomato field 10 BODY SHERD FINEWARE ORANGE INTERIOR / EXTERIOR SLIP
PA A 1 3b Cultivated tomato field 11 RIM FRAGMENT FINEWARE
PA A 1 3b Cultivated tomato field 12 BODY SHERD FIEWARE ORANGE INTERNAL / EXTERNAL SLIP
PA A 1 3b Cultivated tomato field 13 HANDLE FRAGMENT
PA A 1 3b Cultivated tomato field 14 BODY SHERD OPEN BOWL GREEN INTERIOR/ EXTERIOR GLAZE
PA A 1 3b Cultivated tomato field 15 HANDLE BASE
PA A 1 3b Cultivated tomato field 16 BODY SHERD WHITE EXTERNAL SLIP
PA A 1 3b Cultivated tomato field 17 BODY SHERD WHITE FABRIC
PA A 1 3b Cultivated tomato field 18 BODY SHERD CLOSED VESSEL
PA A 1 3b Cultivated tomato field 19 BODY SHERD CLOSED VESSEL EXTERNAL LINE DECORATION
PA A 1 3b Cultivated tomato field 20 RIM FRAGMENT 20TH CENTURY CHINA
PA A 1 3b Cultivated tomato field 21 TILE FRAGMENT 20TH CENTURY CONTINENTAL
PA A 1 3b Cultivated tomato field 22 RIM FRAGMENT OPEN BOWL FINEWARE
PA A 1 3b Cultivated tomato field 23 RIM FRAGMENT COARSEWARE
PA A 1 3b Cultivated tomato field 24 BODY SHERD FINEWARE
PA A 1 3b Cultivated tomato field 25 BODY SHERD OPEN BOWL BROWN / YELLOW INTERIOR GLAZE
PA A 1 3b Cultivated tomato field 26 HANDLE FRAGMENT
PA A 1 3b Cultivated tomato field 27 BODY SHERD BROWN FABRIC
PA A 1 3b Cultivated tomato field 28 BODY SHERD CLOSED VESSEL?
PA A 1 2a Panagia ALMOND ORCHARD (ABOVE ROAD)
PA A 1 2a Panagia 1 BODY SHERD AMPHORA, HANDLE BASE
PA A 1 2b Panagia ALMOND ORCHARD (BELOW ROAD)
PA A 1 2b Panagia 1 RIM FRAGMENT OPEN BOWL FINEWARE
PA A 1 2b Panagia 2 RIM FRAGMENT OPEN BOWL OR PLATE
PA A 1 2b Panagia 3 RIM FRAGMENT OPEN BOWL
PA A 1 2b Panagia 4 BODY SHERD VESSEL NECK
PA A 1 2b Panagia 5 RIM FRAGMENT PLATE?
PA A 1 2b Panagia 6 RIM FRAGMENT / BASE FRAGMENT FINEWARE

Appendix 11: Pottery Register

PA A 1 2b Panagia 7 BASE FRAGMENT CLOSED VESSEL
PA A 1 2b Panagia 8 RIM FRAGMENT
PA A 1 1 Panagia MANDRA AREA AND VINEYARD
PA A 1 1 Panagia 1 BASE FRAGMENT OPEN BOWL PURPLE / CREAM INTERNAL GLAZE
PA A 1 1 Panagia 2 BODY SHERD OPEN BOWL BROWN INTERIOR GLAZE
PA A 1 1 Panagia 3 BODY SHERD FINEWARE WHITE EXTERNAL GLAZE?
PA A 1 1 Panagia 4 RIM FRAGMENT INCISED PATTERN ON RIM EDGE
PA A 1 1 Panagia 5 RIM FRAGMENT LARGE OPEN VESSEL
PA A 1 1 Panagia 6 RIM FRAGMENT
PA A 1 1 Panagia 7 RIM FRAGMENT OPEN BOWL BROWN INTERIOR GLAZE
AD A 16 FORD AREA - ALMOND ORCHARD / RIVER BED
AD A 16 1 TILE? (register reads TITLE?)
AD A 16 2 PITHOS? TILE? FRAGMENT
AD A 16 3 BODY SHERD
AD A 16 4 RIM FRAGMENT
AD A 16 5 HANDLE FRAGMENT
AD A 16 6 BODY SHERD FINEWARE
AD A 16 7 BODY SHERD PLATE? FINEWARE
AD A 16 8 BODY SHERD LARGE OPEN BOWL?
AD A 16 9 BODY SHERD CLOSED VESSEL WHITE EXTERIOR SLIP
AD A 20 MANDRA ON SPUR LINE
AD A 20 1 BODY SHERD LARGE CLOSED VESSEL
AD A 20 2 BODY SHERD COOKING POT?
AD A 20 3 BODY SHERD OPEN VESSEL WHITE INTERIOR / EXTERIOR SLIP
AD A 20 4 BODY SHERD CLOSED VESSEL BLUE CORE FABRIC
AD A 20 5 BODY SHERD CLOSED VESSEL NECK
AD A 20 6 BODY SHERD OPEN BOWL WHITE INTERIOR GLAZE
AD A 20 7 RIM FRAGMENT OPEN BOWL GREEN INTERIOR GLAZE
AD A 20 8 BODY SHERD CLOSED VESSEL?
AD A 20 9 BODY SHERD CLOSED VESSEL
AD A 19 BIG PILE OF STONES
AD A 19 1 PITHOS FRAGMENT RAISED RIM DECORATION
AD A AD23 T18 1 AMPHORA NECK AND HALF BODY
AD A AD23 T18 2 BODY SHERD CLOSED VESSEL
AD A AD23 T18 3 BODY SHERD CLOSED VESSEL
AD A 26 1 HANDLE FRAGMENT
AD A 26 2 HANDLE FRAGMENT
AD A 26 3 BODY SHERD OPEN BOWL OR PLATE BROWN INTERIOR GLAZE
AD A 26 4 BASE FRAGMENT OPEN BOWL BROWN INTERIOR GLAZE
AD A 26 5 RIM FRAGMENT CLOSED VESSEL FINEWARE
AD A 26 6 BODY SHERD CLOSED VESSEL HANDLE BASE
AD A 26 7 BODY SHERD OPEN VESSEL BROWN INTERIOR SLIP
AD A 26 8 BODY SHERD OPEN VESSEL EXTERIOR RAISED DECORATION
AD A 7 10 HANDLE FRAGMENT
AD A 7 11 RIM FRAGMENT PIE CRUST DECORATION
AD A 7 12 RIM FRAGMENT OPEN VESSEL COARSEWARE
AD A 7 13 BODY SHERD CLOSED VESSEL
AD A 7 14 BODY SHERD CLOSED VESSEL
AD A 7 15 BODY SHERD CLOSED VESSEL?
AD A 7 16 BODY SHERD OPEN BOWL CREAM GLAZE WITH DESIGN
AD A 25 ABANDONED VILLAGE AND INDUSTRIAL COMPLEX
AD A 25 1 PITHOS BASE
AD A 25 2 TILE FRAGMENT
AD A 25 3 RIM FRAGMENT PITHOS
AD A 25 4 TILE FRAGMENT
AD A 25 5 BODY SHERD CLOSED VESSEL HANDLLE BASE
AD A 25 6 BODY SHERD OPEN VESEL WHITE INTERIOR / EXTERIOR SLIP
AD A 25 7 BASE FRAGMENT
AD A 25 8 BODY SHERD CLOSED VESSEL
AD A 25 9 BASE FRAGMENT CLOSED VESSEL

Appendix 11: Pottery Register

AD	A	25	10	BODY SHERD	CLOSED VESSEL	WHITE EXTERIOR SLIP
AD	A	25	11	BODY SHERD	CLOSED VESSEL	WHITE EXTERIOR SLIP
AD	A	25	12	BODY SHERD	CLOSED VESSEL	WHITE EXTERIOR SLIP
AD	A	25	13	HANDLE BASE		
AD	A	25	14	LAMP FRAGMENT?		
AD	A	25	15	BODY SHERD	BROWN INTERIOR / EXTERIOR SLIP	
AD	A	25	16	RIM FRAGMENT	CLOSED VESSEL	
AD	A	25	17	RIM FRAGMENT	OPEN VESSEL	COARSEWARE
AD	A	25	18	RIM FRAGMENT	OPEN BOWL	BROWN SLIP
AD	A	25	19	BASE FRAGMENT	CLOSED VESSEL	
AD	A	25	20	BASE FRAGMENT	CLOSED VESSEL	
AD	A	25	21	BODY SHERD	WHITE EXTERNAL SLIP	
AD	A	25	22	HANDLE FRAGMENT		
AD	A	25	23	BODY SHERD	CLOSED VESSEL	
AD	A	25	24	RIM FRAGMENT	OPEN BOWL	WHITE EXTERIOR SLIP
AD	A	25	25	RIM FRAGMENT	CLOSED VESSEL	
AD	A	25	26	HANDLE FRAGMENT		
AD	A	25	27	RIM FRAGMENT	WIDE SHALLOW DISH	
AD	A	25	28	HANDLE FRAGMENT		
AD	A	25	29	BASE FRAGMENT		
AD	A	25	30	HANDLE FRAGMENT		
AD	A	25	31	BODY SHERD	BLACK EXTERNAL SLIP	
AD	A	25	32	HANDLE FRAGMENT	MOULDED STRAP DECORATION	
AD	A	25	33	BODY SHERD	CLOSED VESSEL	VESSEL NECK
AD	A	25	34	RIM FRAGMENT		
AD	A	25	35	RIM FRAGMENT		
AD	A	25	36	RIM FRAGMENT		
AD	A	25	37	BASE FRAGMENT?		
AD	A	25	38	BODY SHERD	NECK?	
AD	A	25	39	BASE FRAGMENT	RED / BROWN EXTERNAL SLIP	
AD	A	25	40	RIM FRAGMENT		
AD	A	25	41	LAMP FRAGMENT		
AD	A	25	42	TILE FRAGMENT?		
AD	A	25	43	TILE FRAGMENT		
AD	A	25	44	TILE FRAGMENT		
AD	A	11		OCCUPATION AND INDUSTRIAL (KILNS) SITE. (FOUNTAIN)		
AD	A	11	1	HANDLE FRAGMENT		
AD	A	11	2	RIM FRAGMENT		
AD	A	11	3	BODY SHERD	RIM BROKEN OFF	

Appendix 12: Pottery Analysis

Pottery Analysis Report from Timothy Gregory.

MKA 8

5-2-1 CP rim, similar to 12-10, Roman?
5-2-2 Slipped rim, Ottoman
5-2-3 PHW Form 3, in CRS fabric, 5-6AD
5-2-4 TR (thin red) WR BS, 6-7AD
5-2-5 Glazed W-1 BS, Ottoman
5-2-7 Glaze painted rim, Ottoman, 18AD
5-2 Modern glass frg.
5-18 Glazed rim, Ottoman, 18-19AD
59 Coarse BS with resin (?)

MKA 8

8-1 Relatively well-preserved Group III bowl bottom, 13-14AD
8-2 Lemba RF with incised decoration (graffito?) 13-14AD
8-3 Group IVC RF, 14AD
8-5 Yellow glazed RF, medieval
8-6 Sgraffito BS, medieval
8-6 (duplicate?) Glazed RF
8-7 sgraf rim, medieval
8-8 Glazed rim, medieval
8-10 Sgraffito BS, medieval
8-11 Glazed BS
8-19 Sgraffito BS, medieval
8-22 Glazed BS
8-25 Glazed rim, UNK
8-26 Sgraffito BS, medieval
8-27 Sgraffito BS, medieval
8-35 Sgraffito BS, medieval
8-40 Flat tile
8-49 CP rim/handle, LR
8-51 Prob. CRS Form 10, 6-7AD
8-53 ARS RF, LR

MKA 12

1. Flat tile with ridge, M/M
2. Flat tile with ridge, M/M
4. Flat tile with ridge, M/M
5. Flat tile with ridge, M/M
6. Pithos rim, black glaze, CL/HE
7. Well-preserved tile with narrow inset ridge, M/M
8. Pithos rim, UNK
10. CP rim, Roman?
11. Combed W-1 BS, M/M
12. Coarse strap handle
13. Flat tile with inset ridge, medieval
22. Round weight
Round weight
Round weight

MKA 18 A

1-8-1 W-3 Pithos BS with circular stamp, M/M, probably 19-20AD
1-8-2 W-3 Pithos BS with circular stamp, M/M
1-8-3 W-3 Pithos BS with circular stamp, M/M
2-1 CGBG Base, Ottoman
2-2 CGBG Base, Ottoman

Appendix 12: Pottery Analysis

2-3 CGBG BS, Ottoman
2-6 CGBG BS, Ottoman
2-7 CGBG BS, Ottoman
2-8 W-3 Handle, M/M
2-11 W-3 BS
2-12 Combed W-3 BS
2-13 W-3 BS
2-14 CR SG SB, UNK
3-1 Combed W-1 BS, M/M
3-2 Combed W-1 BS, M/M
3-3 CGBG RF, Ottoman
3-4 CGBG BS, Ottoman
3-5 ARS RF, LR
3-6 CGBG BS, Ottoman
3-7 Green glazed BS
3-8 CGBG BS, Ottoman
3-9 CGBG BS, Ottoman
3-10 Red-painted glazed BS, Ottoman
3-11 CGBG BS, Ottoman
3-12 Green glazed BS
3-15 CGBG RF, Ottoman
3-16 CR RS BS
3-18 W-3 pithos BS
3-27 W-3 pithos rim
4-1 W-1 moulded pithos rim
4-2 W-1 BS
5-2 CR handle, Ancient?
5-3 TR KW WR BS, LR
5-4 CL BS
8-1 W-3 pithos BS with circular stamp, prob. Ottoman-Modern
8-2 W-3 pithos BS with circular stamp, prob. Ottoman-Modern
8-3 W-3 pithos BS with circular stamp, prob. Ottoman-Modern

MK-73-MK118 3 W-3 BS, Ottoman/medieval
T5-MK66 CL Handle, OT/MOD
T6-MK16 combed W-3 BS, OT/MOD
T6-MK30 many sherds of CL Pitcher/jug with flat base, OT
MK-138-139 W-3 rim, OT
MK-110 4 W-3 tile (?) frgs
MK-73-MK75 1 W-1 bowl rim, medieval
T8 C8-C45 1 worn W-3 BS
T3 3 W-3 BS

MKA-18-B

6-1 Coarse rim, CL?
6-4 TR Rim, IA/CL?
9-1 PHW Form 3 rim, 6-7AD
9-2 PHW Form 3 rim, 6-7AD
9-5 CL RF, Iron Age
9-8 CRS, Form 9 rim
9-13 Fine RF, UNK
9-17 PHW Rim, LR
9-20 Fine rim, UNK
9-21 TR KW Rim, IA?
9-24 Fine Rrim, IA-HE?
9-27 W-3 pithos tim
9-29 W-3 vat rim, OT/MOD
9-30 Tile

Appendix 13: Gazetteer

MAKHERAS FOREST ARCHAEOLOGICAL SITES: Preliminary Report.

A systematic, scientific survey to assess the current biodiversity of the Makheras Forest so as to ascertain the ecological shifts as a result of anthropogenic impact was completed in late February 1995. The survey, Operation Spinosum, was undertaken with the assistance of Dr. Michael Given with occasional help rendered by Dr. Stuart Swiny, Mrs. Laina Swiny, Dr. Edvard Polak, Cameron Lownds and Miss Alassandra Swiny. A total of 101 30m^2 sample plots or quadrants were established on north-south transect lines one kilometre apart.

An important aspect of the data recorded was both modern and ancient anthropogenic evidence. A total of 19 archaeological sites were located, the majority of these being mandras or agricultural centres. These, however, appear to have long histories as ascertained by the range of pottery dating, in some cases, from the medieval period.

Two important sites requiring rescue archaeology were located. The first of these coded MK.A.8, (T7, MK21) "Bishop's Retreat" has been exposed by Forestry Dept. reforestation techniques following the 1986 wildfire. It appears to be a single occupation, early 12th century site, and may be associated with court residence in the vicinity of the Makheras Monastery during the plague of 1393.

The second site coded MK.A.9, (T5, MK13) "Ayios Onouphrios" has been exposed by road construction. It is located in a bank opposite the Ayios Onouphrios Church on the Kaphades to Makheras Monastery road.

All other sites appear to be in a stable condition although several have been completely destroyed by bulldozer activity.

A brief synopsis of the sites is presented here in preliminary report form and a full analysis of the archaeological importance of the forest together with a description of the pottery associated with each site, site plans and photographs, will be presented in final report form when the analysis is completed.

Appendix 13: Gazetteer

MK.A.I: Mine Site. October 2, 1994.

Location: T1; MK 139; Elevation: 830 m.
Access: Along ridge line to boundary cairn 156.
Area description: Mine shaft located about 5 meters from the north-west corner of MK 139.5. Shaft appeared about 3.5 meters deep; 2.5 m long and 1 to 1.5 m wide. It seems to have been cut through a dike of altered diabase with a heavy quartz vein. Six other shafts were then located in the immediate area between this ridge and one adjacent to the west - two with associated platforms.

Note: In total seven shafts were located down the spur, each following a quartz seam. The dominant site aspect was westerly. The condition of the site was good although some in filling had occurred on the floor. The sides of the shaft appeared sound and stable. Although it has not yet been possible to have a geologist investigate the site, it is known that gold was mined in the Makheras Forest and the quantity of quartz associated with the shafts may suggest this activity.

Site photographed. Geological samples collected. No ceramics.

MK.A.II: Forest Hut/Byzantine Church? October 29, 1994.

Location: T2; MK 99; Elevation: 680 m.
Access: Forest road from Kaphades.
Area description: Bulldozed. Byzantine and/or Medieval tile present together with native rounded tile. Some solid 20th century brick and a little 4 cm. Medieval or Ottoman brick present. Site situated on ridge line with a north-west aspect. Water was located on site.

Tiles photographed. No ceramics collected.

MK.A.III: Prophetis Elias Monastery. September 18, 1994.

Location: T2; MK 111; Elevation:
Access: Kaphades to Lythrodondas road.
Area description: The road bank and cut terraces in front of the current monastery and church contain a vast quantity of Byzantine material including tile, pithoi, glazed and plain ware sherds. It is possible that earlier foundations of the monastery existed at a slightly lower altitude than the present structure, thus accounting for the quantity of material observed. The material is wide spread and extends down a gully line below the current spring head.

Ceramics collected.

MK.A.IV: Mandra tou Klosmatou. October 10, 1994.

Location: T3; MK 129; Elevation: 460 m.
Access: Along narrow dike from Kyprovasa Chiftlik to Prophetis Elias road.
Area description: A single roomed mandra with three stone wall enclosures. Access to a series of deep water holes in creek line into which a permanent spring flows. Good defensive position. Quantity of medieval pottery located on area above creek line and below the house where the ground was slightly eroded. The building roof has collapsed raising the floor level over a meter. No windows were observed in the structure although remnant wooden fabric, presumably from the door, were noted. A sheep or goat fold appeared to be located between two walls. A range of domestic and semi-domestic trees and herbs were located in the area.

Site mapped, photographed. Ceramics collected.

MK.A.V: Mandra tou Kambiou. November 9, 1994.

Location: T4; MK 89; Elevation: 670 m.
Access: Main Kaphades to Makheras road; Mandra tou Kambiou Forest Station sign posted.
Area description: Original mandra building completely destroyed by construction of the present Forest Station structure, with the exception of an oven or similar installation. Scatter of sherds and tile very dense to north and east of Forest Station. Four fields were defined for the area.
Field 1: Lying below the road line and trending toward the eastern side of the Forest Station. To the south-west a greater concentration of thicker pithos type sherds were located together with some large tap slag pieces. These decreased toward the Forest Station. The field was systematically walked and the recently ploughed surface afforded recognition of a dense sherd scatter. Finer sherds and those of a diagnostic nature appeared to be concentrated in the northern sector, that is, closer to the Forest Station.
Field 2: Tiles and pithos were concentrated in the north-east corner of this field, due perhaps, to bulldozer activity. Sherds were more concentrated in the south-eastern area. The north-western sector of the field is a recently constructed terrace but remains of a small square structure with a bottom opening remains. It is built of plastered masonry. The original pre-Forest Station mandra was probably situated on the platform directly north of the present building.

Appendix 13: Gazetteer

Field 3: Following the path from behind the forest station to the creek below where one large rock pool lies with a wider shallow wade below. Sherds from plain ware vessels (non-diagnostic body sherds) were observed although many may have fallen from the platform above. Diagnostic sherds from two vessels were collected - one green glazed, the other yellow glazed. These sherds were scattered across the rocky gorge face above the rock pool and two were recovered from the pool.

Field 4: Lies adjacent to a capped spring north and west of a line of cedars. A concentration of fine ware (Roman and Hellenistic) was located within ten meters of the spring. The area had been heavily ripped by a bulldozer in recent times. The scatter of sherds extended north to the water stand pipe and south into an olive grove. Numerous pithos sherds were observed but not sampled.

Site mapped, photographed. Ceramics collected.

MK.A.VI: Stone Structure 1. December 31, 1994.

Location: T6; MK 16; Elevation: 950 m.
Access: On spur line above Makheras to Philani roadway.
Area description: Rubble pile from small building perhaps 4m X 3m. Spoil from the structure had collapsed down slope in a southerly direction. The site is strategically placed with excellent views north down the valley and to the Makheras Monastery to the south. A quantity of pithos, tile and ceramics were associated with the site. Distance from a water source appeared to be about 500m.

About 200m further up the same spur another similar, but smaller, rubble pile was located with some bedrock exposed. This area may be entirely natural with the bedrock used for building blocks. A fairly distinct rectangle was discerned on the top of the rock pile, approx. 3m X 2m, with bigger stones around edge and fill of smaller rubble. No visible ceramics.

Site mapped, photographed. Ceramics collected.

MK.A.VII: Mandra tou Petri. November 6, 1994.

Location: T4; MK 117; Elevation:
Access: Forest road leading to the adjacent capped spring.
Area description: Bulldozed platform above spring on a ridge with a southerly aspect. Dense concentration of Byzantine tile or plain ware pottery with about 10 pieces per square meter.

Site photographed, ceramics photographed. Ceramics not collected.

MK.A.VIII: "Bishop's Retreat". January 22, 1995.

Location: T7; MK 21; Elevation: 910 m.
Access: Down spur line from the Makheras Monastery to Kionia road.
Area description: A low density scatter of pottery throughout quadrant was sufficient evidence to record it as an archaeological site. No pottery occurred in the area west of quadrant above road, however, a concentration of early Medieval ottery was located in a cutting exposed by forestry terracing. The site therefore, appeared to extend 20 meters north of the quadrant with a distinct layer of ashy deposit above a level of pottery. A quantity of glazed and decorated pottery was collected for analysis. Portion of a free standing wall on a bedrock foundation had been exposed by the road cutting. The site appears to be a single occupation early Medieval dwelling.

Site photographed. Ceramics collected.

MK.A.IX: Ayios Onouphrios.

Location: T5; MK: 13; Elevation:
Access: Main road from Kaphades to Makheras Monastery.
Area description: In the road cutting opposite the current church structure a floor with two and possible more walls are exposed. A small quantity of medieval pottery is eroding from the bank in the area. There appears to be a distinct floor level of large flat river cobbles making up two building areas. There is no structure visible at the surface level nor are any ceramics located at the surface. The original monastery was founded in the area in 1450 and was dissolved and fell into ruin in 1571. It is possible that the foundations exposed by the road cutting pertain to this early foundation as is borne out by the relevant pottery dates.

Ceramics collected.

MK.A.X: Stone Structure 2. November 2, 1994.

Location: T3; MK 76; Elevation: 765 m.
Access: Spur line from boundary cairn 27A situated on track from Prophetis Elias Monastery.

Site description: Ruins of a small stone building made of rubble masonry. Approximately 5m X 3.5m in size. There was no tiles or pottery visible, however, the site is heavily overgrown.

Site photographed.

MK.A.XI: Mandra tou Jeramais. January 8, 1995.

Location: T6; MK 22; Elevation:
Access: Forest track.
Site description: The abandoned mandra lies along a flattened spur trending south and overlooking the headwaters of the Argaki tou Ayiou Onouphriou. The site is situated in a high and isolated location with excellent views up and down the valley. An old path crosses the mountain opposite. The remaining standing structure consists of a two roomed enclosure with walls about 70 cm thick. Old pottery sherds have been reused in the wall construction. Two large tree stumps have been incorporated into the walls along the eastern and western lengths. Four fields were defined. Field 2 is almost entirely paved while field 3 is partially paved. Here the paving is much better laid with river cobbles carried from the valley below. Similar paving is locally associated with threshing floors used for the separation of millet and barley grains from their husks. Another structure, believed to be the dwelling, has been completely razed and the area planted to seeding pine and cedar. Two areas yielded diagnostic pottery - Field 4 to the west and fields 1 and 3 in the vicinity of the former dwelling. Several large pithoi ruins were located in Field 4 on the eastern side. Part of a stone bowl was recovered near a semi-circular feature in Field 2. The site and immediate area carried a quantity of medicinal herbs.

Site mapped, photographed. Ceramics collected.

MK.A.XII: "Loom Weight Site". February 5, 1995.

Location: T7; MK 3; Elevation: 910 m.
Access: Forest road.
Site description: Bulldozed. The artificially flattened area on a spur seems typical of other mandras located in the forest. Concentrations of pithos and tile were mapped and three loom weights were recovered.

Site mapped. Ceramics, loomweights collected.

MK.A.XIII: Stone Structure 3. February 5, 1995.

Location: T7; MK 3; Elevation: 880 m.
Access: Spur line from "Loom weight site".
Site description: A long, stone built enclosure or structure 19.8m in length and 6m in width. The walls averaged 60cm in width. The bearing of long wall was 100 degrees. One large piece of pithos was observed.

No ceramics collected.

MK.A.XIV: "Fort Site". February 26, 1995.

Location: T1; MK 5; Elevation:
Access: Road from Lazania to Politiko.
Site description: A large four roomed complex on a ridge line with strategic views over the Pedhieos River valley and towards Nicosia on the plain. A steep slope fell sharply below the site and old paths ran on either side. A concentration of pottery occurred on the eastern slope below the site. Fragments of a cup or beaker were recovered and joined giving a full profile. Two other diagnostic sherds were also collected from this area however the sherds observed were predominately non-diagnostic body sherds. Above the structure an area of bedrock was located with cutting lines distinctly marked. The inner most room of the complex was cut into the bedrock with its end wall standing to a height of 1.93 meters. Walls ranged in thickness from between 70 cm and 120 cm, the average being usually between 70-80 cm. No immediate water source was located, although later map investigation marked a spring about 50 meters distant.

Site mapped, photographed. Ceramics collected.

MK.A.XV: Kionia Forest Rest House. January 22, 1995.

Location: T7; MK 62; Elevation: 1300 m.
Access: Forest road from Kionia picnic area.

Site description: The original structure appears to have been two square rooms divided by a stone wall bonded into the western wall. The two doors, set adjacent to each other, opened to the east. A window opened to the north and another - later made into a door opening, to the south. The roof top was of moulded tile and the floor of local brick tile. The structure was later extended to include an additional room to the south with two southern opening windows and a door onto one of the original rooms. Another door opened to the east. A larger room was constructed at the northern end and the original northern window closed and plastered as shelf space. A fire place was constructed on the northern wall with a drainage pit? adjacent. The original roofing and flooring was re-used in the construction of the additional room which used much less stone of a smaller size. The original building still has a heavy white wash coating in parts, whereas the later additions show little white work, paint or plaster. The entire building was shuttered with only the shutter bolts and burnt timber framing around the windows remaining. The walls for both building periods were two feet thick. Two internal roof beams provided a false ceiling in the kitchen. The internal wall between the two original rooms was one foot thick. Second period wall thickness was eighteen inches. The new extention also included a covered verandah. Capping on the original building and continuing around the new extensions was of re-inforced concrete now largely collapsed on the western side. Agregate for the concrete appears to be too large and not well pressed into the scaffolding forms. The masonary above appears contemporary with the extention.

Five Cupresses sempervirens planted along the terrace edge show burn marks similar to lighting strike burns. These extend about 3 meters above ground level to just above ground level for the southern most trees, 1.25 meters to above ground level for the third, The tree situated closest to the kitchen doorway is almost completely destroyed from burn de-barking, and the tree opposite the kitcken window shows about 60 cm of burn scarring.

Site mapped, photographed. No ceramics.

MK.A.XVI: Stone Structure 4. January 29, 1995.

Location: T7; MK 52; Elevation: 1420 m.
Access: Ridge line from Kionia.
Site description: A rough oval enclosure which utilized large rocks already in situ and included a small terrace on the eastern side. There was not sufficient rubble or tumble to suggest a large or complete structure, however, the structure did appear to be dug into the surrounding slope. Surface covered by dense pine litter. The situation appeared sheltered and a very large Arbutus andrachne grew nearby. There appeared no immediate water resource apart from the springs at Kionia picnic area. A second enclosure was located five meters to the north east and was of similar construction. This structure measured approximately 2.70 m (9 feet) square and a quantity of rubble tumble from its collapse extended several meters down slope. One body sherd, not diagnostic, was recovered. The location of these enclosures afforded views to the southern coastline. From the site a bearing on Forest boundary cairn 190 was 142 degrees true.

Site mapped and photographed. One ceramic (non-diagnostic).

MK.A.XVII: Mandra tou Anayiou. March 25, 1995.

Location: T6; MK 8; Elevation: 720 m.
Access: Makheras to Philani road.
Site description: The mandra is situated with main entrances orientated toward the south. Walls are rubble built with a flat, open platform type yard on the southern exterior. The entire complex is roughly rectangular. Some walls include reused pottery or tile material. Five "rooms" were mapped, Room 1 being called the large southern enclosure. The wall height for room 1 on the south, east, and westerly aspects was averaged at 40 cm. The northern wall of Room 2 was constructed of small blocks filled with stones and round tile. It survives to a height of 140 cm. Other walls appear similar, though with less tile filling. The doorway was well constructed. Room 3 appeared to be another enclosure with roughly built walls of round river cobbles and some squared blocks. A shallow pit was located inside. The walls of this room extend to a height of 80 cm. Room 4 also contained a shallow pit of square form in the north-west corner. Traces only of Room 5 were visible, but a curious, elongated semi-circle was located similar in form to one observed at Mandra tou Jeramais.

Site mapped and videoed. No ceramics collected.

MK.A.XVIII: Khelidonia Industrial Site. March 30, 1995.

Location: T3; MK 92; Elevation: 650 m.
Access: Mandra tou Kambiou to Lythrodondas road.
Site description: A large settlement and industrial complex with some associated low grade copper ore seams situated on the confluence of four streams. The area extends over an area of approximately 10 hectares. To facilitate recording the complex was divided into three "site" areas: Site A, a more recent Ottoman occupational area; Site B, an earlier occupation area with Roman and Medieval pottery, and Site C, the slag heap and industrial installation area.

Appendix 13: Gazetteer

Site A was situated on a spur above the main river line with creeks on either side and a steep cliff face above the southern river line. Five "fields" were instituted to describe Site A: Field one covered the extensive construction which was divided into eleven "rooms". Field 2 lay between the structures and the river line to the east. Field 3 covered the steep slope to the south where extensive erosion had occurred. Field 4 lay to the north and incorporated the creek line down to its confluence with the river. Field 5 incorporated the small area of alluvium opposite Site A.

Site B was situated on a plateau to the south of Site A. Seven fields were instituted to describe the area. Field 6 lay between a lightly marked pathway above a visible wall line opposite the slag heap. Several areas of erosion had exposed pottery fragments which were collected. Several geological samples were also collected from this area. Field 7 was an eroded peninsula which proved generally sterile of artifactual material with only a small number of sherds collected. One large piece of vitrified material however was located here. Field 8 covered an area of thick grassy pasture with scattered small fire places presumably built by hunters. One "T" shaped dug-out appeared to be associated with military activity. A considerable amount of finely broken slag occurred in the spoil from the dug-out. Field 9 was a cut bank above the track where a concentration of pottery fragments were observed in situ and some were eroding onto the track. The bank extended for about 30 meters. A quantity of diagnostic sherds were collected for analysis. Field 10 extended down a steep bank from the track to the river line. An old wall appears quite distinctly running parallel to the track. Field 11 included the second building complex whose walls were, to a large extent, standing and intact. This area was measured and photographed. A curtain wall or enclosure wall appeared to extend about 20 meters from the building to the north-east. Field 12 incorporated a platform area defined by a low retaining wall. Some pottery was recovered from under very old olive trees where the ground had been cultivated. The olives were planted in rows approximately ten meters apart. Several piles of rubble were evident to the south but building foundations were not discernable although the area was heavily grassed.

Site C was situated on the end of a long spur. An old wall and possibly old smelting installations were observed on the crown of the spur adjacent to the slag heap. The slag heap extended from the top of the spur end to the creek line below where its width was about 25 meters. Extensive landscape modification had occured along the length of the spur with several deep cuttings observed. These are thought to have facilitated the movement of ore. One small outcrop of low grade ore was observed midway along the spur. The area had been used for military exercises.

Site mapped, photographed, videoed. Ceramics, slag and ore collected.

MK.A.XIX: Stone Structure 5. March 1, 1995.

Location: T5; MK 83; Elevation: 855 m.
Access: Abandoned path from Mandra tou Kambiou to Mandra tou Jeramais.
Site description: Unroofed stone building of squared rubble construction situated on a hilltop with commanding views up the adjacent valleys on either side and to the Messoria. Walls are standing on four sides from a height of 30 cm. to 1.5 cm. One doorway facing south was noted but no windows appeared visible. A quantity of broken tile was apparent but no diagnostic pottery. Several old olive trees were nearby.

Site mapped, photographed and videoed. No ceramics collected.

Appendix 13: Gazetteer

Adelphi Archaeological Sites.

T14 AD97 AD.A.I. Vrysi tou Marrou Tomb/Sanctuary?

A dense concentration of small fine-ware fragments were located in a small saddle along the ridge line dividing forest compartments 93 and 97. The site aspect is westerly with excellent views to the south through a gap in the ranges. The area is approximately two kilometres by foot from the densely settled sites along the river alluvium east of Panagia Bridge. The site is also located approximately 100 meters from the spring Vrysi tou Marrou (Spring of the Black Man) from which the toponym portion of the given name was derived.

The pottery scatter which included rims of ten different vessels was concentrated in an area of 1.30 m in diameter. Fragments about 1 cm in diameter were spread 15.4 meters uphill in a NE direction. This is assumed to be the action of the bulldozer during the construction of the fire break. A total of 290 fragments were collected for preservation.

Samples of rotted wood were collected for analysis. The presence of tree roots may indicate a deeper soil horizon present excavated for a tomb. A small scatter (three sherds) of similar manufacture were located at Ypsitas about one kilometre away.

AD.A.II Panagia Pitch Site

Broken pithos sherds, some with a pitch-like substance adhering to their sides were collected from a bulldozed access road. In winter a natural seep to the NNW of the site was covered with a thick grass growth. Many hare droppings were observed in the area. The pithos fragments were located near a gossan seam exposed by gully erosion. A vegetation transect was drawn across the site from the SSE to NNW to include the gully line and embankment below the fire trail and the slope above the site.

AD.A.III Marrou (Dad & Dave Industrial Site)

The small disturbed industrial area includes eight features. The former slag heap has been removed during the creation of a roadway linking Panagia Bridge and Askas. Slag extends along the roadway for 31 metres. no slag deposit was recorded on the cliff face below the road line. Area or locos ! was a small slag and rubble heap with pithos fragments. Locos 2 included three rubble dumps; locos 3 contained gossans and ore (sample collected); locos 4 was an ash lode suspected to be modern (charcoal collected); locos 5 appeared to be a circular stone placement; locos 6 was a spoil heap containing ore, glass fragment, ash and charcoal; locos 7 was another rubble pile with broken rock, pottery and slag (samples collected); locos 8 appeared to be a preparation area with crushed ore, charcoal, heat fractured rocks and crucible fragments.

The whole site has been heavily disturbed. Plan drawn and samples collected for analysis.

T15 AD70 AD.A.IV Mandra tou Peraskeva Mandra/Monastic? Site

altitude 480m asl; access: old forest road from ford; aspect: south east

A large agricultural or monastic complex situated on a level alluvial terrace above the gorge of the Phterirouni Potamos. Four separate structures of stone rubble construction and a large rough stone platform, possibly a threshing floor, were recorded. The well appeared to be in alignment with a small square room. The older and smaller buildings were of better construction than the larger room (locos 1). Some walls were standing to a height of 30 cms. The area was surrounded by a narrow curtain wall. The site was thickly overgrown with ground cover and plantation pines. Three diagnostic sherds, one pithos sherd and one undiagnostic were collected from the curtain wall area. One piece of slag and several lumps of gossan were observed.

The site is sheltered with first early morning sun exposure. A rich bird life was noted locally. Several ancient olive trees were located within the site perimeter. These had stone wall surrounds. The inner diameter of one olive tree was 2.60 meters. The stone enclosure measured 3.60 meters in diameter. A dense covering of wild garlic was noted.

The cliff face down to the river appears to have previously had a stair cut into it, now eroded out. A large quantity of water worn pottery was located on the pebble beach below the cliff. Two very deep natural water-holes were in the river bed below the site. An old ford crossing was also noted.

The area was mapped, vegetation transect established and photographed.

T15 AD69 AD.A.V Sylitis Pitch Extraction Site?

altitude:

On a slope above a very steep gully line trending north a quantity of pithoi fragments and coarse ware pottery were located. The area appeared to have a small platform or terrace adjacent to the pottery scatter. No structural remains were otherwise visible. An unusual rock formation was nearby and easily seen from a forest road above the site.

Pottery collected.

T16 AD64/51 AD.A.VI Kamitirkou Square Structure

altitude:

A square stone structure with two standing walls of a width of approximately 55 cm. The northern and western wall foundations were well constructed and preserved measuring 4.50 m and 4 m respectively. A large amount of rubble obscured the remaining walls. A small level platform behind a terraced retaining wall was situated just above the structure. Two pieces of pithoi were recorded. Not collected.

Site photographed and sketched.

T16 AD64/51 AD.A.VII Mandra tou Kamitirkou

altitude:
A medieval? mandra was recorded on the spur line about 100 meters below the watch tower or lookout recorded as AD.A.VI. The structure consisted of one large room with a narrow smaller, or secondary room built in one corner. The inner room is built against an earth bank cut into the spur line. Both rooms have doorways facing west down the spur line. A large court yard extends 18 meters to the west and the stone foundations are visible along its entire length. Evidence of wall robbing was apparent and rubble tumble extended down the scarps on the northern and southern sides. Pottery sherds were located in the gully line on the southern scarp.
The structure lies between two small knolls on a narrow saddle. Four gully lines on the south east have low terrace walls just below their confluence suggesting a spring or well site. The stones in this area carried a heavy moss cover. On the northern gully line some modification to the catchment area was suggested and a large earth spoil heap was recorded.
Pottery collected. Area drawn and photographed.

T16 AD59 AD.A.VIII Voudhokreanou Square Structure (Quadrat)
altitude
A small square stone structure was located in the corner of quadrat AD59 on a spur line descending from Muti tou Voudhokreanou. From a distance the area appears to be a slightly raized knoll. A pit like depression occurs within the rubble fill within the structure. One corner and the outline of the western wall is visible. One pottery sherd, jar handle, was recovered near this corner area. A well defined path crosses the quadrat and leads directly to the structure. Two rows of rough terraces were located above the path and the soil appears to be a deeper red loam with a higher moisture content.
Area photographed.

AD59 AD.A.IX Square Structure

T16 AD62 AD.A.X Voudhokreanou Square Structure
altitude: 800m; aspect: west-south-west

A stone structure consisting of rough rubble walls, cut mid way into a spur line. A deep gully was located to the south of the site with water access to the Row Potamos about 200 meters distant. The area appeared sheltered The structure which was mostly collapsed was approximately five metres square.
One pithos sherd collected. area photographed

T15 AD69 AD.A.XI Sykioes Industrial Kiln Site
altitude 580 m asl aspect south-east

A small stone building with agricultural terraces, adjacent footpath leading to a natural spring and a double pitch kiln emplacement comprised this industrial site. A small pit for charcoal, ash dump and vitrified earth was also recorded. The building had collapsed although the foundations were partially visible. The kilns were intact and represent a typical example of the pitch extraction process. The site was situated within 25 metres of water. It was located on the southern side of the Phterirouni Potamos above the confluence of a creek line from the spring and the river.
Site mapped and photographed.

T15 AD66 AD.A.XII Kamnel tou Dhespoti Porto (Settlement at Bishops Leap)
altitude :520m asl; aspect: north and east.

A series of abandoned terraces have been cut by recent road works to reveal walls from an ancient settlement located above the medieval bridge replaced in 1956 by a modern construction. A quantity of pottery was recovered from the cutting and the terraces adjacent. Ground visibility was poor. A path led to the top of a small hill where a rubble pile suggested the presence of another small stone structure. The area possibly covers approximately half an acre.

T15 AD.A.XIII Mandra tou Kimitirin Settlement / Industrial
altitude: 600m asl ; aspect: north and south

An abandoned mandra located on a spur overlooking the Phterirouni Potamos consisted of two rooms and a large walled enclosure. The site extended over an area of approximately four acres with pottery from the road line to the river escarpment. A quantity of fine ware, very dense pithoi and tile fragments were recorded.
The structure has standing walls with an older foundation wall visible in room 2. Some slag was located in the almond orchard on the adjacent southern slope. Very dense ground cover obscured most of the area with pottery recovered only from exposed areas under cultivated trees. The old olive trees have individual terraces or walls for protection.

T15 AD65 AD.A.XIV Kimitirin Industrial / Kiln.
altitude: 580m asl; aspect: east

A fire pit or kiln was located approximately 500 meters from Mandra tou Kimitirin. The kiln had been exposed by road works and partially destroyed. It measured 2.35 metres in diameter and appeared to be surrounded by stones. Samples of ash, charcoal and vitrified material were collected for analysis.
One pottery sherd was collected from the road verge above the kiln.
Site drawn and photographed.

T16 AD43 AD.A.XV Circular Stone Arrangement

altitude: 700m asl; aspect: west

A circular stone arrangement with a door way or insert was located above a modern forest track. A long stone wall or terrace ran north from the structure. A small platform appeared to be adjoining the structure. No pottery was evident. The site was at least 400 metres from the Row Potomas.

AD.A.XVI Adelphi *Poro* Settlement / Mandra above ford.
A two roomed stone building or enclosure dug into the ground above the Potomos. Extensive old terraces and cultivated fields were adjacent. An old path descended from the building to the river where some modification to the river bed was observed.
Pottery collected.

T16 AD44 AD.A.XVII Circular Stone Arrangement
altitude:

A circular stone arrangement or pit was located on the ridge line within the quadrat. No pottery was observed. Site photographed.

AD43 AD.A.XVIII Square Stone Structure
altitude: 540m asl; aspect: west

A small stone structure consisting of two rooms was located on a ridge line in apparent alignment (north-south) with the previously recorded circular stone arrangement (AD.A.XV). The building had foundation only in part with three large rubble heaps. Visible walls were 80 cms thick. The structure may have consisted of one room and an enclosure.

AD45 AD.A.XIX Vateron Square Stone Structure / Mandra
altitude 530 m asl aspect westerly
A two roomed stone structure cut into a steep spur was located on a spur descending from Muti tou Vaterou. Only one foundation wall was standing, the others were collapsed. Good stone masonry was however, apparent. A scatter of pithoi sherds was recorded and one sherd collected.
Site photographed.

AD64 AD.A.XX Kamitirkou Mandra (mansion house)
altitude 590m asl aspect westerly
A four or five roomed stone dwelling or mandra was cut into a very steep spur above the river line and between two steep gullies. Very thick pine litter obscured visibility. Two pieces of rough tile, several large pieces of flat stone (roofing?) were recorded and a few sherds. The stonework was very well laid and no re-use of material was noted. The site was approximately 50 meters from the river where several large natural pools occurred.
Area mapped, photographed

AD45 AD.A.XXI Kiln (Pitch?)
altitude 650m asl aspect west
A double kiln similar to those traditionally described for pitch extraction was located on a steep ridge. The kiln walls were carefully made split stone 1m in thickness. The diameter of the larger pit was 1.70 metres. A smaller pit below had been largely infilled.
Site photographed.

AD43 AD.A.XXII Mandra - Early Modern?
altitude 650m asl aspect west
An early modern mandra possibly sited on a much older foundation was recorded. The wall material contained re-use pithoi. Several unusual pitted stone boulders were located beside the building. The adjacent terraces appeared to be very old.

AD.A.XXIII Laxia tou Papa Yianni Settlement?
An area within the forest enclave appeared to have a quantity of pottery suggesting a small settlement site. Pottery extended over five terrace levels overlooking the ford crossing into forest Beat 3.

AD45 AD.A.XXIV Mandra - Abandoned Early Modern?
altitude 580m asl aspect west.
An abandoned mandra was located on an adjacent ridge line opposite the kiln site. The remains of the building appear consistent with other foundations in close proximity to kiln placements.

T18 AD19 AD.A.XXV Adelphi *Lithosouri* Abandoned Village + Industrial
altitude 540m asl aspect easterly.
The site of a possible village extends over a large area, probably in excess of 6 acres with dense concentrations of rubble piles and large quantities of pithoi, tile and pottery. The site extends from a ridge line along which a forest fire break has been bulldozed, to the creek line trending north. By following the ridge line for approximately 400 meters the village of Kato Moni is clearly visible.
No building foundations were observed above ground level although visibility was very poor due to a dense pine litter cover. Areas cut for pit planting of trees and along the road cuttings exposed quantities of white slipped pottery.

Above the creek line one kiln site was observed intact. Others may lie adjacent to it. A quantity of slag and vitrified material was concentrated near the kiln. The structure and the type of material associated with it does not suggest a pitch or charcoal kiln. It is very similar to one (or more) destroyed in the Makheras forest during post wild fire terracing.
Area photographed, sketched and pottery collected.

T18 AD16 AD.A.XXVI Lakxioes tou Seraphim Settlement / Mandra?
altitude 480m asl aspect south
A series of rubble piles cut by road construction have exposed pottery and vitrified material. The quantity of material and the extent of the complex is difficult to ascertain due to the condition of the site.
Pottery collected. Vitrified material sampled.

Ti9 AD10 Dhiplaera Square stone structure
897m asl aspect east.
The foundations of a small stone structure are located beside an old path linking Ayia Marina Xyliatos with Stavros Dhiosmatou. No pottery was observed. The location afforded excellent views to Muti Zizitias and the valleys trending north.

PANAGIA BRIDGE PA.A.I Settlement / Mandra
altitude 400m asl aspect east
Locos 1: An extensive area of abandoned agricultural land and eight roomed mandra were located on a fluvial terrace adjacent to the medieval bridge of Yephyri tis Panayias. The area covered included vineyards to the north and almond groves to the south. The mandra has solid walls of broken stone and pithoi fragment construction. One room was dug below the ground level and covered with iron sheeting. The building appears structurally sound. An old pithoi house was recorded about 25 meters up hill from the main building. Although only the foundation of the pithoi house remains the holding places for six pithoi were evident. Three different broken pithoi were photographed on the slope below the out house.
Other stone walls or foundations show in the road cutting below the mandra. Pottery was collected and the area photographed.
Locos 2:orchard area with dense grass covering. Rat middens observed under the almond trees. One large body sherd was recovered.
Locos 3a: abandoned and cultivated vineyards with some diagnostic sherds, however, visibility was very poor. One red jasper threshing sled flint collected.
Locos 3b: roadside plantation of almonds with pottery visible under trees. Two cultivated areas with good visibility, however, manure dumps contained broken pottery.

PANAGIA BRIDGE PA.A.II Mandra / Settlement
altitude 510m asl aspect west

An extensive settlement site was located on the fluvial terrace on the northern bank of the Phterirouni Potamos. The area had been bulldozed to create modern agricultural terraces. A large quantity of fine ware, pithoi and tile were recovered.

www.ingramcontent.com/pod-product-compliance
Lightning Source LLC
LaVergne TN
LVHW071205210926
848482LV00015B/730